Threatened Swallowtail Butterflies of the World
The IUCN Red Data Book

INTERNATIONAL UNION FOR CONSERVATION OF
NATURE AND NATURAL RESOURCES

Threatened Swallowtail Butterflies of the World
The IUCN Red Data Book.

N. MARK COLLINS
IUCN Conservation Monitoring Centre, Cambridge, U.K.

MICHAEL G. MORRIS
Chairman, IUCN/SSC Butterfly Specialist Group
Institute of Terrestrial Ecology, Wareham, U.K.

With the help and advice of the network of IUCN
Commissions and other experts throughout the world.

Photographic plates by
N. MARK COLLINS

PUBLISHED BY IUCN, GLAND, SWITZERLAND
AND CAMBRIDGE, U.K. 1985.

IUCN

IUCN (International Union for Conservation of Nature and Natural Resources) is a network of governments, non-governmental organisations (NGOs), scientists and other conservation experts, joined together to promote the protection and sustainable use of living resources.
Founded in 1948, IUCN has more than 450 member governments and NGOs in over 100 countries. Its six Commissions consist of more than 700 experts on threatened species, protected areas, ecology, environmental planning, environmental policy, law and administration, and environmental education.

IUCN

> monitors the status of ecosystems and species throughout the world;
> plans conservation action, both at the strategic level through the World Conservation Strategy and at the programme level through its programme of conservation for sustainable development;
> promotes such action by governments, inter-governmental bodies and non-governmental organisations;
> provides assistance and advice necessary for the achievement of such action.

Published by IUCN, Gland, Switzerland and Cambridge, U.K.
Prepared with the financial assistance of the World Wildlife Fund, the United Nations Environment Stamp Conservation Fund, the United Nations Environment Programme, Exxon Corporation and Citibank. A contribution to the Global Environment Monitoring System.

ISBN No.2 88032–603–6

Citation: Collins, N.M. and Morris, M.G. (1985). *Threatened Swallowtail Butterflies of the World. The IUCN Red Data Book*. IUCN, Gland and Cambridge. vii + 401pp. + 8 pls.

Cover illustration: *Papilio homerus* by Sarah Anne Hughes.
Book Design by James Butler.

Typeset by Text Processing Ltd., Clonmel, Ireland.
Printed by Unwin Brothers Limited, The Gresham Press, Old Woking, Surrey, U.K.

Contents

Appendices

Plates

1

How to use this book

Introduction

Work on 'Threatened Swallowtail Butterflies of the World' began in 1983. Information has been obtained from a very wide variety of sources including published and
unpublished material, and from an extensive worldwide correspondence with
amateur and professional lepidopterists, taxonomists, conservationists and
government officials. We hope that all have been acknowledged at the end of this
section. For those not mentioned, please accept our apologies for the omission and
our thanks for your help. Whilst we have tried to make the book as up-to-date as
possible and have sought advice worldwide, any mistakes and opinions are, of course,
the responsibilities of the authors.

Swallowtail butterflies comprise the family Papilionidae of the insect order
Lepidoptera. They were chosen for this study because 1) they are relatively well
known and are familiar to laymen and amateurs as well as professional entomologists,
2) they are large, often spectacular, insects, relatively easy to see, identify, and
therefore monitor, and 3) they have a worldwide distribution and have adapted to
many habitats and foodplant families, particularly in the tropics.

Layout and content of the book

The contents of the book have been listed in the preceding pages. The main body of
the book is divided into a number of sections, each of which will be described and
explained briefly below. These notes explain the plan of the book, the inter-relationships of the sections, and the terms and categories utilized.

**The swallowtail butterflies: an introduction to the family and its conser-
vation** This section is a general introduction to the swallowtails. A
description of their biology is followed by sections summarizing their distribution,
classification, foodplants and origins. Finally the conservation of swallowtails is
discussed, including current threats and efforts that are being made, or are needed, to
combat those threats.

**Swallowtails of the world: their nomenclature, distribution and conservation
status** This is a comprehensive list of the whole family in taxonomic order.
Each species has been numbered sequentially to permit easy cross-reference with the
geographical index (see below). Published synonyms have been included where they
may ease confusion, but no attempt has been made to list all synonyms. All names
mentioned are indexed at the back of the book. The complete known distribution of
each species is given, along with notes on taxonomy, conservation status, threats and
a list of references. Threatened subspecies have been noted in the list but, with three

1

particularly important exceptions, not given a full review. Common names are given where known, but no new names have been suggested. To find the species occurring in a particular country or geographical region, consult the geographical index to species (section 3, page 124). Under each country, or in certain cases each regional heading, there is a list of numbers corresponding to the species in the annotated taxonomic list which occur in that country or region.

The annotated taxonomic list was also used as 1) the basis for a critical fauna analysis to indicate those countries which have faunas of particular interest (section 4, p. 137), 2) the basis for an analysis of trade in swallowtail butterflies (section 5, p. 155) and 3) to permit a preliminary assessment of threats to swallowtails and the choice of a list for full review (section 6, p. 181). The status of species chosen for review in this final section is indicated in the taxonomic list and the name with status is also printed in full under the appropriate country or regional heading in the geographical index.

Analysis of critical faunas This is a computer-assisted analysis of the distribution of the swallowtails, which has enabled us to pinpoint those faunas that are critical to conservation effort. The analysis is based on endemic species within political boundaries. It is emphasized that this is a preliminary and coarse-grained analysis. It may be used for guidance in international efforts for swallowtail conservation but in no way affects national or local considerations.

Trade in swallowtail butterflies This section describes the extensive worldwide trade in butterflies with particular reference to the swallowtails. The main kinds of trade are outlined and the areas concerned are pinpointed. An assessment is made of the impact on wild populations, with attention given to the various ranching projects and their potential as a form of conservation. Appended to the text is a complete list of the family Papilionidae showing those species advertised for sale during the last five years and the range of prices demanded by dealers. Notes on legislation and captive breeding are also included in this table.

Reviews of threatened species This section contains reviews of the 78 taxa which were chosen from the taxonomic list as being threatened or in need of special conservation attention. Three important subspecies have been included because they are of particular interest for legislative or other reasons. The choice of species for review is always problematical and for swallowtails it has been made more difficult by the dearth of information for many species (see Appendix B).

Each review follows a similar format to that of previous IUCN Red Data Books. The following sections and headings are used:

Name and taxonomic position The name comprises generic (and sometimes subgeneric), specific (and sometimes subspecific) names, authority for the description of the taxon and the date when the description was published. The family Papilionidae is divided between subfamilies and tribes which are also indicated.

Red Data Book (RDB) status categories The traditional IUCN RDB status categories have been used throughout this volume. It is emphasized that the application of these categories to threatened organisms of any kind poses many problems and inevitably involves subjective judgements. They should always be regarded as a working tool, subject to change at any time as new information is received. The categories are defined as follows:

Extinct (Ex)
Species not definitely located in the wild during the past 50 years (criterion as used in the Convention on International Trade in Endangered Species of Wild Fauna and Flora—CITES).

Endangered (E)
Taxa in danger of extinction and whose survival is unlikely if the causal factors continue operating. Included are taxa whose numbers have been reduced to a critical level or whose habitats have been so drastically reduced that they are deemed to be in immediate danger of extinction. Also included are taxa that are possibly already extinct but have definitely been seen in the wild in the past 50 years.

Vulnerable (V)
Taxa believed likely to move into the Endangered category in the near future if the causal factors continue operating. Included are taxa of which most or all of the populations are decreasing because of over-exploitation, extensive destruction of habitat or other environmental disturbance; taxa with populations that have been seriously depleted and whose ultimate security has not yet been assured; and taxa with populations which are still abundant but are under threat from severe adverse factors throughout their range.

Rare (R)
Taxa with small world populations that are not at present Endangered or Vulnerable, but are at risk. These taxa are usually localized within restricted geographical areas or habitats or are thinly scattered over a more extensive range.

Indeterminate (I)
Taxa known to be Endangered, Vulnerable, or Rare but where there is not enough information to say which of the three categories is appropriate.

Insufficiently Known (K)
Taxa that are suspected but not definitely known to belong to any of the above categories, because of lack of information.

Out of Danger (O)
Taxa formerly included in one of the above categories, but which are now considered relatively secure because effective conservation measures have been taken or the previous threat to their survival has been removed.

N.B. In practice, Endangered and Vulnerable categories may include, temporarily, taxa whose populations are beginning to recover as a result of remedial action, but whose recovery is insufficient to justify their transfer to another category.

Threatened is a general term to denote species that are Endangered, Vulnerable, Rare, or Indeterminate, and should not be confused with the use of the same word by the U.S. Office of Endangered Species.

The information needed to fulfil the criteria for these categories is often lacking for swallowtails, even though they are among the best known of all insects. There is generally little difficulty in assessment of the threats to well-known species, particularly when they have restricted distributions, but the status of species whose habitat requirements are poorly understood is more difficult to assess, particularly where their distribution is wide.

The category Insufficiently Known could have been very widely applied in this volume, but such action was deemed to be of little conservation purpose. Fourteen examples of species reviews in this category are given in section 6, but it will be seen that the quality of data is often poor and the review may add relatively little to what has already been stated in the species list (section 3). Appendix A (p. 369) is a list of those 78 species assigned to RDB categories, followed in Appendix B by a further list of almost 100 species for which more data on distribution, ecology and conservation status are required before an assignment of RDB category can be made.

Description As the colour and size of most species of swallowtails will be unfamiliar, a brief description is included. In the case of Endangered, most Vulnerable and a selection of other interesting threatened species this description is supported by a colour plate of a museum specimen. The size of the species is in most cases indicated by the length of the forewing. The wings are the main distinguishing feature of butterflies and the terms used in describing them are explained in Figure 1.1.

Distribution Distributional data are the most recent available and are generally only given to the level of region, state or province. Precise localities are occasionally mentioned.

Habitat and Ecology Notes include data on life-cycles and foodplants (which are all too often unknown), flight times and seasonal cycles. In some cases the biotope is described in a general way, including data on climate, vegetation and altitude where appropriate.

Threats This section includes any known or surmised threats to the survival of the species or its populations. In some cases the threats are well defined, in which case a higher conservation category may be expected. Often the threats will not be known and a species may be known from very few specimens. In such cases holding categories such as Indeterminate or Insufficiently Known may be expected. Most of the listed species are threatened by loss of habitat. In many reviews where this is so, a general account of habitat loss and its causes may be given. Clearly such problems affect a wide range of wildlife.

Conservation Measures Herein are included measures already taken to conserve the species, including protection by legislation or gazetting of protected areas. Proposed measures are also listed, including legislation, appeals for protective measures and further scientific study to establish population sizes, distribution and status.

Index The general index includes all scientific and common names used in the book and a selection of key words.

Plates Finally, colour plates of all but four Endangered and Vulnerable swallowtails are presented. A number of species in other threatened categories are also included, mainly to demonstrate the range of form to be found within the swallowtail family. All photographs have been reduced to a standard size because in life many of the species are larger than the page upon which they are illustrated. Their actual size is clearly indicated in the legend and in the appropriate review in section 6.

Figure 1.1. Explanation of terms used in the description of butterfly wings. In general the wings of the two sexes are similar in form but the males of some genera (e.g. *Parides*) have the anal margin of the hindwing folded over to contain secondary sexual organs. This fold is variously called the abdominal fold, hair-pouch or scent-organ.

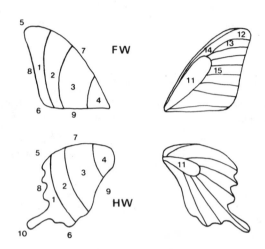

UFW:	upper forewing	LFW:	lower forewing
UHW:	upper hindwing	LHW:	lower hindwing

1.	submarginal	8.	outer margin
2.	postdiscal	9.	anal or inner margin
1&2.	distal	10.	tail
3.	discal	11.	cell
4.	basal	12.	apical
5.	apex	13.	subapical
6.	anal angle or tornus	14.	subcostal
7.	costa or costal margin	15.	median

The following terms apply particularly to *Ornithoptera* males

16. radial band
17. cubital band
18. anal band

Acknowledgements

We are very grateful to the many institutions and people who have given generous assistance in the preparation of this volume and without whom the task could not have been attempted. We would particularly like to thank the following:

Members of the IUCN/SSC Butterfly Specialist Group.

The staff of the British Museum (Natural History), in particular the Entomology librarian Miss P. Gilbert and the staff of the Rhopalocera section of the Entomology Department, particularly R.I. Vane–Wright, P.R. Ackery and R.L. Smiles.

The staff of the Royal Entomological Society of London and particularly the librarian Mrs B. Leonard.

Sally Anne Hughes for the cover drawing.

All our colleagues at the IUCN Conservation Monitoring Centre and the Institute of Terrestrial Ecology, Furzebrook, who gave enthusiastic encouragement and helpful advice on many matters.

Michael Morton of Brunel University and David Webb of Hatfield Polytechnic, sandwich students who gave invaluable assistance in data-processing during their time at the Centre.

And the following individuals for providing information and refereeing various sections of the book:

B. d'Abrera	P.R. Ackery	S.A. Ae	H.S. Barlow
M.J.C. Barnes	A. Bedford–Russell	M.J. Braga	F.M. Brown
K.S. Brown Jr.	J.P. de Carvalho	P.B. Clark	Sir C.A. Clarke
M.P. Clifton	M.J. Coe	S. Collins	T.C.E. Congdon
G.A. Dacasin	E.W. Diehl	P. Dollinger	F.M. Dori
M.J. Dourojeanni	H. Dove	K. Eltringham	R.D. Estes
P. Feeny	T.L. Fenner	R.W. Garrison	M. Ghilarov
A.K. Ghosh	S.K. Ghosh	H.C.J. Godfray	R. Guilbot
N. Gunatilleke	D.L. Hancock	J. Haugum	T. Hidaka
J.D. Holloway	W. Hsu	A.F. Hutton	M. Ishii
J.N. Jumalon	J. Kielland	J. Kingdon	Kim Hon Kyu
G. Lamas	T.B. Larsen	R.R.B. Leakey	R.E. Lewis
D. Lowe	R. Macfarlane	D.K. Mandall	A.G. Marshall
J.de la Maza E.	O.H.H. Mielke	J.S. Miller	L.D. Miller
T. Milliken	J. Nabhitabahata	T.R. New	D. N'Sosso
A.U. Oboite	S.N. Okiwelu	P.A. Opler	G.W. Otis
D.F. Owen	M.J. Parsons	R. Paulian	N.D. Penny
M.en C.H. Pérez	A. Pinratana	R.M. Pyle	C.D. Quickelberge
T. Racheli	J. Riley	A.H.B. Rydon	T. Saigusa
R.L. Smiles	D. Spencer–Smith	T.T. Struhsaker	B. Turlin
T.W. Turner	A. Tye	the late H.Tyler	R.I. Vane–Wright
L. Vazquez G.	P. Viette	M.V. Walker	B.M. Wankhar
D. Wankhar	P.G. Waterman	J.D. Weintraub	J. G. Williams
M.L. Winston	A. Youdeowei		

This book was prepared and published with the support of the World Wildlife Fund, the United Nations Environment Stamp Conservation Fund, the United Nations Environment Programme, Exxon Corporation and Citibank.

2

The swallowtail butterflies:
an introduction to the family
and its conservation

The butterfly family Papilionidae includes some of the most spectacular and magnificent of all insects. When compared with other insects, all the swallowtails have large wings, ranging from the dainty, 50 mm wingspan of the dragontails of India and Malaysia (*Lamproptera* species), to the giant birdwings (*Trogonoptera*, *Troides*, *Ornithoptera*) of Indonesia and New Guinea. The female Queen Alexandra's Birdwing (*Ornithoptera alexandrae*) is the largest butterfly in the world, attaining a wingspan of over 250 mm. Because of their large wings many swallowtails are powerful fliers, the apollos being an exception with their rather feeble flight. The colours and patterns of swallowtails are astonishingly rich and diverse. They range from sombre black or brown (e.g. *Papilio acheron*, *Troides dohertyi*, Plate 3) to the lustrous sheen of the gloss swallowtails (e.g. *Papilio chikae*, Plate 8). Some of the birdwings are iridescent and highly changeable in colour; *Troides prattorum* can appear golden-orange, green, pale blue or pink from different angles of view (11).

In many species the hindwings are extended into tails, superficially resembling the long tails of swallows, hence the popular name for these butterflies. The precise function of the tails has never been examined and it is not clear whether they assist flight or have some other purpose. The term swallowtail is used very loosely at both the familial, generic and specific level. Other common names include apollos (*Parnassius* spp), festoons (*Allancastria* and *Zerynthia*), gorgons (*Meandrusa*), kites (*Eurytides*), dragontails (*Lamproptera*) swordtails or jays (*Graphium*), windmills or clubtails (*Atrophaneura*), birdwings (*Trogonoptera*, *Troides*, *Ornithoptera*) and fluted swallowtails (*Papilio*). Common names are given in the species list (section 3, p. 33) and the general index (p. 375).

Biology of swallowtails

The adult butterfly is only the last of four distinct phases in the life cycle. The early stages comprise the egg, larva or caterpillar, and chrysalis or pupa. The process of development, or metamorphosis, takes several weeks unless the species over-winters in an immature phase. For example, *Papilio machaon* eggs hatch after about two weeks, the caterpillars feed for 30 days and may then pupate for as little as 14 days or may last through the winter. Adult Papilionidae may survive for up to four months, but 20–30 days is more usual. Predation and parasitism of butterflies are high at all stages of development and only a small proportion will survive to adulthood (47). The eggs of many species, from the primitive *Baronia brevicornis* of Mexico to the *Troides* and *Ornithoptera* of New Guinea, are heavily attacked by minute parasitic (or, more correctly, parasitoid, as they kill their host) chalcid wasps (Hymenoptera: Chalcidoidea) (48, 49). Studies on a range of butterflies, including the British race of *Papilio machaon* in the Norfolk Broads, have indicated high predation of the first two larval instars by arthropods, particularly spiders (16). Later instars may be too big to be taken by arthropods but suffer heavy mortality due to birds. Pupae are severely predated by small mammals (16).

The eggs of swallowtails are spherical, lack any sculpturing and in the birdwings may be up to 4 mm in diameter, the largest butterfly eggs of all. Most species can lay several hundred eggs, but predators usually prevent such a potential from being realized. The genus *Ornithoptera* is again unusual in generally laying no more than 30 eggs per brood. Eggs are usually laid on the caterpillar's foodplant, often on the underside of leaves where they are inconspicuous and shaded. After hatching, caterpillars immediately begin feeding and growing, passing through five moults before the quiescent pupal stage.

The caterpillars of Papilionidae may have fleshy spines or tubercles but they are never hairy and always have a curious structure called an osmeterium. This is a forked scent gland which the larva can extrude through a slit in the thorax when it is disturbed or attacked. The gland secretes a powerful-smelling liquid which consists, in the species studied so far, of two aliphatic acids (isobutyric and 2-methyl butyric acids) (22, 23). When attacked, the larva tries to wipe the gland against its attacker, a procedure that is certainly successful against ants (22) and may deter not only other predators, but also parasites (47). Mature caterpillars are green in *Baronia* and the Leptocircini, black spotted with red or yellow in the Parnassiinae and Troidini, and variable in the Papilionini. Some caterpillars, like those of *Papilio polymnestor*, present a curious 'eyed' appearance and may adopt a threat posture as a deterrent to would-be predators (57). Others, such as *Papilio clytia*, mimic the appearance of distasteful species in the caterpillar as well as the adult (13). Species that feed on Aristolochiaceae and other poisonous plants are all believed to store the poison and themselves become poisonous. This fact may be advertized in a gaudy range of patterns and hues that acts as a reminder to predators, particularly birds. Many swallowtail larvae change their patterns between instars. *Papilio machaon*, for example, resembles a bird dropping in its first two instars, but thereafter becomes conspicuous in a bright pattern of black, orange and green (35). It is fairly commonplace for the final instar larva to leave the foodplant before pupating. For example, the caterpillar of *Ornithoptera paradisea* leaves its foodplant, the *Aristolochia* vine, and pupates about 1 m above the ground on shrubs and trees (11).

The pupae of Papilionidae are generally brown, sometimes green, with prominent keels or tubercles in the Troidini and a single dorsal protuberance in the Leptocircini (30). A characteristic of the Papilionidae is that the fully grown larva spins a silk tail pad and girdle prior to pupation. The chrysalis is thus suspended at an angle from its substrate, often resembling a broken twig. An exception is the genus *Parnassius*, in which the caterpillar spins a fine silken web and pupates among leaves near the ground. Further details of the early stages of Papilionidae are given in Igarashi's recent study (34).

Adult swallowtails have a number of distinctive characteristics, apart from the tails already described. In the mouthparts the palps are small, the proboscis well-developed. Adults frequently take nectar from flowers whilst still in flight, the fluttering stance being quite distinctive. All six legs are fully developed and fit for walking (unlike the Nymphalidae), and have a spur-like projection on the tibia of each foreleg. The cell is closed on both the fore- and hindwings and only a single anal vein is present on the hindwing (except in *Baronia*). In the males of some genera (e.g. *Parides*, *Atrophaneura*) the anal margin of the hindwing is folded over and contains scent brushes used in courtship and mating.

In the majority of swallowtails the sexes are broadly similar in appearance. Nevertheless, there are many species that are sexually dimorphic, often, but not invariably, as a result of mimicry. The males and females of birdwings are very different (Plates 3,4,5) but are not known to be mimetic except in the case of the female *Ornithoptera croesus lydius* (11) (Plate 6). Females of other swallowtails are often slightly larger and less colourful than males, as in the birdwings, but again this is not always so. For example, the female *Papilio antimachus* is smaller than the male (Plate 8). Females are generally more retiring in their habits and in forests females often keep to the tree-tops while males congregate at drinking spots or fly rapidly along rides and streams. Males may be highly aggressive, jostling and fighting each other for a spot on a hilltop or defending their territory along a stretch of river.

In some species, such as *Graphium idaeoides* (Plate 2), *Papilio agestor* and *P. slateri*, both sexes are mimics. In others, such as *Papilio jordani* and *P. phorbanta* (Plate 8), only the female is a mimic and extreme sexual dimorphism is apparent. A further variation is seen in *Eurytides euryleon*, a dimorphic species in which each sex appears to mimic the same sex of the dimorphic model, *Parides iphidamas phalias* (12).

Mimicry is a form of protection against predators and is a widespread phenomenon amongst butterflies. In classical definitions it consists either of a palatable mimic species gaining protection from predators by resembling a distasteful and unpalatable model (Batesian mimicry), or of a number of distasteful models joining a mimetic ring in which each gains protection from the other (Müllerian mimicry). Recent research has shown this to be an over-simplification and many new variations and interactions are now recognized (1, 60).

Table 2.1 lists a small selection of mimics and their models. Models are generally unpalatable species from the Danainae, Heliconiinae, Acraeinae or *Aristolochia*-feeding Papilionidae, but other groups may also be used. *Papilio laglaizei* is unusual in being a mimic of a day-flying moth, *Alcidis agarthyrsus* (Uraniidae).

In numerous species the females (and less often the males, *Papilio paradoxa* is an example) are polymorphic, i.e. they occur in a number of different forms. The African Mocker swallowtail, *Papilio dardanus*, is yellow and tailed in the male but the female is extremely variable and may appear mainly white, yellow or orange, often tail-less. The females mimic a variety of distasteful model species (Table 2.1) (4).

Distribution, classification and origin of swallowtails

The Papilionidae is a pre-eminently tropical family (ca 20°N–20°S), although some species can reach latitudes 60–70°N and 40–50°S. In an analysis of the distribution of Munroe's (42) list of 538 species, Scriber (53, after Slansky, 56) demonstrated the great increase in species richness at tropical latitudes. In Figs. 2.1–2.4 we have repeated Scriber's analysis using the 573 species in our list (section 3). As expected, the pattern is much the same, with the richest areas in the equatorial rain forest zones. South East Asia has the highest number of species but interestingly, species richness in the eastern Asian region between 20 and 40°N is also high, even exceeding the richness of the American and African tropics. This trend is reflected in the analysis of critical faunas (section 4, p. 137) where Indonesia, the Philippines and China alone are shown to include well over a third of the world's swallowtails.

The classification of the Papilionidae has been a source of great difficulty to

Table 2.1 A selection of possible models and mimics in the swallowtail family Papilionidae (1, 9, 12). Key: m = male, f = female.

Mimics	Models
Indo–Australia	
Graphium idaeoides	*Idea leuconoe*
Papilio memnon f.	*Atrophaneura varuna* f.
Papilio memnon f.	*Atrophaneura nox* f.
Papilio memnon f.	*Atrophaneura coon*
Papilio polytes f.	*Atrophaneura aristolochiae*
Papilio clytia	*Tirumala septentrionis*
Papilio agestor	*Parantica sita*
Papilio slateri	*Euploea algea*
Papilio paradoxa	*Euploea mulciber* m.
Papilio clytia clytia	*Euploea klugii*
Papilio paradoxa aegialus	*Euploea diocletianus*
Papilio jordani f.	*Idea blanchardii*
Papilio laglaizei	*Alcidis agarthyrsus*
Africa and its Islands	
Papilio phorbanta f.	*Euploea goudotii*
Papilio dardanus f.	*Amauris albimaculatus*
Papilio dardanus f.	*Danaus chrysippus*
Papilio dardanus f.	*Amauris niavius*
Papilio dardanus f.	*Bematistes poggei* m.
Papilio cynorta f.	*Bematistes epaea* f.
Papilio cynorta f.	*Amauris niavius*
Papilio echerioides f.	*Amauris echerius*
Papilio rex	*Danaus formosa*
South America	
Eurytides thymbraeus aconophos	*Parides alopius*
Eurytides thymbraeus	*Parides photinus*
Eurytides branchus	*Parides arcas mylotes*
Eurytides belesis	*Parides polyzelus*
Eurytides euryleon	*Parides iphidamas phalias*
Eurytides euryleon clusoculis	*Parides lycimenes*
Eurytides pausanias	*Heliconius wallacei*

taxonomists. The main problem has been that the wide diversity of colouring, life cycle and habits is not reflected in clear morphological differences. This has been particularly problematical at the generic level where great variations in colour and pattern often overlie an essentially similar structural design. In the genus *Papilio* for example, neither Munroe in 1961 (42) nor Hancock in 1983 (30, p. 30) could find morphological differences useful in creating generic groupings with a smaller number of species. Fortunately however, morphological, biological and chemical studies have met with success in developing our understanding of the taxonomy of the Papilionidae at the suprageneric level (26, 42, 43).

The living Papilionidae are divided into three subfamilies, the Baroniinae, Parnassiinae and Papilioninae. Each of these will be discussed only briefly here; the reader should refer to the literature for further details (see the bibliographies in 30, 42) and to section 3 (p. 33) for information on particular species.

The Baroniinae includes a single species, **Baronia brevicornis** from Mexico (see

Fig. 2.1. Latitudinal gradients in the species richness of the New World Papilionidae (adapted from Scriber (53) and Slansky (56)).

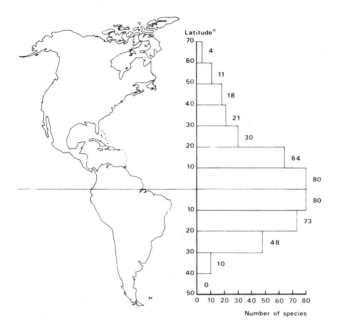

Fig. 2.2. Latitudinal gradients in the species richness of the Eurasian and African Papilionidae (adapted from Scriber (53)).

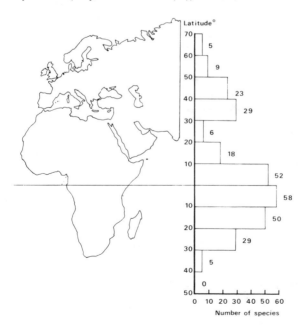

Fig. 2.3. Latitudinal gradients in the species richness of the Indo–Australian and Eastern Asian Papilionidae (adapted from Scriber (53)).

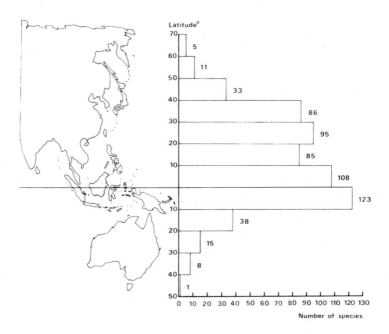

Fig. 2.4. Latitudinal gradients in the species richness of the Papilionidae of the world.

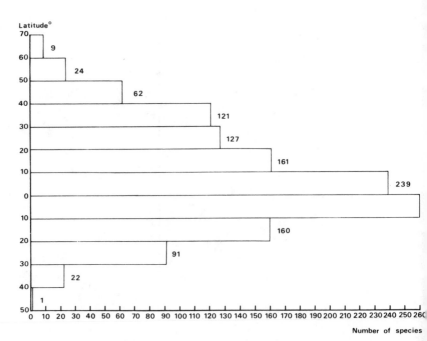

p. 182), considered to be the most primitive living papilionid. Although the adults have the fore-tibial spur and osmeterium which are characteristic of the Papilionidae, the wing venation is very uncharacteristic in having both anal veins present on the hindwing, like Pieridae and Nymphalidae. *Baronia* is therefore believed to be a descendant of primitive papilionid ancestors close to the evolutionary stem shared by the Papilionidae and Pieridae (43). *Baronia* is also most unusual in feeding on *Acacia* (Leguminosae). The position of *Praepapilio colorado*, a fossil papilionid from the middle Eocene deposits of Colorado, U.S.A., is open to question. It shares some of the primitive features found in *Baronia*, but has recently been elevated to a monotypic subfamily of its own, the Praepapilioninae (17).

The subfamily Parnassiinae is divided into two tribes, Parnassiini and Zerynthiini. The former includes the monotypic genera *Archon* and *Hypermnestra* and the holarctic and species-rich genus *Parnassius*. *Archon* and *Hypermnestra* are confined to Asia Minor while *Parnassius* species are primarily butterflies of the montane areas of Europe, northern and central Asia and western North America. Primitively the larval foodplant is the lowland vine *Aristolochia*, as in *Archon*, but a switch within the Aristolochiaceae to the perennial herb *Asarum* by *Luehdorfia*, to the family Zygophyllaceae by *Hypermnestra* and to the families Saxifragaceae, Fumariaceae, Crassulaceae and Dipsaceae by *Parnassius* has permitted a great expansion in alpine regions. Tectonic upheavals in central and southern Asia have encouraged extensive speciation by *Parnassius*; many species in the Asian mountains and plateaux remain very poorly known. The Zerynthiini have remained more conservative and retained Aristolochiaceae as larval foodplants. The tribe includes the genera *Sericinus*, *Allancastria*, *Zerynthia*, *Bhutanitis* and *Luehdorfia*, also found predominantly outside the tropics in southern Europe and Asia. *Bhutanitis* species such as *B. lidderdalei*, the Bhutan Glory, are highly prized and spectacular insects with a wingspan of over 10 cm and three tails to the scarlet-spotted hindwing.

The third subfamily, Papilioninae, is by far the richest (90 per cent of all swallowtails) and is divided into three tribes, the Leptocircini, Troidini and Papilionini. The Leptocircini includes seven genera which fall into two groups, *Iphiclides*, *Teinopalpus* and *Meandrusa* (sometimes called the subtribe Teinopalpiti) and *Eurytides*, *Protographium*, *Lamproptera* and *Graphium* (sometimes called the Leptocirciti) (30). The first group is primarily Palaearctic, although *Meandrusa* extends to Indonesia. The larval food plants of *Iphiclides* and *Teinopalpus* are Rosaceae and Thymeleaceae respectively, but the foodplants of *Meandrusa* are unknown. The second group has a very wide distribution, believed to be the result of a spread through Gondwanaland from a North American/Laurasian ancestor (30). *Eurytides* from the Americas is a large genus of over 50 species using Annonaceae, Lauraceae and other Magnoliales as larval foodplants. Munroe's subgenera *Protesilaus* and *Eurytides* (42) are retained here but were raised to generic level by Hancock (30). *Protographium* is a monotypic genus that feeds on *Rauwenhoffia* (Annonaceae) in Australia. The two species of *Lamproptera* fly in South East Asia and breed on Hernandiaceae and Combretaceae. Finally, the genus *Graphium*, with over 90 species, occurs in Africa, Asia and Australia. Munroe (42) adopted a number of subgenera that are retained in essence here, but adapted to comply with more recent work by Hancock (30) and others. Larval foodplants are variable and include Lauraceae for *Graphium* (*Pazala*), Aquifoliaceae for *Graphium* (*Pathysa*), Annonaceae, Malpighiaceae, Apocyanaceae, Annonaceae and Anacardiaceae for *Graphium* (*Arisbe*), and Annonaceae, Apocyanaceae, Lauraceae, Magnoliaceae, Monimiaceae, Hernandiaceae and Piperaceae for *Graphium* (*Graphium*) (30).

The second tribe in the Papilioninae, the Troidini, includes eight genera almost all species of which feed on Aristolochiaceae. *Battus* is primitive and occurs in the Americas. In South America *Euryades* is a genus of temperate grasslands while *Parides* is found in tropical forests. *Cressida* is a monotypic genus from Australia, closely related to *Euryades*. *Atrophaneura* is mainly eastern Palaearctic and Indo–Australian, but with the extraordinary relict species *Atrophaneura antenor* in Madagascar. Finally, the three birdwing genera are *Trogonoptera* from the forests of Sundaland and Palawan, and *Troides* and *Ornithoptera* from Indo–Australia.

The third tribe, Papilionini, includes over 200 species in the single genus *Papilio*, divided into a series of numbered sections by Munroe (42) and into several smaller genera by Hancock (30), Miller and Brown (39) and others referred to in those works. Here we adopt a conservative nomenclature of subgeneric divisions based on Hancock's phylogenetic studies (30) but Munroe's nomenclature (42). The larval foodplants of *Papilio* are extremely variable. The American and Caribbean subgenus *Papilio* (*Pterourus*) feeds primitively on Rutaceae, but has spread to include at least 11 other plant families, notably the Lauraceae, Magnoliaceae and Piperaceae (30). *Papilio* (*Heraclides*), also from the Americas and Caribbean, uses mainly Rutaceae and Piperaceae. *Papilio* (*Eleppone*) is a monotypic subgenus from Australia that feeds on Rutaceae. *Papilio* (*Chilasa*) species often mimic distasteful milkweed butterflies (Danainae) and they are distributed in eastern and South East Asia, mainly on Lauraceae or Magnoliaceae. *Papilio* (*Papilio*) includes only the 14 species in the *machaon* group. Some of these butterflies are very familiar to entomologists in the northern hemisphere; *machaon* itself has a vast range stretching from Great Britain to Saudi Arabia, North America and Japan. The foodplants are Umbelliferae, Compositae or Rutaceae. The final subgenus, *Papilio* (*Princeps*), is a vast assemblage of over 130 species that feed almost entirely on Rutaceae and inhabit Africa, Europe, Asia and Australia. They have evolved a stunning variety of form and pattern, from the iridescent blue of *Papilio zalmoxis* and the giant wings of *Papilio antimachus* in Africa, to the delightful hues of the Gloss or Peacock Swallowtails (*paris* group) and the realistic mimics found in the *memnon* group of Asia.

In this volume we have adopted a conservative and utilitarian approach to nomenclature simply to serve the best interests of our readership. The message of our book is conservation rather than taxonomy and our aim is to help the many politicians, legislators, scientists, conservancies and individual campaigners whose task it is to protect and preserve wild fauna and flora. Those specialist entomologists who have a deeper understanding of the inter-relationships and evolution of the swallowtails will surely not be misled by such an approach. Readers interested in the arguments for and against the alteration of long-established systems of nomenclature and classification will not be disappointed by the literature, in which conflicting views are regularly aired (19, 20, 30, 38, 39, 42).

The present patterns of diversity of the three Papilionidae subfamilies may not necessarily reflect their centres of origin. Although Shields and Dvorak (55) proposed a Neotropical or Indo–Australian origin for butterflies (including the Papilionidae), Hancock has recently argued convincingly against this (30). He concludes that in the case of the Papilionidae it is inappropriate to use diversity as an indication of centre of origin. The South East Asian fauna contains elements derived from the Palaearctic, Australian, Oriental and Afrotropical regions. No doubt it owes its richness to speciation patterns on the continental shelves and island arcs that comprise the present geography of the area. Similarly, speciation in the neotropics, which also has a fairly rich fauna, reflects the great diversity of habitats and perhaps the influence of Pleistocene forest refugia. Hancock contends that the centre of origin of the Papilionidae was Laurasia, one of the two Tertiary super-continents. According to the hypothesis, during the Cretaceous period Laurasia divided into three sections, each with a subfamily of swallowtails: Baroniinae in the western North America block, Papilioninae in the eastern North America–Europe block and Parnassiinae in the Asian block (30). Dispersal to Gondwanaland (which gave rise to South America, Africa, Australia, Madagascar, India and Antarctica) was supposedly facilitated via an island arc between North and South America. The theory is supported by the present distribution of the primitive genera from the three subfamilies, *Baronia* (Baroniinae) in Mexico (with the closely related fossil *Praepapilio* in Colorado), *Archon* and *Sericinus*/*Bhutanitis* (Parnassiinae) in western and eastern Asia respectively, *Eurytides* (*Protesilaus*), *Papilio* (*Pterourus*) and *Battus* in the Americas, with the most primitive groups in North and Central America (30). With *Baronia* and *Praepapilio* in the western North America block, this may be the most likely area for the origin of the Papilionidae. It must be admitted, however, that these ideas are still contentious and there is a need for further research. The fossil record for papilionid butterflies is poor and a few new finds could greatly alter the present interpretation of patterns.

Swallowtails and man

Swallowtails, and indeed all other butterflies, are of value to us in a number of different ways. Without doubt the most widely enjoyed aspect is the aesthetic one. With their fabulous colours and delicate flight, the beauty of swallowtails can be appreciated by all ages and races of people.

The enjoyment of swallowtails can take a number of forms, divided broadly between appreciation of the living animal and of the dead, dried and set specimen. The most natural and rewarding pleasure is to observe them in their native setting, flying across a forest glade or sipping water from the edge of a muddy pool. There is no need to know the scientific names and life history of swallowtails in order to appreciate the extra dimension of richness they can add to our environment. Most people around the world may not voice their enjoyment of butterflies, but experience in northern industrialized countries is showing that the response is more vociferous when the butterflies are suddenly no longer around.

Swallowtails, being a largely tropical group, are generally not well known to the general public in countries with a temperate climate. However, there is a growing interest in "butterfly houses", where exotic species fly in large greenhouses amongst tropical vegetation. Visitors can walk through to watch, photograph and even handle the butterflies. Some of the more common birdwings and swallowtails are often well represented in such establishments. The stock is usually imported from the native country either as adults or as pupae which emerge in captivity. However, there is generally some research going on behind the public galleries and captive breeding is possible for an increasingly wide range of species. Established and responsible butterfly houses can do much to advance the cause of butterfly conservation, in terms of both scientific research and educating the public. So long as wild-caught material is harvested in a sustainable manner there will be no threat to the species exhibited.

The appreciation and use of dead swallowtails and other butterflies is a more contentious and complex issue. It can take many forms, from the Papuan farmer who decorates his hair with a birdwing, to the Taiwan industrialists that mass produce table mats made of butterfly wings set in clear plastic. The collecting method also varies widely in approach, from the private collector seeking a specimen or two for study, to the disinterested local labourer who is paid by the specimen and employed by dealers in cities remote from the habitat of the species. The moral issues are tortuous and perhaps best guided by a single consideration. Butterflies are living beings that fly, breathe and breed; they are not lifeless objects like stamps or match-books. They should not be taken merely to satisfy pointless acquisitiveness, but should only be collected with restraint and without unnecessary wastage. Examples of the types of trade in swallowtail butterflies are given in more detail in section 5 (p. 155). As will be seen, small-scale butterfly collecting generally does no appreciable harm to the population. The conservation of butterflies is much more seriously affected by habitat destruction (see below). Nevertheless, there is an unmistakable trend away from collecting and towards monitoring and photographing butterflies (51, 62). This fashion towards appreciation of the living animal in its natural or semi-natural setting is greatly to be applauded and encouraged.

Swallowtails and science

Swallowtails have been used as a tool for biological research in a number of different fields. In the early 1940s E.B. Ford carried out novel research using chemical analysis as a taxonomic tool (26). He found that pigments could be used to give insight into the phylogenetics of swallowtail butterflies. The presence of yellow anthoxanthins was believed to be a primitive character shared with certain Pieridae, supporting the hypothesis of a common evolutionary stem. The red pigments are of two types, A and B. Type A is found in all genera except *Battus* and *Papilio*, which have type B, not known anywhere else in the Lepidoptera. Type A is widely distributed in butterflies and moths and is taken to be the primitive red pigment of the Papilionidae.

Biochemical analysis has also proved to be central to our understanding of the foodplant relationships of the Papilionidae and other phytophagous insects. Chemical similarity of larval foodplants is a widespread phenomenon that frequently cuts across botanical affinities (21). Correlations have been found between feeding habits and a number of natural products, including alkaloids, coumarins, essential oils, glucosinolates, acetylenic fatty acids and phenolic glycosides. The role of these chemicals is to act as physiological barriers to feeding by animals (27), but they have been variously overcome by butterfly larvae in the course of evolution (21). Once an animal species has evolved the capacity to metabolize or sequester the poison, any plant that uses that particular chemical deterrent becomes a potential foodplant.

Feeny and his research team have drawn together the foodplant data for the Papilionidae and studied the ways in which the butterflies detect the correct plant (25). It is now known that host-plant recognition by swallowtails is the result of a complex series of tests, including visual searching, contact chemoreception and possibly olfaction of plant defence chemicals. It has been demonstrated that the families of the order Magnoliales (subclass Magnoliidae, including the Annonaceae, Lauraceae and Magnoliaceae) are linked with the Aristolochiaceae and Rutaceae by their content of benzylisoquinoline alkaloids. Hydroxycoumarins and furanocoumarins are likewise associated with larval foodplant families, particularly those used by *Papilio*. However, the most consistent chemical link of all is provided by the essential oils, which are reported from virtually all families of swallowtail food plants (25). Essential oils are volatile organic mixtures, usually containing terpenes and esters, that have the odour or flavour of the wide variety of plant families in which they are found. The structure of the oil-producing cells in the Rutaceae are similar to those in the Aristolochiaceae, the Piperaceae and the Magnoliales families mentioned above. Larvae that normally feed on either Rutaceae or Umbelliferae have been shown to be attracted to the essential oils of both families, explaining the numerous observations of atypical egg-laying behaviour (25).

Research in this field is continually giving rise to fresh challenges and insights that are applicable to a wide range of pure and applied biological sciences. In particular, there is a growing emphasis on the search for natural plant compounds with insect-repellant qualities. The hope is that they will be as effective as the man-made chemicals now generally used for plant protection, but less environmentally hazardous. A well-known example of a natural product already widely used to combat insects is pyrethrum, extracted from the plant *Tanacetum cinerariifolium* (Compositae).

As has already been mentioned, many species of swallowtails are polymorphic in the female, most notably the African Mocker Swallowtail, *Papilio dardanus*. C.A. Clarke and P.M. Sheppard of Liverpool University chose first this species and later the South East Asian *P. memnon* for their examination of the genetic basis of mimicry, a study that continued for 30 years and produced a result of unexpected and exceptional importance. The story is a complicated one, but will be explained fully in order to demonstrate the way in which a seemingly unrelated piece of research may finally have a great impact on human well-being.

The work began with the discovery that *Papilio* species could be hand-paired quite easily, by opening the male's claspers with a fingernail and introducing the male to a female. Clarke and Sheppard then set about obtaining specimens of *P. dardanus* from many parts of Africa and specimens of *P. memnon* from the Far East. By controlled cross-breeding of the mimetic and non-mimetic forms in each species they found that the genetic coding for the mimetic form was not contained in a single gene. This is best explained with an example. *Papilio memnon* form *achates* is a mimic of *Atrophaneura coon*, while form *agenor* is non-mimetic. Breeding experiments showed that mimetic *achates* is dominant to *agenor* and the various features of the mimic all appear together. Such features include yellow body, presence of tails and a small white 'window' on the hindwing. Very rarely, however, the features were found to be split up and although, for example, tails were present and the wing window small, the body was black. This suggested that more than one gene locus was involved and that random crossing-over of the chromosomes may occasionally split up the several closely linked loci of what is now known to be a mimicry 'supergene'. Furthermore, Clarke and Sheppard found that breeding of hybrids between races produced a great increase in phenotypic variability. This indicated that each race had interacting and co-adapted gene complexes which insured a low degree of variability in the phenotypic expression of the major genes. These interactions were markedly disturbed in the race hybrids (6).

As a result of studies on swallowtail races in which two forms are common but intermediates caused by genetic rearrangements are also occasionally found, the attention of the scientists was directed to the human Rhesus blood group system, where crossovers probably account for the rare genotypes and where there is a marked interaction between the Rhesus (Rh) and ABO systems (5). Here, as in the mimetic butterflies, the various Rhesus phenotypes are controlled by a series of closely linked genes in a supergene. The most important interaction is that ABO incompatibility between a mother and her baby often confers protection against Rhesus immunization of the mother (5).

Rhesus immunization is a problem caused by the movement of red blood cells from an Rh-positive baby into the blood system of its Rh-negative mother. This generally happens at or near to the time of delivery. The mother subsequently develops an immune response to Rh-positive blood cells and although the first child is safe, having already been born, subsequent children may suffer severe rejection problems in the womb. The baby may be killed before birth, or die shortly afterwards. Complete blood transfusion immediately after birth of the so-called "yellow baby" may save the child's life.

Clearly the protection against Rh-immunization conferred by ABO incompatibility of mother and child is greatly to the advantage of second, third and later children. The reason for this protection is not clear but Clarke suggested that any Rhesus-positive foetal cells which have crossed into the maternal circulation are destroyed by the mother's naturally occurring anti-A or anti-B before they have had time to stimulate the production of anti-Rh antibodies (5). This protection against Rhesus immunization is highly desirable, but it only occurs in the 20 per cent of cases in which the foetus is ABO incompatible with the mother. After a great deal of research it proved to be possible to imitate the protection by giving suitable antibodies to Rh-negative mothers in the 80 per cent of cases where mother and baby are compatible on the ABO systems. As a result, nearly all cases of maternal Rh immunization can now be prevented (5). The success of the research programme is due, in large part, to the usefulness of certain swallowtail butterflies for scientific study.

Threats to swallowtails

Four main factors threaten swallowtails, as well as other terrestrial forms of wildlife: habitat destruction, pollution (which might be considered as a special form of habitat destruction), the introduction of exotic species and commercial exploitation. In general these threats are either the direct result of increasing human population pressure or else are enhanced by it (7). It is now internationally recognized that conservation efforts can only reach long-term fruition in partnership with policies for population planning and control (32). A full consideration of the global problem of human populations outstripping their resources, and of the political, economic and social causes, are beyond the scope of this book and other publications should be consulted for more detailed studies (3, 24, 33).

Each of the four types of threat will be briefly reviewed, with examples drawn from the swallowtails.

Habitat alteration and destruction As demonstrated in the companion IUCN Red Data Books for plants, mammals, amphibians, reptiles and invertebrates, the primary threat to wild species throughout the world is the destruction and alteration of the biotopes and habitats in which they live (29, 37, 58, 61). All species depend for their survival upon the integrated network of physical and biological factors that make up their environment. Most are able to withstand a small degree of disturbance and manipulation, but human activities have steadily modified wilderness areas into man-made landscapes of settlement, agriculture and industry that are altered to an extent that precludes coexistence with many wild creatures.

Habitat alteration has a universal impact on swallowtails and other wildlife, but the effects are more noticeable in biomes and ecosystems that are restricted in area. Hence the extensive destruction and alteration of African savannas has certainly caused great reductions in distribution patterns, but no swallowtail species are regarded as threatened there. Even in the more restricted rain forests of the Zaire basin only *Papilio antimachus* is recognized as threatened (category Rare), and this species is still widespread. A similar situation prevails in the massive forest block of the Amazon basin. Conversely, islands are particularly prone to extinctions because they tend to have rich endemic faunas in very restricted areas. Mauritius for example, has suffered the extinction of many species of animals and plants, largely because of deforestation followed by extensive cultivation of sugar cane. The endemic swallowtail *Papilio manlius* has been fortunate in successfully adapting to *Citrus* for its foodplant, but *Libythea cinyras* (Libytheinae) is now believed to be extinct there and *Cyclyrius mandersi* (Lycaenidae) may be seriously threatened (14). Similarly, *Papilio phorbanta* has adapted to *Citrus* on Réunion, but the subspecies *nana* from the Seychelles has not been seen since 1890. A number of endemic taxa in Taiwan are under threat, including *Papilio maraho* (Vulnerable) and *Troides aeacus kaguya*. In Jamaica *Eurytides marcellinus* is now extremely rare and populations of the Homerus Swallowtail (*Papilio homerus*) have been split up and endangered as a result of coffee production and timber extraction. *Graphium epaminondas* may be under similar pressures in the Andaman Islands, as may *Graphium levassori* on Grande Comore. Many other examples of island-living swallowtails at risk will be found in section 6, including amongst others the birdwings of South East Asia, *Papilio hospiton* from Corsica and Sardinia and *Papilio neumoegeni* from Sumba.

Montane species suffer many of the same problems as those on islands. In the tropics there are many swallowtails restricted to a habitat that may have been more extensive during the cooler climate of the last Ice Age, but is now found only on mountains. In Sabah and Sarawak *Papilio acheron* and particularly *Graphium procles* have restricted ranges in montane regions. Mt Kinabalu, the only place where *G. procles* flies, is a protected area, but economic considerations have taken high priority there and the region has suffered incursions by mining companies and recreational developments. In Indonesia the little known species *Graphium stresemanni* is possibly confined to the Manusela range in Seram and *Atrophaneura luchti* has a very restricted montane range in eastern Java. *Graphium sandawanum* is found only on Mt Apo in the Philippine island of Mindanao, and both *Papilio chikae* and *P. benguetanus* are relict species in the cordilleras of Luzon.

In Africa the effect of the last Ice Age on vegetation patterns is still contentious (63), but whatever the historical causes, there is no question that the string of relict forests running from Mt Kulal in northern Kenya to Mt Mulanje in Malawi now all have their own endemic subspecies of butterflies, if not species, and all are under threat from deforestation for agriculture and plantation forestry (8). Of particular note is the now highly Endangered Taita Blue-banded Swallowtail (*Papilio desmondi teita*) from the Taita Hills in southern Kenya. Other Kenyan montane papilios of interest include *Papilio dardanus flavicornis* from Mt Kulal and *P. d. ochracea* from Mt Marsabit and Mt Nyiru. In Tanzania *Papilio sjoestedti* flies only in montane forests in the regions around Mt Kilimanjaro, Mt Meru and the Ngorongoro range, much of which is fortunately protected inside national parks. Many more butterflies from East Africa's montane forests could be listed here, particularly from the nymphalid genus *Charaxes*.

The forms of habitat alteration of most importance to swallowtail butterflies are 1) deforestation, 2) agricultural conversion and intensification, 3) alteration of pastures and 4) industrialization and urbanization.

Deforestation As a result of the ecology and distribution of swallowtail butterflies, forestry practices in temperate latitudes have little impact on their populations. An exception is the Japanese *Luehdorfia japonica*, which is discussed below.

In 1981, following an extensive survey of 76 tropical countries, the Food and

Agriculture Organization (FAO) and the United Nations Environment Programme (UNEP) jointly published an assessment of the global tropical forest resources (24, 28). These documents have supplied an extremely valuable framework within which to discuss the conflicts between conservation and exploitation of forest resources, conflicts that must be resolved if sustainable economic growth is to be achieved (33). When combined with data on protected areas and the distribution patterns of animals, plants and, in the present case, swallowtail butterflies, it becomes possible not only to assess current conservation problems, but also to predict which regions and species are likely to be of particular concern in the future.

Tropical forests in the broadest sense are estimated to cover 2970 million hectares, i.e. 20 per cent of the land surface of the globe (28). The main forest formations are closed broadleaved forest (including rain and monsoon forests, 40 per cent), open forest types (including the cerrados of South America and the wooded savannas of Africa, 25 per cent), coniferous forests (1 per cent) and shrubland (21 per cent) (28). These ecosystems include by far the most important swallowtail habitats, as is clearly demonstrated by the distribution patterns in Figs 2.1–2.4. The majority of swallowtail species use larval foodplants that are vines, shrubs or trees, although there are important exceptions in the Parnassiinae (see 1.3 below). Exploitation of tropical forests by man will inevitably disturb swallowtail populations, but the extent of the impact depends on the level of disturbance, the extent of the biotope and its ability to recover, and the willingness of people to allow it to do so.

Tropical forests may be altered in two ways: 1) through complete deforestation followed by non-forestry usage such as agriculture, or 2) by selective felling for industrial use (mainly rain and monsoon forests) or fuel production (mainly drier forest formations). However, it is usual for logged over forests eventually to become completely deforested so that for all practical purposes deforestation becomes the major factor altering tropical forests (28).

The most important single cause of deforestation is shifting cultivation, accounting for 70 per cent, 50 per cent and 35 per cent of the total areas that have been deforested in Africa, Asia and the Americas respectively (24, 28). Thus the division of threats between deforestation and agricultural conversion in this discussion is to some extent artificial. Some of the conservation problems associated with shifting cultivation will be found in section 1.2 below, but might equally well have been discussed here under deforestation. Other significant causes of deforestation are clearance to create grazing lands (particularly in South and Central America) and settlement of agriculturalists along new logging roads with subsequent expansion into the forest (particularly in Africa and Asia).

Logging can damage forests to varying extents. In South East Asia for example, the 'volume actually commercialized' (VAC), an index of production from the point of view of commercial exploitation, ranges from 15 cu. m/ha in Burma to 90 cu. m/ha in Sabah, i.e. from less than 10 per cent to more than 30 per cent of the corresponding standing volume of timber (24). This may mean that whereas Burmese swallowtails like the Kaiser-I-Hind (*Teinopalpus imperialis*) could survive logging, native birdwing species in Sabah would be unable to withstand the rapid degradation and complete loss of habitat which can result from such intensive logging rates. Another variable is the extent to which logging roads are subsequently used as access roads to new farms by land-hungry people. This problem is most apparent in Africa and Asia where 70 per cent of the deforestation occurs in areas of closed forests that have been previously logged. In the Americas the comparable figure is only 44 per cent because most logs are transported along rivers or existing roads (28).

In the drier regions of the tropics, particularly of Africa and Asia, the accelerating demand for fuelwood and charcoal is causing severe deforestation of open woodland. The technology is available to meet these demands through plantation forestry (36) and Brazil produces 29 per cent of its fuelwood needs from *Eucalyptus* plantations (28). In Africa and Asia however, the low levels of investment in plantations to meet these basic demands of the rural poor must eventually have dire consequences in socio-economic and resource conservation terms.

The global rate of clearance of tropical closed broadleaved forest is now believed to be lower than was previously feared. It has been estimated that by the end of the century 88 per cent of the present world cover of closed tropical broadleaved forests will still remain intact (24, 28). However, this is no cause for complacency, it simply means that there is a little more time in which to implement sound conservation and management strategies. A large proportion of those remaining forests will be disturbed to some degree.

There are considerable differences between the rates at which countries are utilizing and altering their closed forest resources. Whereas Zaire and Brazil will still have extensive forest resources in the year 2000, the forests of Peninsular Malaysia, Ivory Coast, Nigeria, Sri Lanka and Java will be virtually gone. If present trends continue, nine countries will have used practically all their closed broadleaved forests within 30 years, and a further 13 will join them within 55 years (28). All 22 of these countries between them only account for 11 per cent of the world's tropical closed forest, but some, such as Jamaica and Sri Lanka have important swallowtail faunas.

What impact is tropical deforestation having on swallowtail butterflies? Considering in the first place the loss of tropical closed forests, the Asian sector is of greatest concern. About one quarter of the world's closed tropical forest area is found in Asia, 61 per cent of this being in the islands of South East Asia including New Guinea. As will be demonstrated in section 4 (p. 137), this is the world's richest area for swallowtail butterflies. Indonesia has 121 species of swallowtails, over 20 per cent of the world total. It has also emerged as the world's most important producer and exporter of tropical hardwoods (24, vol. 3). Rain forest species from the South East Asian islands that have been given threatened status include *Graphium idaeoides*, *G. megaera*, *G. sandawanum*, *Papilio osmana* and *P. carolinensis* in the Philippines, *Graphium stresemanni*, *Ornithoptera aesacus* and *O. croesus* in the Moluccas, *O. rothschildi*, *O. chimaera*, *O. paradisea* and particularly the Endangered Queen Alexandra's Birdwing, *O. alexandrae*, in New Guinea and *Graphium meeki*, *G. mendana* and *Papilio toboroi* in Bougainville and the Solomon Islands. In continental Asia the threats from deforestation are less serious because most swallowtails are more widespread in their distribution. However, deforestation throughout its range, but particularly in India and Nepal, is placing the Kaiser-I-Hind, *Teinopalpus imperialis*, at risk. Numerous species are locally at risk in Peninsular Malaysia, including *Troides helena*, *T. aeacus*, *T. amphrysus*, *T. cuneifer*, *Papilio agestor*, *P. mahadeva*, *P. palinurus*, *Graphium empedovana*, *Meandrusa payeni*, *Lamproptera meges* and *L. curius* (2).

African rain forests are still poorly studied and the two Insufficiently Known species, *Graphium weberi* and *G. aurivilliusi* from the main Cameroon–Congolese block, require further research. There are few endemics in the much-depleted forests of the western block, but *Graphium maesseni* is only known from the Ghanaian type series. The Giant African Swallowtail, *Papilio antimachus*, is now very scarce in this western part of its range but although it is confined to primary forests, its future may be assured in the forests of the Zaire basin.

In the South American rain forests the swallowtail fauna also remains poorly known. There are few seriously threatened species but the narrow ranges of *Parides hahneli, P. pizarro, P. steinbachi, P. coelus, P. klagesi, P. burchellanus* and *Papilio garleppi* are cause for concern. Many other poorly known rain forest species are listed in Appendix B.

Turning to the loss of tropical open woodlands, a number of threats to swallowtail butterflies can also be identified. Of the Asian fauna, *Papilio jordani* is a Rare species from possibly open formations of semi-deciduous forest in northern Sulawesi and *Troides dohertyi* is Vulnerable in the nearby Talaud and Sangihe Islands. Two Vulnerable species, *Papilio neumoegeni* from Sumba and *Atrophaneura schadenbergi* from Luzon, probably fly mainly in open woodlands. *Eurytides* is a genus of kite swallowtails that fly in the open woodlands of the Neotropics. Both the Yellow Kite, *Eurytides iphitas*, and Harris' Mimic Swallowtail, *E. lysithous harrisianus*, are in serious decline in Brazil, and the Jamaican Kite, *Eurytides marcellinus* is now a great rarity in Jamaica. As already noted, there are few threatened species in the extensive savanna formations on the African mainland. A number of threatened island species have already been mentioned; two others that are confined to open woodland, *Graphium levassori* from the Comores and *Graphium pelopidas* (possibly a subspecies of *G. leonidas*) from Pemba Island and Zanzibar, are at risk.

Agricultural conversion and intensification The spread of subsistence farming has been cited as the biggest single cause of forest clearance, soil degradation and loss of wildlife in the world today (18). Landless labourers, sharecroppers and marginal farmers now form a majority in rural areas of Latin America and Asia, and in Africa the recent and rapid rise in human population levels is imposing severe pressure on land unsuitable for settlement (18). Approximately 3.8 million hectares of open woodland are being cleared every year, mainly as agriculture extends into marginal lands (24). Much of Africa is too infertile and dry for crops or even grazing, but farmers are moving further into semi-arid areas in their search for land and food. Pastoralists are being squeezed into desert margins, causing overstocking and overgrazing which in turn lead to degradation of soils and an increasing proportion of plants unpalatable to domestic animals. Inevitably the wildlife suffers too and land in national parks throughout Africa, and to a lesser extent other continents, is under pressure for grazing and development. In Kenya for example, where the population growth rate of about 4 per cent is one of the highest in the world, sections of the Masai Mara, Tsavo and Lake Nakuru reserves have already been excised.

Many examples of threats to swallowtails from deforestation followed by agriculture have already been cited. A few more that specifically concern agricultural pressure may also be mentioned. In the central African countries of Rwanda and Burundi agricultural land is so overcrowded that pressures on the few relict forests are intense. Encroachment into these forests is depleting forest wildlife such as the Vulnerable Cream-banded Swallowtail, *Papilio leucotaenia*. *Papilio morondavana*, *P. grosesmithi* and *P. mangoura*, all endemic to Madagascar, are threatened by uncontrolled burning, the spread of shifting agriculture and the failure to sustain adequate fallow periods. In Sri Lanka the few remaining forests are under severe pressure from agriculturalists and logging companies. The Vulnerable clubtail *Atrophaneura jophon* has its last stronghold in the island's Sinharaja Forest Reserve, until recently severely encroached by agriculturalists and fuel-gatherers but now under full legal protection.

In temperate latitudes where agricultural conversion is already extensive, emphasis in recent decades has been on intensification of farming practices. Many traditional forms of land use are compatible with butterflies, e.g. coppicing and moderate grazing, but the demand for higher food production and living standards has resulted in much more intensive and exhaustive exploitation of the fertile areas. In Great Britain, the U.S.A. and many other parts of the northern hemisphere, this has caused severe reductions in butterfly populations (41, 52, 61). Factors detrimental to virtually all forms of wildlife are involved in these processes of intensification, including use of insecticides and herbicides, destruction of hedgerows, drainage, short fallow periods and heavy use of fertilizers. With the very limited swallowtail fauna in the north temperate regions, the family has suffered less than many other forms of wildlife. Nevertheless, drainage has severely reduced the range of the British Swallowtail (*Papilio machaon britannicus*) (16), and in Japan *Luehdorfia japonica* is becoming more scarce as its previously lightly managed open woodland haunts are destroyed in favour of agriculture or intensive forestry.

In tropical regions problems associated with intensification of agriculture on the main continental blocks are of much less significance to conservation than the extension of agriculture into virgin lands. However, in eastern Asia human population levels are such that intensive agriculture is already very widespread and incompatible with wildlife, notably in the subtropical parts of China, parts of Indochina and Thailand, Peninsular Malaysia and Java. In Brazil drainage and development of coastal swamps and thickets is threatening the Vulnerable species *Parides ascanius*. Its mimic *Eurytides lysithous harrisianus* is now seriously Endangered and has been proposed for Brazil's protected list, an action which will only be successful if suitable habitat for the mimic and its model can be set aside and maintained.

Alteration of pastures Permanent grasslands are important habitat for many butterflies, particularly in the families Lycaenidae and Hesperiidae, but they are of less significance to the Papilionidae. The swallowtail foodplants are generally shrubs, vines or trees and most species fly in woodlands or forests. An important exception is the north temperate subfamily Parnassiinae, many of which use herbaceous species that grow in open situations. *Luehdorfia* feeds on the aristolochiaceous herb *Asarum*, which grows in open woodland, and the foodplant families used by *Parnassius* are all herbs of alpine meadows. In the mountains of the Hindu Kush, the Himalaya and western China there are very many species of *Parnassius* that are poorly known and may be at risk from pasture degradation (see appropriate countries in section 3 and the discussion of Chinese *Parnassius* on p. 151). Some are in such inaccessible regions that their future may be assured, but others, such as *Parnassius autocrator*, are certainly coming under new pressures. This particular species lives in the Hindu Kush of Afghanistan and the Pamir mountains of Tadzhikskaya S.S.R. In certain parts of the southern Himalaya, particularly in Nepal, there are reports of heavy tourist foot traffic resulting in pollution, trampling, erosion and degradation of vegetation through firewood collecting. Such impacts are not in the best interests of either the wildlife or the tourist industry.

Also Endangered is *Papilio hospiton* from Corsica and Sardinia, primarily as a result of alteration and destruction of pastures. The umbelliferous foodplants are poisonous to sheep and are destroyed by fires started by local farmers. Pressures from recreational developments are reducing the butterfly's range, particularly on Corsica.

Industrialization and urbanization Since the Papilionidae is a mainly tropical group, there are relatively few examples of conservation problems resulting from these factors. However, the range of *Papilio himeros* in south–eastern Brazil is declining as a result of development in Rio de Janeiro and coastal regions in general.

Schaus' Swallowtail (*Papilio aristodemus ponceanus*) from the Florida Keys is Endangered because of demand for building land and the consequent destruction of the hardwood hammocks in which the butterflies live.

Pollution The effects of atmospheric pollution on butterflies are rather poorly understood. An assessment of Europe's threatened butterflies has reported that a number of widespread species have suffered severe declines in Fennoscandia, northern Germany, Poland, Czechoslovakia, Netherlands, Switzerland, Austria, Hungary and the Italian Alps (31). All these areas are to the north and east of the principal industrial zones in western Europe from which atmospheric pollution is carried on the prevailing south–westerly winds. The decline of *Parnassius apollo* in Norway has, for example, been attributed to 'acid rain' - a weakly acidic solution of sulphur and nitrogen oxides in rainwater. It seems likely that acid rain has deleterious effects on a wide variety of butterflies and possibly also their foodplants, as well as other invertebrates and wildlife. Ozone has been implicated as a particularly toxic component of acid rain but further research is needed to demonstrate its effects. In Europe there are widespread demands to instigate and increase international efforts to improve anti-pollution measures.

Pesticides are not generally hazardous to butterflies when applied in the recommended manner directly to the target species. However, there is a significant risk in the effects of wind drift, particularly with the growing trend towards aerial application of pesticide to rice, wheat and other extensive monocultures. When spraying is carried out in a cross-wind pesticide may drift into adjacent natural habitats. Although no serious conservation problems have been reported directly from this cause, there is still a need for vigilance in the responsible use of pesticides (40).

While in the industrialized countries of the northern hemisphere pollution control is showing valuable results, there is little evidence of similar efforts in the developing world. Here, serious pollution may result from forest and savanna fires (10, 54, 64) as well as industrial effluent and incorrect or excessive use of pesticides. Serious threats to all forms of wildlife may be expected from such sources in the future.

Exotic introductions Introduction of exotic animals and plants inevitably upsets the balance of natural communities to some extent. There are no documented cases of severe effects on swallowtail butterflies but the spread of *Papilio demodocus* (Orange Dog) from the African mainland to Madagascar, Mauritius and Réunion is a matter for concern. Apart from being a minor pest of *Citrus*, *P. demodocus* is an aggressive species that is capable of ousting less vigorous native species. Research is needed concerning the impact *P. demodocus* has had on the Indeterminate Mauritius endemic *P. manlius* during the three-quarters of a century since *demodocus* was introduced. No effects on the closely related Madagascar endemics *Papilio morondavana* (Vulnerable) and *P. grosesmithi* (Rare) have been noted, nor on the Vulnerable Réunion endemic *Papilio phorbanta*.

Commercial exploitation The level of commercial exploitation of swallowtails and other butterflies has never been estimated globally. With butterfly trade in Taiwan worth about US $20–30 million per year, a speculative figure in the order of $100 million per year worldwide may not be unrealistic. As has been noted in the above discussion of swallowtails and man, exploitation can be anything from the capture, sale or barter of a single specimen for the sake of interest or decoration, to large-scale commercial ventures employing hundreds of local collectors. Invertebrates can frequently, but not invariably, withstand a considerable level of harvest because of their high reproductive capacities. However, heavy exploitation can have serious effects under three circumstances: 1) if the population is already critically depleted by other factors like habitat destruction, e.g. *Papilio hospiton* may be seriously at risk in Corsica and Sardinia, 2) if the population is small and has a high value per individual, e.g. the highly prized species *Papilio chikae* from Luzon, 3) if the species has a low reproductive rate and low juvenile recruitment, e.g. the *Ornithoptera* birdwings of Indonesia, New Guinea and the Solomon Islands.

The need for self-restraint in the collecting of butterflies and other insects was first formally recognized by the Royal Entomological Society of London, which set up a Protection Committee in 1923. In 1968, the Society replaced this with the Joint Committee for the Conservation of British Insects, which includes representatives from all parts of Britain and all the major entomological societies. In 1972 the Committee published "A Code for Insect Collecting" which has been reproduced and adopted not only in Britain, but also in many parts of Europe and throughout the world. The Xerces Society, a U.S. organisation devoted to conservation of insects, responded with its own policies in 1975, and the Lepidopterists' Society of America published guidelines in 1982. The only code adopted by commercial entomologists is the Entomological Suppliers Association of Great Britain "Code of Conservation Responsibility", published in 1974. In this document the trade in a number of British and exotic species was restricted to specimens already in circulation (31).

For a full consideration of the value of rational, sustainable exploitation of swallowtails and of the dangers of unsustainable levels of commercialization refer to section 5, Trade in swallowtail butterflies (p. 154).

Conservation of swallowtails

Documentation and education The documentation of threats to wildlife is the first step in any effective conservation programme. This book is an attempt to fulfil that requirement for the swallowtails. It is the first worldwide assessment of the conservation needs of any invertebrate group and the first Red Data Book (RDB) to be based upon a published consideration of every species in the taxon under review, in this case the family Papilionidae.

A number of national RDBs and official or unofficial lists have documented the threats to local Papilionidae. The majority of these are for European countries and consider only the very limited swallowtail fauna of that region (61, p. 325). Such lists are referenced under individual species in section 3 (p. 33). Although the Papilionidae is one of the best-known of all insect groups, documentation of the conservation and biology of rare swallowtails is relatively sparse because of their mainly tropical distribution. Much of the work that has been done is referred to in the reviews of threatened species in section 6 (p. 181).

The IUCN Invertebrate Red Data Book (61) included seven species of swallowtails and a number of other butterflies, but drew attention to the fact that destruction of habitat is putting whole invertebrate communities at risk. The category Threatened Community, coined in the Invertebrate RDB, has not been utilized here because it was considered inappropriate and artificial to describe communities based on a single family of butterflies. Nevertheless, the principal remains valid since swallowtails comprise a small, but significant and often highly visible, part of innumerable Threatened Communities worldwide. As described above, the most severe threat is destruction of entire biotopes and habitats, which destroys plants and animals indiscriminately.

The species list in section 3 shows that many swallowtails still remain very poorly known. This book is only a preliminary attempt to document what we do know and to draw attention to important areas for future research. Hopefully it may act as a catalyst for further studies on the biology and conservation of swallowtails.

Protected areas The protection of swallowtail butterflies in wild and natural areas designated as national parks and reserves is a vital priority that is emphasized throughout the species reviews in section 6 of this book. Two main needs have become evident.

Firstly, very few existing protected areas have been surveyed for any invertebrate groups. The swallowtails and other butterflies provide an opportunity for extending the traditional species lists of birds and mammals to include a spectacular and popular insect group that is of interest to an increasing number of tourists and visitors to protected areas. Once the local species are known, fine opportunities exist for tropical countries to emulate the success of the British butterfly houses in more natural surroundings. Habitat enrichment to attract butterflies after the fashion of the Papua New Guinea ranching programme (see below) could do much to conserve swallowtails at the same time as interpreting the value of swallowtails to the general public in a pleasantly assimilable fashion.

Secondly, there are certain countries that have important swallowtail faunas but very limited allowances for protected areas. In section 4 of this book, Analysis of critical faunas, the distribution of the family Papilionidae is analyzed in detail, demonstrating that just five countries, Indonesia, the Philippines, China, Brazil and Madagascar, between them contain over half of the world's species of swallowtails. A further five countries, India, Mexico, Taiwan, Malaysia and Papua New Guinea, brings the total to more than two thirds (p. 141). All of these countries, and many others, could use their swallowtail faunas as an extra yardstick in judging the value of existing or proposed protected areas. An analysis of the Indonesian fauna, given in section 4, demonstrates how this could be done.

Certain of the world's tropical forest areas are known to be under particularly severe threat of total deforestation (44), many of them with valuable endemic swallowtails already listed above. An exhaustive account cannot be given here, but a few illustrative examples may be cited. Because of economic and population pressures most of the Philippines' lowland forests may be gone by the end of the century. The national parks system is said to be undergoing review, but there is evidence of wholesale abuse of reserve boundaries by logging companies and agriculturalists (e.g. see review of *Graphium sandawanum*). Similarly, the population profile and extent of logging in Indonesia indicate that the loss of huge areas of forest cover is also now inevitable. In recognition of the need for conservation planning as part of their development programme, the Indonesian government has prepared a thorough conservation plan (59). Malaysia is also drafting a state by state conservation strategy that should rationalise the parks system throughout that country, but there is a need for haste as the rate of logging in eastern Malaysia is growing rapidly, threatening to cause deforestation on the scale seen in Peninsular Malaysia before long (24).

Other priority areas for the protection of swallowtails in reserves include western China, the southern foothills of the Himalayas, the western forests of India and the forests of Sri Lanka. In the Caribbean, Jamaica's system of reserves is in need of revision and there is opportunity for more effective measures in the Dominican Republic and possibly Cuba. Brazil's rain forests are still extensive despite increasing settlement in Pará, Mato Grosso and Rondonia, but the Atlantic seaboard forests are poorly protected and include many highly endangered species. In Africa there is increasing attention being paid to the protection of the relict montane forests, and in the unique biotopes of Madagascar more has been done to strengthen the national park system in recent years.

The Tropical Rain Forest and Primates Campaign, launched by the World Wildlife Fund in 1982, is currently promoting conservation measures in 14 countries. Many of the projects focus on the need to protect forested areas not only as a genetic resource for the future, but also as a valuable tool in watershed and soil protection (50). World Wildlife Fund's Plants Campaign, launched in 1984, emphasizes the value of protected forests as a source of plants useful to mankind as crops and medicines.

Where natural forests are already severely depleted, the value of plantation forestry as a conservation tool should not be underestimated. Plantations can provide timber for fuel, decoration and building, thus alleviating the pressures on the natural woodlands and forests that remain. They can also serve as buffer zones around natural forest refuges, as has been proposed for the Nyungwe forest in Rwanda (see review of *Papilio leucotaenia*). However, the aims of resource conservation are largely defeated when natural forest areas are cleared to make way for plantations. In practically every part of the tropics there are adequate areas of already cleared and partially degraded land that could be focal points for afforestation and regeneration. International co-operation is needed in a global programme of conservation-orientated investment in plantation forestry.

Legislation and international conventions International agreement on wildlife trade control is contained in the Convention on International Trade in Endangered Species of Wild Fauna and Flora (CITES), which controls and monitors import and export of listed species. A total of 87 countries have so far become party to CITES, and the European Community has adopted a Regulation (3626/82) to enforce the Convention uniformly throughout the Community. Appendix I, which contains no insects, is a list of species in which trade is subject to strict regulation and commercial trade is virtually prohibited. Appendix II lists species in which trade is regulated for the purpose of monitoring. The butterflies listed on Appendix II are the birdwings (*Ornithoptera*, *Troides* and *Trogonoptera*) and the Apollo butterfly (*Parnassius apollo*), all in the Papilionidae. These species may be commercially traded, but an export permit from the country of export is required before specimens may be removed from that country or imported into another state which is party to CITES. There are inconsistencies in the present listings of swallowtails on CITES and it is perhaps time for a reassessment of their past effectiveness and future value. A case could be made for delisting certain birdwings supplied by the ranching trade in Papua New Guinea, while possibly raising Queen Alexandra's Birdwing to Appendix I (48). Many other threatened butterflies are traded quite heavily and ideally could be monitored. However, it must be recognized that there are serious problems in the implementation of controls on insects that are difficult to identify (compared with many vertebrates) and very easily transportable.

The European Community Regulation 3626/82 implementing CITES includes an Annex of species listed on Appendix II of CITES that the Community treats as though they were on Appendix I, i.e. preventing virtually all trade in those species. The EC has unexpectedly added the CITES Appendix II Papilionidae to this list, a move that was carried out without consultation with the CITES Secretariat or the IUCN/SSC Butterfly Specialist Group. The Regulation will severely jeopardize the Papua New Guinea birdwing ranching programme and may cause the demise of proposed ranching programmes in Indonesia and the Solomon Islands, all of which

27

rely heavily on European markets. As will be stated on a number of occasions throughout this book, butterfly ranching is seen as an important tool for conservation. International efforts are being made to have the Papilionidae removed from Annex C Part I in order to bring the EC Regulation more into line with the requirements of CITES..

National legislation to protect local swallowtails is now quite commonplace. A full consideration of these laws cannot be given here, but protected species are noted in section 3; section 5 includes extensive notes on legislation. General restrictions on trade in Lepidoptera apply in both East and West Germany (GDR and FRG), Kenya, Madagascar, Mexico and Turkey. One or more species of swallowtail are protected or banned in trade under national legislation in Austria, Czechoslovakia, Finland, France, Germany (GDR), United Kingdom, Greece, Hungary, Luxembourg, Netherlands, Poland, Switzerland, U.S.S.R., Brazil, India, Indonesia, Malaysia, Papua New Guinea and the U.S.A.

In recognition of the importance of conserving the habitats of threatened animals and plants, all northern European countries give greater emphasis to protected areas than to legislation concerning individual species. No amount of protective legislation will succeed if a species' habitat is permitted to be destroyed. Nevertheless, legislation and international conventions do have the added advantage of drawing public attention to the plight of particularly rare or threatened species. The U.S. Endangered Species Act is probably one of the most effective forms of wildlife protection in the world, but is not without its problems. Taxa listed are required to be thoroughly studied and a recovery plan drawn up. If so recommended, the Act allows for designation of complete protection for habitats critical to the survival of threatened taxa, as well as for protection of the taxa themselves. An example of the working of the Act is described in the review of Schaus' Swallowtail (p. 301), a case-history which also demonstrates that problems still exist in the practical application of the Act.

Management and research Research on the management of threatened insects began in Great Britain, where it is now the responsibility of the Institute of Terrestrial Ecology. The centres at Furzebrook and Monks Wood have been particularly active in management and conservation of rare insects, one of which, the British Swallowtail (*Papilio machaon britannicus*), has been the subject of long-term research and is now one of the world's best-known papilionid taxa.

The Swallowtail is a very widespread Holarctic species that in Britain has become specialized on a single foodplant found only in the fens (marshes) of East Anglia, Milk Parsley (*Peucedanum palustre*). At one time the Swallowtail occurred throughout the East Anglian fens and possibly in marshes along the River Thames and River Lea (16), but extensive drainage in the early 19th century destroyed most suitable habitat. The butterfly survived at Wicken Fen until the early 1950s when that population became extinct, leaving the species confined to marshes around the Norfolk Broads, notably Hickling Broad (16). In 1975, 228 artificially reared adults were released at Wicken Fen in an attempt at re-introduction and it was estimated that over 2000 individuals pupated that year, but by 1980 the population was once again extinct (15). The failure was attributed to a gradual lowering of water levels in Wicken Fen, which had a deleterious impact on the foodplants. The Swallowtail is one of about 15 target species in a new campaign of research and conservation of British butterflies currently being run by World Wildlife Fund (U.K.).

In the U.S.A. Schaus' Swallowtail (*Papilio aristodemus ponceanus*) is now confined to the Florida Keys. It is listed as Threatened but has been proposed for re-listing as Endangered under the Endangered Species Act. Schaus' Swallowtail has been the subject of intensive biological research but despite these efforts the species is still declining, mainly as a result of habitat destruction. For a full account of research on this swallowtail refer to section 6, p. 301.

These attempts at swallowtail management have perhaps not met with the success that was hoped for, but important lessons have been learnt. It has been recognized that recovery programmes are expensive on resources and may be very risky. Long-term planning and adequate provision for protected areas are undoubtedly preferable. Nevertheless, as the threats to swallowtails and other butterflies become more intense, there will be a growing need for careful management studies, particularly in tropical regions. Research on the birdwing butterflies has demonstrated that an intimate knowledge of breeding biology and general ecology can pay dividends in terms of both conservation and rational exploitation (45, 48). Further carefully directed research could be of great benefit to the conservation of the family as a whole.

In 1984 the Lepidoptera Specialist Group of the Species Survival Commission of IUCN was divided into a Butterfly Specialist Group and a Moth Specialist Group. The Butterfly Group inherited certain long-standing priorities such as the conservation of the Monarch Butterfly over-wintering grounds in Mexico and the protection of Queen Alexandra's Birdwing in Papua New Guinea (61), but now has the opportunity to re-examine butterfly conservation problems, particularly those of the swallowtails. With their worldwide connections the Specialist Groups are of great value in centralizing conservation data and in advising on priorities for conservation attention. Attracting funds for conservation projects is more difficult for invertebrates than vertebrates, but the recommendations of the Group should be recognized as being of international priority and importance.

Ranching and farming of swallowtails The term 'farming' in this context means that young are reared in captivity from parents that are also held in captivity. 'Ranched' swallowtails are captured as young stages of wild parentage and reared to the adult stage in captivity. Farming of swallowtails on a large scale is relatively rare but perfectly possible, particularly since the technique of hand-mating has been perfected for *Papilio*. As noted above, the hybrids used in the study of the genetics of mimicry were farmed, as were the Swallowtails that were released at Wicken Fen. Farming has also been used as a conservation measure for the Apollo, *Parnassius apollo*. Normally this species breeds one generation per year, over-wintering in the egg stage. By artificially rearing two generations per year, material for recolonization of depleted areas can be rapidly accumulated (46).

The finest example of successful butterfly ranching is without question the development of the birdwing industry in Papua New Guinea. The programme is described in the trade review in section 5 so only a brief summary is needed here. In Papua New Guinea the Insect Farming and Trading Agency (I.F.T.A.) of the Department of Primary Industry sells high quality specimens of birdwings ranched locally, returning three quarters of all profits to the ranchers (45). The I.F.T.A. demonstrates ways to enrich the habitat of the birdwings by planting foodplants and nectar plants around the gardens of the ranchers, and provides basic equipment for rearing the pupae in cages, killing the adults and storing them safely for later setting and sale. Imperfect and unwanted specimens are returned to the wild in order to keep up the stock of individuals visiting the gardens. The main species ranched are

29

Troides oblongomaculatus and *Ornithoptera priamus*, but recent research into foodplant requirements and conservation status has suggested that *O. goliath, O. victoriae* and *0. chimaera* could also be ranched and traded (48). The butterfly ranching project in Papua New Guinea has demonstrated that trade and conservation can be of mutual benefit. Careful biological studies of other large and spectacular swallowtails, with a view to replacing the trade in wild-caught specimens with ranched specimens, is greatly to be encouraged. Although a thorough analysis of the potential for butterfly ranches around the world has never been made, there are certainly innumerable opportunities throughout tropical Africa, South America and Asia (45).

References

1. Ackery, P.R. and Vane–Wright, R.I. (1984). *Milkweed Butterflies: Their Cladistics and Biology*. British Museum (Natural History), London, and Cornell University Press. 420 pp.
2. Barlow, H.S. (1983). Butterfly protection in Peninsular Malaysia. Manuscript *in litt.*, 4 pp.
3. Brandt Commission (1980). *North–South: A Programme for Survival*. Pan, London. 304 pp.
4. Carcasson, R.H. (1981). *Collins Handguide to the Butterflies of Africa*. Collins, London. 109 pp.
5. Clarke, C.A. (1967). Prevention of Rh-haemolytic disease. *British Medical Journal* 4: 7–12.
6. Clarke, C.A. and Sheppard, P.M. (1963). Interactions between major genes and polygenes in the determination of the mimetic patterns of *Papilio dardanus*. *Evolution* 17: 404–413.
7. Collins, N.M. (1984). The impact of population pressure on conservation and development. *Research in Reproduction* 16: 1–2.
8. Collins, N.M. and Clifton, M.P. (1984). Threatened wildlife in the Taita Hills. *Swara* 7(5): 10–14.
9. Corbet, A.S. and Pendlebury, H.M. (1978). *The Butterflies of the Malay Peninsula*. (Third edition, revised by J.N. Eliot). Malayan Nature Society, Kuala Lumpur. 578 pp.
10. Crutzen, P.J., Heidt, L.E., Krasnec, J.P., Pollock, W.H. and Seiler, W. (1979). Biomass burning as a source of atmospheric gases CO, H_2, N_2O, NO, CH_3Cl and COS. *Nature* 282: 253–256.
11. D'Abrera, B. (1975). *Birdwing Butterflies of the World*. Landsdowne Press, Melbourne. 260 pp.
12. D'Abrera, B. (1981). *Butterflies of the Neotropical Region. Part 1. Papilionidae and Pieridae*. Lansdowne Editions, Melbourne, in association with E.W. Classey, Faringdon. 172 + xvi pp.
13. D'Abrera, B. (1982). *Butterflies of the Oriental Region. Part 1. Papilionidae and Pieridae*. Hill House, Victoria, Australia. xxxi + 244 pp.
14. Davis, P.M.H. and Barnes, M.J.C. (in press). The butterflies of Mauritius. *Journal of Research on the Lepidoptera*.
15. Dempster, J.P. and Hall, M.L. (1980). An attempt at re-establishing the swallowtail butterfly at Wicken Fen. *Ecological Entomology* 5: 327–334.
16. Dempster, J.P., King, M.L. and Lakhani, K.H. (1976). The status of the swallowtail butterfly in Britain. *Ecological Entomology* 1: 71–84.
17. Durden, C.J. and Rose, H. (1978). Butterflies from the Middle Eocene: the earliest occurrence of fossil Papilionidae (Lepidoptera). *Pearce–Sellards Serial of the Texas Memorial Museum* 29: 1–25.
18. Eckholm, E. (1982). *Down to Earth*. Pluto Press, London. 238 pp.
19. Ehrlich, P.R. and Murphy, D.D. (1981 (82)). Butterfly nomenclature: a critique. *Journal of Research on the Lepidoptera* 20:1–11.
20. Ehrlich, P.R. and Murphy, D.D. (1981 (83)). Nomenclature, taxonomy and evolution. *Journal of Research on the Lepidoptera* 20:199–204.

21. Ehrlich, P.R. and Raven, P.H. (1964). Butterflies and plants: A study in coevolution. *Evolution* 18: 586–608.
22. Eisner, T. and Meinwald, Y.C. (1965). Defensive secretion of a caterpillar (*Papilio*). *Science* 150: 1733–1735.
23. Eisner, T., Pliske, T.E., Ikeda, M., Owen, D.F., Vazquez, L., Pérez, H., Franclemont, J.G. and Meinwald, J. (1970). Defense mechanisms of arthropods. XXVII. Osmeterial secretions of papilionid caterpillars (*Baronia, Papilio, Eurytides*). *Annals of the Entomological Society of America* 63: 914–915.
24. FAO/UNEP (1981). *Tropical Forest Resources Assessment Project*. (3 volumes). FAO, Rome.
25. Feeny, P., Rosenberg, L. and Carter, M. (1983). Chemical aspects of oviposition behaviour in butterflies. In: *Herbivorous Insects: Host-seeking Behaviour and Mechanisms*. Ed. S. Ahmad. Pp. 27–76.
26. Ford, E.B. (1944). Studies on the chemistry of pigments in the Lepidoptera, with reference to their bearing on systematics. 4. The classification of the Papilionidae. *Transactions of the Royal Entomological Society* 94: 201–223.
27. Fraenkel, G.S. (1959). The *raison d'être* of secondary plant substances. *Science* 129: 1466–1470.
28. Global Environment Monitoring System, UNEP (1982). The global assessment of tropical forest resources. *GEMSPAC Information Series* No. 3, 14 pp.
29. Groombridge, B. and Wright, L. (1982). *The IUCN Amphibia–Reptilia Red Data Book. Part 1*. IUCN, Gland. xliii + 426 pp.
30. Hancock, D.L. (1983). Classification of the Papilionidae (Lepidoptera): a phylogenetic approach. *Smithersia* 2: 1–48.
31. Heath, J. (1981). *Threatened Rhopalocera (Butterflies) in Europe*. Nature and Environment Series No. 23. Council of Europe, Strasbourg.
32. IUCN Commission on Ecology (1984). *Population and Natural Resources*. Commission on Ecology Occasional Paper Number 3, 12 pp.
33. IUCN/UNEP/WWF. (1980). *World Conservation Strategy*. IUCN, Gland.
34. Igarashi, S. (1979). *Papilionidae and Their Early Stages*. Vol. 1 219 pp., Vol. 2 102 pp. of plates. Kodansha, Tokyo. (In Japanese).
35. Larsen, T. (1984). *Butterflies of Saudi Arabia and its Neighbours*. Stacey International, London. 160 pp.
36. Leakey, R.R.B. and Last, F.T. (in press). Deforestation in the tropics: how to mount a counter attack. *Spectrum*. Central Office of Information.
37. Lucas, G. and Synge, H. (1978). *The IUCN Plant Red Data Book*. IUCN, Gland. 540 pp.
38. Miller, L.D. and Brown, F.M. (1981 (83)). Butterfly taxonomy: a reply. *Journal of Research on the Lepidoptera* 20: 193–198.
39. Miller, L.D. and Brown, F.M. (1981). A catalogue/checklist of the butterflies of America north of Mexico. *Memoirs of the Lepidopterists' Society* 2: vii + 280 pp.
40. Moriarty, F. (1975). *Pollutants and Animals. Chapter 6: Where have all the butterflies gone?* Pp. 81–97. Allen and Unwin, London.
41. Morris, M.G. (1981). Conservation of butterflies in the United Kingdom. *Beiheft Veröffentlichung Naturschutz Landschaftspflege Baden–Württemberg* 21: 35–47.
42. Munroe, E. (1961). The classification of the Papilionidae (Lepidoptera). *Canadian Entomologist* Supplement 17: 1–51.
43. Munroe, E. and Ehrlich, P.R. (1960). Harmonization of concepts of higher classification of Papilionidae. *Journal of the Lepidopterists' Society* 14: 169–175.
44. Myers, N. (1979). *Conversion Rates in Tropical Moist Forests*. Report to National Academy of Sciences. National Research Council, Washington D.C. 205 pp.
45. National Research Council (1983). *Butterfly Farming in Papua New Guinea*. Managing Tropical Animal Resources Series. National Academy Press, Washington D.C. 30 pp.
46. Nikusch, I. (1981). Die Zucht von *Parnassius apollo* Linnaeus mit jährlich zwei Generationen als Möglichkeit zur Erhaltung bedrohter Populationen. *Beiheft Veröffentlichung Naturschutz Landschaftspflege Baden–Württemberg* 21: 175–176.
47. Owen, D.F. (1971). *Tropical Butterflies*. Clarendon Press, Oxford. 214 pp.

Troides (Lepidoptera: Papilionidae) in Papua New Guinea. Final report to the Department of Primary Industry, Papua New Guinea. 111 pp.

49. Pérez R., H. (1971). Algunas consideraciones sobre la poblacion de *Baronia brevicornis* Salv. (Lepidoptera, Papilionidae, Baroniinae) en la region de Mezcala, Guerrero. *Anales del Instituto de Biologia Universidad de México* 42, Ser. Zool. (1): 63–72.

50. Poore, D. (1976). *Ecological Guidelines for Development in Tropical Rain Forests*. IUCN, Gland. 39 pp.

51. Pyle, R.M. (1984). *The Audubon Society Handbook for Butterfly Watchers*. Charles Scribner's Sons, New York. 274 pp.

52. Pyle, R.M., Bentzien, M. and Opler, P. (1981). Insect conservation. *Annual Review of Entomology* 26: 233–258.

53. Scriber, J.M. (1973). Latitudinal gradients in larval feeding specialization of the world Papilionidae (Lepidoptera). *Psyche* 80: 355–373.

54. Seiler, W., and Crutzen, P.J. (1980). Estimates of gross and net fluxes of carbon between the biosphere and the atmosphere from biomass burning. *Climatic Change* 2: 207–247.

55. Shields, O. and Dvorak, S.K. (1979). Butterfly distribution and continental drift between the Americas, the Caribbean and Africa. *Journal of Natural History* 13: 221–250

56. Slansky, F. (1972). Latitudinal gradients in species diversity of the New World swallowtail butterflies. *Journal of Research on the Lepidoptera* 11(4): 201–218.

57. Smart, P. (1975). *The Illustrated Encyclopedia of the Butterfly World*. Hamlyn, London. 275 pp.

58. Thornback, J. and Jenkins, M. (1982). *The IUCN Mammal Red Data Book. Part 1*. IUCN, Gland. xl + 516 pp.

59. UNDP/FAO National Parks Development Project (1981/1982). *National Conservation Plan for Indonesia*. Vols 1–8. FAO, Bogor.

60. Vane–Wright, R.I. (1976). A unified classification of mimetic resemblances. *Biological Journal of the Linnean Society* 8: 25–56.

61. Wells, S.M., Pyle, R.M. and Collins, N.M. (1983). *The IUCN Invertebrate Red Data Book*. IUCN, Cambridge and Gland. L + 632 pp.

62. Whalley, P. (1980). *Butterfly watching*. Severn House, London.

63. White, F. (1981). The history of the Afromontane archipelago and the scientific need for its conservation. *African Journal of Ecology* 19: 33–54.

64. Woodwell, G.M., Hobbie, J.E., Houghton, R.A., Melillo, J.M., Moore, B., Peterson, B.J. and Shaver, G.R. (1983). Global deforestation: contribution to atmospheric carbon dioxide. *Science* 222: 1081–1086.

3

Swallowtails of the world: their nomenclature, distribution and conservation status

This section is divided into two parts. The first is a complete list of swallowtail species arranged in taxonomic order. Details of distribution, conservation status and common names are given. The second part is a geographical analysis, listing swallowtail species by region and country (p.124).

Swallowtails of the world: an annotated species list

The arrangement of this list is a slightly amended version of Hancock's recent taxonomic re-assessment of the Papilionidae (109), which itself relies heavily on the earlier work of Munroe (188), Rothschild and Jordan (144, 145, 233, 234) and many others. These references give basic background information on all groups of species in the list that follows. Important regional works include those of Berger (16), Pennington (208), Van Son (273) and Carcasson (34) for Africa , D'Abrera for Australia (48), the Afrotropics (50), the Neotropics (51) and the Orient (52), Corbet and Pendlebury for Malaysia (45), Common and Waterhouse for Australia (40), Tsukada and Nishiyama (262) for the South East Asian islands, Rothschild and Jordan (234) and D'Almeida (55) for South America and Chou (35) for China. Many other valuable volumes are cited in the references, which follow the species list. Important systematic works include Ackery (1) and Eisner (70–74 and numerous others) for the Parnassiinae, Jordan (144, 145) for the Papilionidae and Haugum and Low (119) and D'Abrera (49) for the birdwings. Ford (88) gives a useful account of general systematics in the Papilionidae. Useful illustrations are given by Lewis (169) and Smart (250), who also gives a species list. Authorities and dates of description are taken mainly from Bryk (28–31) and other authors, and have not necessarily been checked back to the original publication.

The use of this list has been described in section 1. It is essentially a vehicle for notes on the distribution and conservation of all swallowtail butterflies. Our main aim has been to produce a document of utilitarian value to conservation planners in wildlife and national park authorities, most of whom may not be professional entomologists. Common names have been given where known. These have been taken from dealers' lists and the literature.

Occasionally it has been necessary to note conflicting opinions where species taxonomy and nomenclature may require further study or revision. However, we must emphasize that the list is not an account of primary taxonomic research.

We have not attempted to examine the anatomy and morphology of the species listed, such a work being beyond our brief and capacity. Instead we have been guided by an extensive literature search combined with correspondence from about 100 expert lepidopterists worldwide. With 573 species listed, differences of opinion were inevitable and in such cases we have given all aspects of the argument in an attempt to present a balanced document.

This section of the book will need to be constantly up-dated. All errors and omissions remain the responsibility of the authors, but in our attempts to reduce these to a minimum we would particularly like to acknowledge the advice and help of four people well-versed in the taxonomy and nomenclature of the swallowtails, David L. Hancock, Jan Haugum, Tomaso Racheli and Bernard Turlin.

Order: LEPIDOPTERA (Butterflies and moths)
Family: PAPILIONIDAE (Swallowtail butterflies)

Subfamily: BARONIINAE

Genus: *Baronia* Salvin

1 *Baronia brevicornis* Salvin, 1893
Rare—refer to section 6, p. 182.
Restricted to a very small area of Mexico and particularly important because of its 'relict' nature as the sole member of the primitive subfamily Baroniinae. Common name: Baronia. Refs: 10, 51, 62, 69, 75, 128, 129, 141, 189, 209, 210, 211, 212, 240, 269, 276, 277, 278, 290.

Subfamily: PARNASSIINAE
Tribe: Parnassiini

The evidently sedentary nature of parnassiine populations and consequent description of many subspecies, particularly of European species, has tended to obscure the conservation status of species in this tribe (1, 31, 141). We are grateful to B. Turlin for allowing us to quote from his unpublished list of *Parnassius* subspecies (265).

Genus: *Archon* Hübner

2 *Archon apollinus* (Herbst, 1798)
A narrow range including Greece, Bulgaria, Romania, Turkey, Syria, Iran, Iraq, Lebanon, Jordan, Israel and the U.S.S.R. (Armenia and Turkmenistan). Fairly rare in collections, although this may be because it flies very early in the season. Not known to be threatened as a species. Listed as Vulnerable in Europe (123). Vulnerable and protected in Greece. Six subspecies. Common name: False Apollo (125). Other refs: 1, 141, 169, 264.

Genus: *Hypermnestra* Ménétriés

3 *Hypermnestra helios* (Nickerl, 1846)
A narrow range in Afghanistan, Pakistan (Baluchistan), Iran and U.S.S.R. (Uzbekistan, Kirghizia and possibly Turkmenistan). Rather poorly known and often considered to be a rarity, but it is not known to be threatened and is apparently rather common in the Maimana Province of Afghanistan (118). The taxon *Hypermnestra* was transferred to an independent tribe by Hiura in 1980 (126) but this was not followed by Hancock (109). Seven subspecies. Common name: Desert Apollo (80). Other refs: 1, 71, 141, 169, 259.

Genus: *Parnassius* Latreille

Species-group: *szechenyii*

4 *Parnassius szechenyii* Frivaldszky, 1886
Western China (north-eastern Xizang Zizhiqu (Tibet), Qinghai, Gansu,

Sichuan (Szechwan) and Yunnan). No threats known but apparently rather rare (264). Possibly eleven distinguishable subspecies (265). Other refs: 1, 31, 169.

5 *Parnassius cephalus* Grum–Grshimailo, 1891
Western China (Xizang Zizhiqu (Tibet), Gansu, Sichuan (Szechwan) and Qinghai). Not known to be threatened but apparently rather rare (264). Eleven distinguishable subspecies, not including *maharaja* (265). Other refs: 1, 31.

6 *Parnassius maharaja* Avinoff, 1916
North-western India (Jammu and Kashmir: Ladakh Range). Not known to be threatened. Treated only as a subspecies of *P. cephalus* by Ackery (1), Eisner (73) and Hancock (109). A full species in Bryk (31), Munroe (188), Smart (250), Talbot (259) and Turlin (265).

Species-group: *delphius* Eversmann

7 *Parnassius delphius* (Eversmann, 1843)
Afghanistan, U.S.S.R. (Tadzhikistan, Kirghizia and Uzbekistan), northern Pakistan, northern India (including Jammu and Kashmir, Himachal Pradesh and Uttar Pradesh), Bhutan and western China (Xizang Zizhiqu (Tibet), Xinjiang Uygur (Sinkiang) and Qinghai). Widely distributed, generally rare but locally common, not known to be threatened, but requiring further research. Protected by law in India (182) and included in the U.S.S.R. Red Data Book, category Vulnerable (8). Up to forty-four subspecies (265). Common name: Banded Apollo (80). Also Astor Banded Apollo (ssp. *nicevillei*), Chitral Banded Apollo (ssp. *chitralica*), Hunza Banded Apollo (ssp. *hunza*), Kafir Banded Apollo (ssp. *kafir*), Pir Panjal Banded Apollo (ssp. *atkinsoni*), Sikkim Banded Apollo (ssp. *lampidius*) and Tibet Banded Apollo (ssp. *latonius*) (80). Other refs: 1, 4, 31, 71, 74, 169, 259, 264.

8 *Parnassius stoliczkanus* C. & R. Felder, 1864
Afghanistan (Badakhshan and possibly Nuristan) (118), northern India, (Jammu and Kashmir, Himachal Pradesh and Uttar Pradesh), Pakistan, and China (Xizang Zizhiqu (Tibet)). Stated by Evans (80) to be very rare and given only subspecific status by him, by Eisner (73) and by Ackery (1), but listed as a full species by Munroe (188) with nine subspecies by Bryk (31), with seven subspecies by Talbot (259) and with fifteen subspecies by Turlin (265). Not mentioned by Hancock (109). It is protected by law in India (182). Further information needed on this species. Common name: Ladak Banded Apollo (80).

9 *Parnassius patricius* Niepelt, 1911
U.S.S.R. (Kirghizia); a rather narrow range but not known to be threatened. Five subspecies (265). Other refs: 1, 31.

10 *Parnassius acdestis* Grum–Grshimailo, 1891
U.S.S.R. (Kirghizia), Nepal, northern India (Jammu and Kashmir, and Sikkim), Bhutan and western China (Xizang Zizhiqu (Tibet), Xinjiang Uygur (Sinkiang) and Sichuan (Szechwan)). No threats known but apparently very local and rather rare (264). Nineteen subspecies (265). Other refs: 1, 31.

Species-group: *imperator* Oberthür

11   *Parnassius imperator* Oberthür, 1883
North-eastern India (Sikkim) (80), western China (Xizang Zizhiqu (Tibet), Qinghai, Gansu, Sichuan (Szechwan) and Yunnan). Not known to be threatened as a species, but *P. i. augustus* is protected by law in India (182). It seems to be a common subalpine species in China (264). Twenty subspecies (265). Common name: Imperial Apollo (80). Other refs: 1, 4, 31, 207.

Species-group: *charltonius* Gray

12   *Parnassius charltonius* Gray, 1853
Afghanistan, U.S.S.R. (Kirghizia, eastern Uzbekistan and Tadzhikistan), Pakistan, northern India (including Jammu and Kashmir, Uttar Pradesh and Himachal Pradesh) and China (Xizang Zizhiqu (Tibet)). Not known to be threatened as a species and is common in Kashmir, but the nominate subspecies is protected by law in India (182). Twenty subspecies (265). Common name: Regal Apollo (80, 282). Other refs: 1, 4, 31, 72, 74, 259.

13   *Parnassius inopinatus* Kotzsch, 1940
Found only in mountain ranges in north-western Afghanistan where it is very localised (264). Little is known about it, but it has recently appeared for sale on dealers' lists. No threats recognized, but possibly a candidate for Rare status. More information is required. Two subspecies. Other refs: 1, 74, 265.

14   *Parnassius loxias* Püngeler, 1901
U.S.S.R. (Kirghizia and Tadzhikistan) and China (Xinjiang Uygur (Sinkiang)). A narrow range, possibly extremely rare but little information available from these inaccessible regions (264). Not known to be threatened but more data required. Two subspecies (265). Other refs: 1, 288.

15   *Parnassius autocrator* Avinoff, 1913
Rare—refer to section 6, p. 185.
Afghanistan and U.S.S.R. (Tadzhikistan). Originally described as a subspecies of *P. charltonius* Gray. Bryk (31) agrees with this and it is omitted by Munroe (188) but thought to be a full species by recent authors (1, 109, 250). Narrow range in the Pamir mountains and Hindu Kush and a rarity according to Smart (250). An extremely Rare species in the U.S.S.R., where it is threatened by the degradation of high mountain pastures (260). It is included in the U.S.S.R. Red Data Book, category Rare (8). Careful monitoring and further study are required. Only one subspecies has been described (221) but up to four have been listed. Other ref: 74.

Species-group: *tenedius* Eversmann

16   *Parnassius tenedius* Eversmann, 1851
Eastern U.S.S.R. (Tuvinskaya, Chitinskaya and Yakutskaya), Mongolia and China (Nei Monggol (Inner Mongolia)). Not known to be threatened as a species, but included in the U.S.S.R. Red Data Book, category Vulnerable (8). Very wide range in tundra and mountains (264). Five subspecies (265). Other refs: 1, 31, 169.

Species-group: *acco* Gray

17   *Parnassius acco* Gray, 1853

Pakistan, northern India (Jammu and Kashmir and further east in Sikkim), Nepal and China (Xizang Zizhiqu (Tibet)). Possibly also Bhutan if *P. hannyngtoni* Avinoff (below) is regarded as a subspecies. Stated by Antram (4), Evans (80) and Turlin (264) to be exceedingly, or very rare (in India (?)) but present status not known. Flies at high altitudes (5500–6200m) so it is very rarely captured (264). The subspecies *P. a. geminifer* is protected by law in India (182). More information is needed, particularly from the Indian highlands and Tibet. Ten subspecies. Common name: Varnished Apollo (80). Other refs: 1, 31, 169, 259.

18   *Parnassius przewalskii* Alpheraky, 1887
Western China (including Xizang Zizhiqu (Tibet), Sichuan (Szechwan), and Yunnan). Four subspecies (265). Although recognized by some authors (31, 188, 191) as a good species, it is not mentioned by Ackery (1) or Hancock (109) and was placed as a subspecies of *P. acco* by Eisner (73). Not known to be threatened but apparently very rare (264).

19   *Parnassius hannyngtoni* Avinoff, 1916
North-eastern India (Sikkim), China (Xizang Zizhiqu (Tibet)) and Bhutan (?). Very rare, according to Evans (80) and Talbot (259) and protected by law in India (182), but specific status uncertain. Flies at a very high level (6000m) (264). Ackery (1) and Eisner (73) treat it as a subspecies of *P. acco* as, more doubtfully, does Smart (250). Hancock (109) does not mention it, but Turlin accepts it as a good species with two subspecies (265). More information is required on this species. The spelling of the trivial name varies greatly with the authors quoted but Avinoff's original 'hunnyngtoni' was amended to the above form by Bryk (31), Munroe (188) and Talbot (259) and is presumably correct since the species was dedicated to Hannyngton (259). Common name: Hannyngton's Apollo (80).

Species-group: *simo* Gray

20   *Parnassius simo* Gray, 1853
U.S.S.R. (Kirghizia and Tadzhikistan), Pakistan, northern India (including Jammu and Kashmir, and Sikkim), Nepal, western China (Xizang Zizhiqu (Tibet), Xinjiang Uygur (Sinkiang) and Gansu) and Mongolia. Not known to be threatened. Thirty-four subspecies (265). Common name: Black-edged Apollo (80). Other refs: 1, 4, 31, 71.

Species-group: *hardwickii* Gray

21   *Parnassius hardwickii* Gray, 1831
Himalayas: northern India (including Jammu and Kashmir, and Sikkim), Pakistan, Nepal, Bhutan and China (Xizang Zizhiqu (Tibet)). Not known to be threatened. Five subspecies (265). Common name: Common Blue Apollo (80, 251). Other refs: 1, 4, 31, 169, 251, 264.

Species-group: *mnemosyne* Linnaeus

22   *Parnassius orleans* Oberthür, 1890
Southern and western China (Xizang Zizhiqu (Tibet), Xinjiang Uygur (Sinkiang), Qinghai, Gansu, Shaanxi (Shensi), Sichuan (Szechwan) and Yunnan) and Mongolia. Not known to be threatened. Sixteen subspecies (265). Other refs: 1, 31, 169, 207, 264.

23 *Parnassius clodius* Ménétriés, 1855
Western U.S.A. (Alaska, Washington, Idaho, Utah, Montana, Wyoming, Oregon, Nevada and California) and south-western Canada (British Columbia). Usually abundant and not known to be threatened as a species. However, two of the twelve subspecies are in decline. The Californian subspecies *P. c. strohbeeni* formerly occurred in the Santa Cruz Mountains, California, but is now believed to be extinct. *P. c. shepardii* has a restricted distribution in the north-western states. It is now absent from former haunts in the Snake River Canyon on the border of Oregon and Idaho (216) and is listed as a Special Species by the Washington State Department of Game. Many conservationists blame logging and dam-building for the decline of these two subspecies, but others blame drought (216). Common name: Clodius Parnassian (216). Other refs: 1, 31, 55, 137, 169, 264, 269.

24 *Parnassius eversmanni* Ménétriés, 1849
Eastern U.S.S.R, Mongolia to northern China, North Korea, Japan (Hokkaido), Alaska and Canada (Yukon Territory, Northwest Territories and British Columbia). Included in the U.S.S.R. Red Data Book, category Vulnerable (8). Inhabits lowland and tundra. Populations difficult to assess but widespread. Not known to be threatened. Twelve subspecies (265). Common name: Eversmann's Parnassian (216). Other refs: 1, 31, 137, 141, 149, 169, 248, 264, 269.

25 *Parnassius felderi* Bremer, 1861
Eastern U.S.S.R. (Khabarovsk Kray). The species is not mentioned by Hancock (109), but is listed by Turlin with three subspecies (265). No threats known, but little information available.

26 *Parnassius ariadne* Lederer, 1853 (= *clarius* Eversmann, 1843)
U.S.S.R (Altay mountains and Tadzhikistan) and western Mongolia. Not known to be threatened. Localized in distribution (264) and rare in collections. Two subspecies (265). Other refs: 1, 31, 169.

27 *Parnassius nordmanni* (Nordmann, 1849)
U.S.S.R. (eastern Armenia and the Bol'shoy Kavkaz (Caucasus) mountains; Azerbaydzhan and Georgia) and eastern Turkey. Locally distributed (264), sometimes abundant (193), but very rare in collections. Inhabits inaccessible areas above 2000m and has a short flight season (90, 245). Included in the U.S.S.R. Red Data Book (8, 260), but known to be present in at least seven reserves and probably several more (193). Four subspecies described (239, 265), but their distinction is somewhat speculative (193). Other refs: 1, 31, 227.

28 *Parnassius mnemosyne* (Linnaeus, 1758)
Hilly or mountainous regions in Spain (Pyrenees), France, Norway, Sweden, Finland, Switzerland, F.R.G., Liechtenstein, Austria (Endangered/Vulnerable (93)), Italy, Sicily, Albania, Yugoslavia, Greece, Bulgaria, Turkey, Romania, Poland, G.D.R. (?), Czechoslovakia, Hungary, U.S.S.R (Latvia, Lithuania, Estonia and Ukraine to Armenia, Bol'shoy Kavkaz (Caucasus) mountains, Uzbekistan, Tadzhikistan, Kirghizia and the Ural mountains), Syria, Lebanon, Iran, Iraq and Afghanistan. Not known to be threatened at the specific level, but now Rare over the whole of its range in the U.S.S.R. and listed in the U.S.S.R. Red Data Book (8, 260).

Protected by law in Czechoslovakia, Finland, G.D.R., Greece, Hungary, Poland and Lithuania (123). Another species which has been excessively subdivided, 125 subspecies being listed by Bryk (31). Common name: Clouded Apollo (125). Other refs: 1, 71, 72, 74, 169, 264.

29 *Parnassius stubbendorfi* Ménétriés, 1849
Eastern U.S.S.R., China (Xizang Zizhiqu (Tibet), Heilongjiang, Gansu, Sichuan (Szechwan) and Qinghai), Mongolia, North Korea, South Korea and Japan (Hokkaido). Twenty-five subspecies listed by Igarashi (141), 28 by Turlin (265). Not known to be threatened and seemingly quite common (264). Other refs: 1, 31, 149, 169, 247.

30 *Parnassius glacialis* Butler, 1886
Eastern China (Hubei, Shandong, Jiangsu, Anhui and Zhejiang), North Korea, South Korea and Japan (Hokkaido, Honshu and Shikoku). Accepted as a good species by recent authors (1, 141, 169, 247), though not by Bryk (31) (subspecies of *stubbendorfi*) or Munroe (188). The larval stages are quite distinct from those of clearly related species. Apparently common (264). Fifteen subspecies (265). Other ref: 149.

Species-group: *apollo* Linnaeus

31 *Parnassius apollonius* (Eversmann, 1847)
U.S.S.R. (Uzbekistan, Tadzhikistan, Kirghizia) and China (western Xinjiang Uygur (Sinkiang)). Twelve subspecies (265), most of which are difficult to distinguish. No evidence of being under threat. Other refs: 1, 31, 169, 264.

32 *Parnassius honrathi* Staudinger, 1882
Central Asia: U.S.S.R. (western Uzbekistan (?), Tadzhikistan and possibly southern Kirghizia) and north-eastern Afghanistan (Pamir Mountains). Five subspecies (265). No evidence of being in danger. Other refs: 1, 31, 74, 169, 264.

33 *Parnassius bremeri* Bremer, 1864
Eastern U.S.S.R., northern China (Heilongjiang, Hebei and Shanxi), North Korea, South Korea where it is threatened (152), Japan (Hokkaido) and possibly eastern Mongolia. The Japanese population is either extinct or the record is erroneous since no recent specimens are known (264). No information available about the status but declining in Korea due to overcollection and loss of foodplant (*Dicentra perigrina*) on skiing slopes. Conservation measures proposed in Korea include legislation prohibiting collection of the species and a captive breeding programme (153). More information is needed on this species, which has up to sixteen subspecies (265). Other refs: 1, 31, 71, 141, 149, 169.

34 *Parnassius jacquemontii* Boisduval, 1836
North-eastern Afghanistan (including Badakhshan), northern Pakistan, north-western India (Jammu and Kashmir), U.S.S.R. (Tadzhikistan (Pamirs) and Uzbekistan) and south-western China (Xizang Zizhiqu (Tibet), Xinjiang Uygur (Sinkiang), Gansu and Sichuan (Szechwan)). Twenty-five subspecies (265); the nominate subspecies is protected in India (182). Not known to be in danger. Common name: Keeled Apollo (80). Other refs: 1, 31, 74, 169, 259.

35 *Parnassius epaphus* Oberthür, 1879
Afghanistan, Pakistan, northern India (including Jammu and Kashmir, and Sikkim), Nepal, Bhutan, China (Xizang Zizhiqu (Tibet), Xinjiang Uygur (Sinkiang), Sichuan (Szechwan), Gansu and Qinghai) and possibly U.S.S.R. (Tadzhikistan). A fairly broad distribution and not known to be threatened. Up to 37 subspecies (265); *P. e. hillensis* is protected by law in India (182). *P. beresowskyi* Staudinger from China is not mentioned by Munroe (188) or Ackery (1) (even as a subspecies) and is treated as a subspecies of *epaphus* by Bryk (31) and Smart (250). Common name: Common Red Apollo (80). Other refs: 1, 169, 259.

36 *Parnassius actius* (Eversmann, 1843)
U.S.S.R. (Turkmenistan, Uzbekistan (?), Tadzhikistan, Kirghizia, Kazakhstan), north-eastern Afghanistan, northern Pakistan, north-western India (Jammu and Kashmir) and south-western China (Xinjiang Uygur (Sinkiang) and Gansu). Very rare in the U.S.S.R. and declining due to changes in high mountain meadows (260). Included in the U.S.S.R. Red Data Book, category Vulnerable (8). More information is needed on this species, which has up to nineteen subspecies (265). Other refs: 1, 31, 74, 169, 259.

37 *Parnassius phoebus* (Fabricius, 1793)
Palearctic: Alps (France, southern F.R.G., Switzerland, Liechtenstein (?) and Austria), U.S.S.R. (Ural mountains, Kazakhstan, Altay, Siberia and Kamchatka), Mongolia, China (Xinjiang Uygur (Sinkiang)), western U.S.A. (Alaska, Washington, Idaho, Montana, South Dakota, Wyoming, California, Nevada, Utah, Colorado and New Mexico) and Canada (British Columbia and Alberta). Not under threat as a species, although some of the 45 subspecies (265) may be threatened. Included in the U.S.S.R. Red Data Book, category Vulnerable (8) and listed as Vulnerable throughout Europe (123). Protected in France. Common names: Small Apollo (125), Phoebus Parnassian (216). Other refs: 1, 31, 62, 72, 118, 137, 169, 269.

38 *Parnassius tianschanicus* Oberthür, 1879
U.S.S.R. (Uzbekistan, Tadzhikistan and Kirghizia), Afghanistan, Pakistan, India (Jammu and Kashmir) and western China (Xinjiang, Uygur (Sinkiang)). Locally common but generally very rare in the U.S.S.R. where it is declining and is included in the U.S.S.R. Red Data Book, category Vulnerable (8, 118, 260). More information is required. Up to eighteen subspecies (265). Common name: Large Keeled Apollo (80). Other refs: 1, 31, 74, 259.

39 *Parnassius nomion* Fischer de Waldheim, 1823
Eastern U.S.S.R. (Irkutsk, Buryatskaya, Amurskaya, Khabarovsk and Altay), Mongolia, China (Gansu, Qinghai, Shaanxi (Shensi), Heilongjiang, Liaoning and Nei Monggol (Inner Mongolia) (?)), North Korea and South Korea. A doubtful record from North America (Alaska) (141) and another from California (55). Authority wrongly given as Hübner by Munroe (188) and Smart (250). Up to 31 subspecies (265) including *P. nomius* Grum–Grshimailo from China, which is not listed by Munroe (188), Hancock (109) or Ackery (1) (even as a subspecies) and is treated as a subspecies of *nomion* by Bryk (31) and Eisner (73). Other refs: 149, 169.

40 *Parnassius apollo* Linnaeus, 1758
   Rare—refer to section 6, p. 187.
   Among mountains at subalpine levels in France, Andorra, Spain (59), Netherlands (only old and questionable records), Norway (Vulnerable, possibly Endangered (123)), Sweden, Finland, F.R.G., Switzerland, Liechtenstein, Austria (Endangered/Vulnerable (93)), Italy, Sicily, G.D.R. (Extinct), Poland, Czechoslovakia, Hungary (migratory), Yugo-slavia, Greece, Albania, Bulgaria, Turkey (including the border area with Iran and Iraq), Romania, Syria, U.S.S.R. (Latvia, Lithuania, Ukraine, to Armenia, Bol'shoy Kavkaz mountains (Caucasus), Ural mountains and Siberia), Mongolia and China (Xinjiang Uygur (Sinkiang)). Concern for this species seems to be based on local threats to populations, particularly those in 'developed' areas of western Europe. In other parts of Europe *P. apollo* can be very numerous. The inclusion of *P. apollo* on Appendix 2 of CITES is questionable when so many threatened species are excluded. The unnecessary division of the species (over 160 'subspecies' are listed by Bryk (31)) has led to exaggerated fears for particular populations. However, numbers of 'Apollos' are taken every year in western Europe for trade, populations are often very isolated and it is declining, threatened, or rare in many countries (64, 123). It is declining sharply in all parts of the U.S.S.R. and is included in the U.S.S.R. Red Data Book (category Vulnerable) and the Red Book of the Ukrainian S.S.R. (8, 260). There is no information from the eastern section of the Apollo's range. The 87 countries that are party to CITES are obliged to invoke national legislation implementing the convention and all countries in the European Economic Community have already done so. The Apollo is also protected by law in several other countries including Austria, Czechoslovakia, Finland, G.D.R., Greece and Poland. The species should be closely monitored. Common name: Apollo. Other refs: 1, 30, 100, 101, 118, 125, 141.

Subfamily: PARNASSIINAE
Tribe: Zerynthiini

Genus: *Sericinus* Westwood

41 *Sericinus montela* Gray, 1843 (= *telamon* Donovan, 1798)
   Eastern U.S.S.R. (Primorskiy Kray), China (Heilongjiang, Jilin, Liaoning, Hebei, Shandong, Anhui, Jiangsu, Hubei, Hunan, Jiangxi and Gansu), North Korea and South Korea. Threatened by changes in flood plain vegetation in the U.S.S.R., where it is included in the U.S.S.R. Red Data Book, category Rare (8, 260). More information is required on this primitive and important zerynthiine. Eleven subspecies. Other refs: 1, 31, 141.

N.B. The following six species have seen several changes of genus. First described as *Thais* by Fabricius in 1807, this invalid name (homonym) was replaced with *Parnalius* by Rafinesque in 1815, and almost simultaneously with *Zerynthia* by Ochsenheimer in 1816. The latter name achieved popular usage and a recent attempt to revive the validity of the name *Parnalius* was suppressed by the Commission on Zoological Nomenclature (118, 194). The name *Allancastria* was raised by Bryk in 1934 and applied to those species of *Zerynthia* found in Asia Minor. This course has been followed by Hancock (109).

Genus: *Allancastria* Bryk

42 *Allancastria cerisy* (Godart, 1824)
Albania, Cyprus, Crete, Greece, Yugoslavia, Bulgaria, Romania, Turkey, U.S.S.R. (Armenia and Bol'shoy Kavkaz (Caucasus) mountains), Iran, Iraq, Syria, Israel and Lebanon. The species is quite local and often rare (160). Declining rapidly over the whole of its range in the U.S.S.R. Included in the Red Book of the Ukraine S.S.R. and the U.S.S.R. Red Data Book (8, 260). Localised but common in Turkey (264). Protected by law in Greece. Not known to be threatened in the rest of its extensive range, but clearly requires monitoring. Eleven subspecies. Common name: Eastern Festoon (125). Other refs: 1, 31, 141, 157, 158, 169, 194.

43 *Allancastria deyrollei* Oberthür, 1872
Turkey, Syria, Lebanon, Israel, Jordan, Iraq and possibly Iran (158). Common and widespread in the Lebanon (158). Treated as a subspecies of *A. cerisy* by Bryk (28) and Igarashi (141) and not listed by Ackery (1), but Larsen (157, 158) has shown it to be distinct. Other ref: 264.

44 *Allancastria caucasica* Lederer, 1864
U.S.S.R. (Armenia and Bol'shoy Kavkaz (Caucasus) mountains) and northern Turkey (160). Specific status uncertain. Regarded as a subspecies of *A. cerisy* by most authors (31, 141). Larsen considers it to be distinct (160) as does Turlin (264). Conservation status needs clarification.

45 *Allancastria louristana* (Le Cerf, 1908)
West and south-west Iran, mountains of Louristan. Originally described as a subspecies of *A. cerisy*, but recently raised to full species status. Not known to be threatened, but conservation status requires confirmation. Ref: 162.

Genus: *Zerynthia* Ochsenheimer

46 *Zerynthia polyxena* (Denis & Schiffermüller, 1775)
Southern France, Italy, Sicily, Austria, Hungary, Czechoslovakia, Yugoslavia, Albania, Greece, Bulgaria, Romania and south-western U.S.S.R. Widely distributed though local and not under threat as a species. Nevertheless, it is rare and declining rapidly in the U.S.S.R.; in need of habitat protection to ensure its survival. Included in the U.S.S.R. Red Data Book, category Vulnerable (8, 260); listed as Vulnerable throughout Europe and protected by law in Czechoslovakia, Greece, Hungary and some of the Austrian provinces (123). Continuous monitoring of this species is necessary. Nineteen subspecies. *Z. hypermnestra* Scopoli, 1763 is an invalid homonym (125). Common names: Southern Festoon (125), Birthwort Butterfly (253). Other refs: 1, 31, 141, 169, 194.

47 *Zerynthia rumina* (Linnaeus, 1758)
Southern France, Italy, Spain, Portugal, Algeria, Morocco and Tunisia. Often fairly common within its somewhat restricted range but has been

listed as Vulnerable throughout Europe. Threatened to varying degrees in France (where it is more local than *Z. polyxena* (264) and the distinctive form *honoratii* is protected by law), Spain and Italy (123). Five subspecies. Common name: Spanish Festoon (125). Other refs: 1, 31, 141, 169.

Genus: *Bhutanitis* Atkinson

48 *Bhutanitis mansfieldi* (Riley, 1939)
Rare—refer to section 6, p. 192
Known to Ackery (1) only from the female holotype from Yunnan, China, but a contemporary male and a second female have since been found. Recent specimens have also come from Sichuan (Szechwan) and a second subspecies described (238). *B. mansfieldi* was separated as the type species of a monobasic genus *Yunnanopapilio* by Hiura in 1980 (126) but this was reduced to a subgenus by Saigusa and Lee (238). The name is not used by Hancock (109). Other refs: 118, 199, 229, 264.

49 *Bhutanitis thaidina* (Blanchard, 1871)
Rare—refer to section 6, p. 194
A narrow range in China (Yunnan, Sichuan (Szechwan) and Shaanxi (Shensi) provinces). Rare status is justified, at least until more is known about the species. No subspecies described. Refs: 1, 19, 31, 207.

50 *Bhutanitis lidderdalii* Atkinson, 1873
Bhutan, northern India (Assam, Sikkim, Manipur and Nagaland), northern Burma, Thailand and China (Sichuan (Szechwan) and Yunnan provinces) (31). Not so restricted as *B. thaidina* and probably not in danger at the moment, but needs monitoring to ascertain its actual status. Up to three subspecies, the nominate subspecies protected by law in India (182), the third recently described from Thailand. The Thailand population is apparently confined to northern areas around Chiang Mai (141) and is considered to be a relict (7). Hundreds of individuals are exported from Thailand annually to collectors (7). Deforestation may also be a threat and more data are needed. Common name: Bhutan Glory (80). Other refs: 1, 259.

51 *Bhutanitis ludlowi* Gabriel, 1942
Insufficiently Known—refer to section 6, p. 196
Bhutan: Trashiyangsi Valley only. The type-series is unique (1). Other ref: 91.

Genus: *Luehdorfia* Crüger

N.B. The genus *Luehdorfia* has a very confusing taxonomy. There is much variation between authors regarding species and the distribution of the individual species is also inconsistently reported.

52 *Luehdorfia chinensis* Leech, 1893
Insufficiently Known—refer to section 6, p. 197
Eastern China (Anhui, Hubei, Jiangsu and Jiangxi provinces). Status not entirely certain: variously treated as a separate species (109, 141), as a subspecies of *L. japonica* (1) or *L. puziloi* (30, 31) or ignored (188). Lee (163) and Igarashi (141) are followed as authors who have studied the early

stages. The above status seems appropriate until its specific identity and possible threats can be assessed. Two subspecies (141).

53 *Luehdorfia japonica* Leech, 1889
Indeterminate—refer to section 6, p. 198
Restricted to the island of Honshu, Japan (141). Ackery (1) includes *chinensis* within this species (although its distribution is in southern China disjunct from the range of *japonica*) while Bryk includes *chinensis* in *L. puziloi* (30, 31). The subspecies supposed to occur in Taiwan appears to be either a doubtful record or now extinct (141, 246, 264). One or two subspecies. Other refs: 149, 166, 247.

54 *Luehdorfia puziloi* (Erschoff, 1872)
Extreme south-eastern U.S.S.R. (Primorskiy Kray), north-eastern China (Manchuria), North Korea, South Korea and Japan (Honshu and Hokkaido). Appears to be declining in Korea due to overcollecting and pollution (153), and in Japan due to overcollecting and habitat destruction (124). Conservation measures proposed in Korea include legislation prohibiting collecting and a programme of captive breeding (153). It has been included in the U.S.S.R. Red Data Book, category Rare (8). Not known to be threatened over the rest of its range, but clearly requires monitoring. Five subspecies. Other refs: 1, 141, 149, 247.

Subfamily: PAPILIONINAE
Tribe: Leptocircini

Genus: *Iphiclides* Hübner

55 *Iphiclides podalirius* (Linnaeus, 1758)
Europe (excluding the British Isles, Norway, Sweden (?) and Finland), North Africa and the Middle East (possibly excluding Saudi Arabia, Yemen, South Yemen, Oman, United Arab Emirates and Qatar), Afghanistan, Pakistan, India and China. Generally common and not seriously threatened although it is protected by law in Czechoslovakia, G.D.R., Hungary, Luxembourg and Poland, included in the U.S.S.R. Red Data Book (8), listed as Endangered-Rare in the provinces of Austria (93), and protected in seven of them, and Indeterminate throughout Europe (123). Six subspecies including *I. p. feisthamelii* Duponchel from Morocco, Algeria, Tunisia, Spain, Portugal and France, which is variously treated as a doubtfully good species (250) or a subspecies of *I. podalirius* (125). Common names: Scarce Swallowtail (125, 253, 282), Sail Swallowtail (253), Pear-tree Swallowtail (158). Other refs: 141, 202.

56 *Iphiclides podalirinus* (Oberthür, 1890)
China (Xizang Zizhiqu (Tibet) and Yunnan). A good species in Munroe (188) and Hancock (109) but stated by Smart (250) to be possibly a subspecies of *I. podalirius*. Not known to be threatened, but more data needed.

Genus: *Teinopalpus* Hope

57 *Teinopalpus imperialis* Hope, 1843
Rare—refer to section 6, p. 200

Nepal, northern India (West Bengal, Meghalaya, Manipur, Sikkim and Assam), Bhutan, northern Burma and China (Hubei and Sichuan). A prized species that commands high prices on dealers lists and is hunted mercilessly in the Himalayas. It is described as rare by Talbot (259) and is protected by law in India (182). The habitat is in mountainous regions, it keeps to the tops of trees, has a strong flight and is difficult to capture (4). *T. behludinii* (Pen, 1937) from Sichuan is almost certainly referable to *T. imperialis* (185, 207). Two subspecies. Common name: Kaiserihind (80), Kaiser-I-Hind (251). Other refs: 52, 135, 141, 179, 197, 286.

58 *Teinopalpus aureus* Mell, 1923
Insufficiently Known—refer to section 6, p. 204.
Despite its unique appearance, its specific status is uncertain; it is an isolated taxon of *Teinopalpus* from south-eastern China (Guangdong Province). If distinct it is presumably Rare, but more information is required. Common name: Golden Kaiser-I-Hind. Refs: 52, 178.

Genus: *Meandrusa* Moore

59 *Meandrusa sciron* (Leech, 1890)
China, India (Sikkim and Assam), Bhutan, south-western Thailand and Burma. The name *hercules* Blanchard is a synonym of *sciron*. The name *gyas* Westwood is a synonym of *lachinus* Fruhstorfer, which is currently regarded as conspecific with *sciron*. Status and distribution of the species not well known but not known to be threatened. Not uncommon in India, where it is protected by law under the name *gyas* (182). More information needed. Two subspecies. Common name: Brown Gorgon (80). Other refs: 4, 141, 207, 259.

60 *Meandrusa payeni* (Boisduval, 1836)
Northern India (Assam and Sikkim), Bhutan, southern Burma, northern Thailand, northern Vietnam, Laos, China (Hainan (Guangdong prov.)), Peninsular and Eastern Malaysia, Brunei, Indonesia (Sumatra, Java and Kalimantan (?)) (52, 262). Not thought to be threatened across most of its range but considered to be Vulnerable and in need of protection in Peninsular Malaysia (10). Six or seven subspecies. Common names: Yellow Gorgon (80), Outlet Sword (3), the Sickle (3). Other refs: 4, 45, 87, 141, 259.

Genus: *Eurytides* Hübner

N.B. The genus name *Eurytides* is usually applied to new world forms whereas *Graphium* is applied to old world forms. There is controversy over the morphological distinction of the two genera and the traditional division is upheld here.

Species-group: *marcellus* Cramer

61 *Eurytides (Protesilaus) marcellus* (Cramer, 1777)
Canada (southern Ontario) and eastern U.S.A. Local, but not threatened. Several named seasonal forms, but subspecies doubtful. Common name: Zebra Swallowtail (154, 216). Other refs: 55, 62, 137, 234, 269.

62 *Eurytides (Protesilaus) epidaus* (Doubleday, 1846)
Mexico, Guatemala, Honduras, Belize, Nicaragua (?) and Costa Rica.
Seasonally abundant in Costa Rica, flying in open areas associated with
deciduous forest (61). Generally common and not threatened. Three
subspecies. Other refs: 51, 55, 57, 62, 232, 234, 269.

63 *Eurytides (Protesilaus) zonaria* (Butler, 1869)
Confined to Hispaniola. Said to be widespread and not uncommon (243)
though it is not well known. No threatened category has been given but the
status of the species should be better determined if possible. Common
name: Haitian Kite (230). Other refs: 51, 55, 234.

64 *Eurytides (Protesilaus) marcellinus* (Doubleday, 1845)
Vulnerable—refer to section 6, p. 206.
Restricted to Jamaica. The butterfly is not common and because of the
relatively small size of Jamaica, the species' status needs to be carefully
monitored. Common name: Jamaican Kite (21, 230). Other refs:
51, 55, 234, 266, 267.

65 *Eurytides (Protesilaus) celadon* (Lucas, 1852)
Restricted to Cuba but the butterfly is 'apparently widespread' in that
country (230). Common name: Cuban Kite (230). Other refs:
21, 51, 55, 234.

66 *Eurytides (Protesilaus) philolaus* (Boisduval, 1835)
Mexico, Belize, Guatemala, El Salvador (?), Honduras, Nicaragua and
Costa Rica. Common and sometimes very abundant (57, 269); no known
threats. Common name: Dark Zebra Swallowtail (216). Other refs:
51, 55, 62, 175, 234.

67 *Eurytides (Protesilaus) anaxilaus* (C. & R. Felder, 1864)
   (=*arcesilaus* Lucas, 1852)
Northern Venezuela and Colombia. Not known to be threatened (234).
Other ref: 51.

68 *Eurytides (Protesilaus) xanticles* (Bates, 1863)
Northern Colombia and Panama, apparently with a restricted range.
Generally uncommon although it is fairly common in the Panama Canal
Zone where there is a national park (221). No threats known, but further
information is needed for this narrowly distributed species. Other refs:
51, 55, 234.

69 *Eurytides (Protesilaus) oberthueri* (Rothschild & Jordan, 1906)
Honduras and Mexico (disjunct range) (269). Specific status questionable,
possibly only a form of *E. (P.) philolaus*, but accepted by Munroe (188) and
Hancock (109). Only three specimens appear to have been recorded (221).
Other refs: 51, 55, 62, 234, 275.

Species-group: *bellerophon* Dalman

70 *Eurytides (Protesilaus) bellerophon* (Dalman, 1823)
Northern Argentina, south-eastern Brazil and possibly Paraguay. Apparently not common, but not known to be threatened. More information
needed. Refs: 51, 55, 122, 234.

Species-group: *protesilaus* Linnaeus

71 *Eurytides (Protesilaus) agesilaus* (Guérin and Percheron, 1835)
Mexico, Central America and South America (excluding Chile and Uruguay). Rare in Costa Rica, common in Panama (61). Generally quite abundant and not threatened. Five subspecies (also variously treated as forms or good species including *E. (P.) a. autosilaus* (Bates, 1861)). Other refs: 26, 51, 55, 62, 122, 234, 269.

72 *Eurytides (Protesilaus) orthosilaus* (Weymer, 1889)
Paraguay and Brazil. Formerly believed to be quite rare, but now known to occur over a very wide area of cerrado in the Mato Grosso of central Brazil (180), where it can be seen almost any day in the year. Males frequent favoured sandy areas and females seek nectar at cerrado flowers (23). Other refs: 26, 51, 55, 234.

73 *Eurytides (Protesilaus) helios* (Rothschild and Jordan, 1906)
Southern Brazil, northern Argentina and possibly Paraguay. Not uncommon, but not known to be threatened. Refs: 26, 51, 55, 122, 234.

74 *Eurytides (Protesilaus) stenodesmus* (Rothschild and Jordan, 1906)
Paraguay, Brazil and northern Argentina. Common and not threatened. Sometimes confused with *E. (P.) helios*. Refs: 51, 55, 122, 234.

75 *Eurytides (Protesilaus) earis* (Rothschild and Jordan, 1906)
Ecuador (55) and Brazil (26). An uncommon and little known species, but no threats are recognized. More information required. Other ref: 234.

76 *Eurytides (Protesilaus) telesilaus* (Felder, 1864)
Panama, Colombia, Venezuela, Guyana, Surinam, French Guiana, Trinidad, Ecuador, Brazil, Peru, Bolivia, Paraguay and possibly Argentina. Not known to be threatened. Ranched in Brazil (24). Four subspecies. Common name: Southern White Page (9, 282). Other refs: 26, 51, 55, 234, 269.

77 *Eurytides (Protesilaus) aguiari* (D'Almeida, 1937)
Brazil, from Belém (Para) to Benjamin Constant (Amazonas). No known threats but more data required. Refs: 51, 55.

78 *Eurytides (Protesilaus) embrikstrandi* (D'Almeida, 1936)
Brazil. A little known and fairly recently described species. No threats known. Refs: 51, 55.

79 *Eurytides (Protesilaus) travassosi* (D'Almeida, 1938)
Brazil. Another little known and relatively recently described species. Not known to be threatened Refs: 51, 55.

80 *Eurytides (Protesilaus) molops* (Rothschild and Jordan, 1906)
Colombia, Venezuela (?), Guyana, Surinam, French Guiana (?), Brazil, Ecuador, Peru and Bolivia. Not recognized as threatened. Three subspecies (all treated as full species by D'Almeida (55)) but *E. (P.) m. hetaerius* was transferred to *E. (P.) macrosilaus* by Hancock (109) and replaced in this species by *E. (P.) m. leucosilaus* (Zikan) (115). Other refs: 51, 234.

81 *Eurytides (Protesilaus) macrosilaus* (Gray, 1852)
Mexico, Guatemala, Honduras, Belize and Nicaragua. Conspecific with *E. (P.) protesilaus*, according to most authors (188, 234), but listed as a species by D'Almeida (55) and Hancock (109). *E. (P.) penthesilaus* (Felder, 1864)

from Mexico is regarded as a subspecies of *E. (P.) protesilaus* by most authors (188, 234, 269), as a subspecies of *E. (P.) macrosilaus* by Hancock (109), and only by D'Almeida (55) as a good species. It should be included under *E. (P.) macrosilaus* if this is to be accepted as a good species (115). Two further subspecies were placed here by Hancock (109).

82   *Eurytides (Protesilaus) nigricornis* (Staudinger, 1884)
Eastern Paraguay and southern Brazil. No known threats but more data needed. Ranched in Brazil (24). Listed by D'Abrera (51) as a subspecies of *E. (P.) protesilaus*, but given full species rank by Hancock (109). The larvae apparently feed on Lauraceae, whilst *E. (P.) protesilaus* utilizes Magnoliaceae (115).

83   *Eurytides (Protesilaus) protesilaus* (Linnaeus, 1758)
Central America, Trinidad and South America (excluding Chile and Uruguay). Apparently common and not threatened. Three subspecies according to Hancock (109). Common name: Northern White Page (9), Swordtail (282). Other refs: 26, 51, 55, 61, 62, 115, 122, 188, 234, 269.

84   *Eurytides (Protesilaus) glaucolaus* (Bates, 1864)
Panama, South America (excluding Chile, Uruguay and Argentina) and possibly Costa Rica (51). Not known to be threatened. Sometimes difficult to separate from *E. (P.) molops* (51). Four subspecies. Other refs: 55, 234.

Species-group: *asius* Fabricius

85   *Eurytides (Protesilaus) asius* (Fabricius, 1781)
South-western Brazil and eastern Paraguay. Not known to be threatened. Ranched in Brazil (24). Other refs: 51, 55, 234.

86   *Eurytides (Protesilaus) microdamas* (Burmeister, 1878)
Paraguay and adjacent areas of Argentina and Brazil. Little known but no threats recognized. Refs: 51, 55, 122, 234.

87   *Eurytides (Protesilaus) thymbraeus* (Boisduval, 1836)
Mexico, Guatemala, Belize, Honduras and El Salvador. Apparently fairly common and widespread within its range (234, 269) and not known to be threatened. Two subspecies. Other refs: 51, 55, 62.

88   *Eurytides (Protesilaus) belesis* (Bates, 1864)
Mexico, Guatemala, Honduras and Nicaragua. Uncommon in some areas (57), but not rare and not known to be threatened. Dimorphic, but no accepted subspecies. Other refs: 51, 55, 62, 232, 234, 269.

89   *Eurytides (Protesilaus) branchus* (Doubleday, 1846)
Mexico, Guatemala, Honduras, Belize (?), El Salvador (?), Nicaragua and Costa Rica. Apparently common north of Costa Rica, but rare within Costa Rica (61). Not believed to be threatened. Two forms, analogous to those of *E. (P.) belesis*. Other refs: 51, 55, 57, 62, 234, 269.

90   *Eurytides (Protesilaus) ilus* Fabricius, 1793
Colombia, Panama and northern Venezuela. Rare and little known but not recognized as threatened. Refs: 51, 55, 234.

91   *Eurytides (Protesilaus) lysithous* (Hübner, 1821)
Brazil, Argentina and eastern Paraguay. Apparently not uncommon and not threatened as a species. Ranched in Brazil (24). *E. (P.) lysithous* mimics

various species of *Parides* and seven taxa are treated variously as subspecies, species, or merely forms (234). *E. (P.) kumbachi* from Salto Grande is regarded as an aberration of *E. (P.) lysithous* by D'Almeida (55) and Hancock (109).

Eurytides (Protesilaus) lysithous harrisianus (Swainson, 1822)
Endangered—refer to section 6, p. 208.
This Brazilian sub-species mimics the Vulnerable *Parides ascanius* and is itself seriously Endangered (283). Nearly all known colonies have been destroyed by development and only a single known locality remains (283). The subspecies is now on the official list of Brazilian animals threatened with extinction (23). Common name: Harris' Mimic Swallowtail. Other refs: 26, 55, 122.

92 *Eurytides (Protesilaus) ariarathes* (Esper, 1788)
Colombia, Venezuela, Guyana, Surinam, French Guiana, Brazil, Ecuador, Peru and Bolivia. Females resemble the females of *Aristolochia* feeding Papilios and exhibit geographical variation (51). Not uncommon and no known threats. Up to nine subspecies. Other refs: 26, 55, 234.

93 *Eurytides (Protesilaus) harmodius* (Doubleday, 1846)
Colombia, Ecuador, Peru and Bolivia. Common and not threatened. Five or six subspecies. The female of the nominate form is known from only one specimen. Refs: 51, 55, 234.

94 *Eurytides (Protesilaus) trapeza* (Rothschild and Jordan, 1906)
Known only from Ecuador and north-eastern Peru (restricted range), but not uncommon. Common in the Napo province of Ecuador but declining in Pastaza province (221). Two subspecies. Other refs: 51, 55, 234.

95 *Eurytides (Protesilaus) xynias* (Hewitson, 1875)
Ecuador, Bolivia and Peru. Not uncommon and not threatened. Two subspecies. Refs: 51, 55, 234.

96 *Eurytides (Protesilaus) phaon* (Boisduval, 1836)
Mexico, Guatemala, Honduras, Belize, Nicaragua, Costa Rica (likely, but no records yet, 61), Panama (?), Colombia, Venezuela and western Ecuador. Not known to be threatened. Two forms, not thought to be of subspecific status (234). Other refs: 51, 55, 57, 62, 269.

97 *Eurytides (Protesilaus) euryleon* (Hewitson, 1855)
Costa Rica, Panama, western Colombia and Ecuador. Present in most habitats throughout the year, at least in Costa Rica (61). Not known to be threatened. Five subspecies. Other refs: 51, 55, 234.

98 *Eurytides (Protesilaus) pausanias* (Hewitson, 1852)
Costa Rica, Panama, Colombia, Venezuela, Trinidad, Guyana, Surinam, French Guiana, northern Brazil, Ecuador, Peru and Bolivia. Very rare in Costa Rica and Panama (61, 274), generally uncommon and possibly threatened. Little known of its biology and ecology. Mimics the unpalatable *Heliconius wallacei* (Heliconiinae) (51). More data required on conservation status in the main part of its range. Up to four subspecies. Other refs: 9, 55, 234.

99 *Eurytides (Protesilaus) protodamas* (Godart, 1819)
Southern Brazil, Paraguay and Argentina. Not known to be threatened.

Two forms, not regarded as subspecies. Ranched in Brazil (24). Other refs: 51, 55, 122, 234.

Species-group: *thyastes* Drury

100 *Eurytides (Eurytides) marchandi* (Boisduval, 1836)
Mexico, Central America, Colombia and western Ecuador. Inhabits rain forest up to 1000 m in Costa Rica (61). Less common than *E. (E.) thyastes* but not rare or threatened. Two subspecies. Other refs: 51, 55, 57, 62, 234, 269.

101 *Eurytides (Eurytides) thyastes* (Drury, 1782)
Eastern Ecuador, Peru, Bolivia and Brazil. Not uncommon and not threatened. Ranched in Brazil (24). Three subspecies. Other refs: 51, 55, 234.

102 *Eurytides (Eurytides) calliste* (Bates, 1864)
Mexico, Guatemala, Belize, Honduras, El Salvador (?), Nicaragua (?), Costa Rica and Panama (261). Apparently common in Mexico, but rarely seen in Costa Rica (61). Not known to be threatened. Two subspecies. Other refs: 51, 55, 62, 234, 269.

103 *Eurytides (Eurytides) leucaspis* (Godart, 1819)
Ecuador, Colombia, Peru and Bolivia. Common and not threatened. Two subspecies. (222). Other refs: 51, 55, 234.

104 *Eurytides (Eurytides) lacandones* (Bates, 1864)
Mexico, Central America, Colombia, Ecuador, Peru and Bolivia. Possibly conspecific with *E. (E.) dioxippus* (222). Rare in Costa Rica (61). Generally poorly known, but wide-ranging and not believed to be threatened. Two subspecies. Other refs: 51, 55, 62, 234, 269.

105 *Eurytides (Eurytides) dioxippus* (Hewitson, 1855)
Colombia. Possibly conspecific with *E. (E.) lacandones* (222). Restricted range, but not uncommon and not known to be threatened. Other refs: 51, 234.

Species-group: *dolicaon* Cramer

106 *Eurytides (Eurytides) serville* (Godart, 1824)
Colombia, Venezuela, Brazil, Ecuador, Peru and Bolivia. Possibly conspecific with *E. (E.) columbus* (222). Common and not threatened. Two subspecies. Other refs: 51, 55, 234.

107 *Eurytides (Eurytides) columbus* (Kollar, 1850)
Colombia, north-western Ecuador and possibly Venezuela. Possibly conspecific with *E. (E.) serville* (222). No known threats. Other refs: 51, 55, 234.

108 *Eurytides (Eurytides) orabilis* (Butler, 1872)
Guatemala, Costa Rica, Panama and Colombia. In Costa Rica always associated with primary forest; always uncommon and solitary but present throughout the year (61). Not recognized as threatened. Two subspecies. Other refs: 51, 55, 234.

109 *Eurytides (Eurytides) salvini* (Bates, 1864)
Mexico, Guatemala and Belize. Not particularly common but not known to be threatened. Refs: 51, 55, 57, 62, 234, 269.

110 *Eurytides (Eurytides) callias* (Rothschild and Jordan, 1906)
Eastern Ecuador and Peru. Not known to be threatened and not uncommon. Refs: 51, 55, 234.

111 *Eurytides (Eurytides) dolicaon* (Cramer, 1775)
South America (excluding Chile and Uruguay). Ranched in Brazil (24). Seven, possibly eight subspecies. Other refs: 26, 51, 54, 55, 122, 234.

112 *Eurytides (Eurytides) iphitas* (Hübner, 1821)
Vulnerable—refer to section 6, p. 211.
Brazil. Extremely rare and not seen for several decades (23). Believed to be seriously threatened but more details needed. Common name: Yellow Kite. Other refs: 51, 55, 139, 234.

Genus: *Protographium* Munroe

113 *Protographium leosthenes* (Doubleday, 1846)
Australia; restricted to the east coast from Cape York southwards to Sydney and (as a separate subspecies) to a small area of the Northern Territory. Apparently not uncommon and not threatened, but status needs to be monitored. Two subspecies. Refs: 14, 48, 141.

Genus: *Lamproptera* Gray

114 *Lamproptera meges* (Zinken–Sommer, 1831)
North-eastern India (Assam), Burma, Thailand, Laos, Vietnam, southern China (including Hainan (Guangdong prov.)), Kampuchea, Peninsular and Eastern Malaysia, Philippines, Brunei and Indonesia (Sumatra, Babi Is, Nias, Bangka, Java, Sulawesi and Kalimantan (52, 262)). Not known to be threatened in most of its range, but considered to be Vulnerable and in need of some protection in Peninsular Malaysia (10). Ten subspecies (262). Common name: Green Dragontail (45, 80, 282). Other refs: 87, 131, 141, 259.

115 *Lamproptera curius* (Fabricius, 1787)
North-eastern India (Assam), Burma, Thailand, southern China (including Hainan (Guangdong prov.)), Hong Kong, Laos, Vietnam, Kampuchea, Peninsular and Eastern Malaysia, Philippines (Palawan), Brunei and Indonesia (Sumatra, Bangka, Nias, Bunguran, Java and Kalimantan (52, 262)). Not known to be threatened in most of its range, but considered to be Vulnerable and in need of some protection in Peninsular Malaysia (10). Four subspecies. Common name: White Dragontail (3, 80). Other refs: 45, 87, 131, 141, 259.

Genus: *Graphium* Scopoli
Subgenus: *Pazala* Moore

Species-group: *alebion* Gray

116 *Graphium (Pazala) eurous* (Leech, 1892–94)
Northern India (Kashmir and Jammu, Himachal Pradesh, Uttar Pradesh, Assam, Sikkim and Manipur), Nepal, northern Burma, south-western and central China and Taiwan. Common and not threatened. Five subspecies including *caschmirensis* Rothschild from Northern India (Jammu and

Kashmir, and the Himalayas). Common name: Six-bar Swordtail (80, 251). Other refs: 4, 52, 141, 259.

117 *Graphium (Pazala) mandarinus* (Oberthür, 1879) (= *glycerion* Gray)
China, Burma and Nepal. Not common, but status and distribution not well known and further information required. Common name: Spectacle Swordtail (80, 251). Other refs: 52, 207.

118 *Graphium (Pazala) alebion* (Gray, 1853)
China and Taiwan. Status, exact distribution and nature of any threats not known. Two or more subspecies. Refs: 52, 141.

119 *Graphium (Pazala) tamerlanus* (Oberthür, 1876)
China (including Xizang Zizhiqu (Tibet)). Apparently not common, but species little known and more information required. Refs: 188, 250.

Subgenus: *Pathysa* Reakirt

Species-group: *antiphates* Cramer

120 *Graphium (Pathysa) aristeus* (Cramer, 1775)
Northern India (Assam and Sikkim), Burma, south-eastern China (including Hainan (Guangdong prov.)), Thailand, Laos, Vietnam, Kampuchea, Peninsular and Eastern Malaysia, Philippines, (possibly not Panay, Negros and Leyte), Indonesia (excluding Sulawesi, Sula, Bangka and Lombok (?)), Brunei, Papua New Guinea, Bismarck Archipelago (including New Britain) and Australia (northern Queensland) (52, 262). Collected for trade in Papua New Guinea (36). *G. (P.) a. anticrates* is protected by law in India (182). Nine subspecies (262). Common name: Five-bar Swordtail (282), Chain Swordtail (80). Other refs: 4, 40, 45, 48, 87, 131, 176.

121 *Graphium (Pathysa) nomius* (Esper, 1798)
Southern and eastern India (including Assam and Sikkim), Nepal, Sri Lanka, Bangladesh (?), Burma, Thailand, Laos (?), Vietnam (?) and Kampuchea (?) (52). Fairly common and not known to be threatened. Two subspecies. Common name: Spot Swordtail (80, 251, 282). Other refs: 4, 259, 287.

122 *Graphium (Pathysa) rhesus* (Boisduval, 1836)
Indonesia (Sulawesi, Butung, Tanahjampea, Tukangbesi and Sula (52, 262)). Regarded as not uncommon and not threatened. This species appears to be the replacement species for *G. (P.) aristeus* in Sulawesi (52). Four subspecies. Other ref: 121.

123 *Graphium (Pathysa) dorcus* (de Haan, 1840)
Sulawesi (Indonesia). Not a well-known species, possibly confined mainly to mountainous areas. Although no threats are known, this species is a rarity and more information is required. Two subspecies. Refs: 52, 121, 262.

124 *Graphium (Pathysa) androcles* (Boisduval, 1836)
Indonesia (Sulawesi and Sula (262)). Apparently not uncommon and not known to be threatened. Two subspecies. Other ref: 121.

125 *Graphium (Pathysa) epaminondas* Oberthür, 1879
Insufficiently Known—refer to section 6, p. 213.

Confined to the Andaman Is; distribution within the group poorly known. Although not particularly rare in its chosen haunts, the butterfly has a very restricted range and requires further study before its status can be verified. It is sometimes treated as a subspecies of *G. (P.) antiphates* (5, 80, 188, 259) but seems to be distinct (250, 262).

126 *Graphium (Pathysa) euphrates* (C. & R. Felder, 1862)
Philippines (Luzon, Mindoro, Palawan and Balabac, possibly other islands too), Malaysia (only Banggi I., off Sabah) and Indonesia (Sulawesi, Halmahera and Obi). A rather disjunct distribution, but not uncommon. Not known to be threatened. Three or perhaps more subspecies (48, 141, 262).

127 *Graphium (Pathysa) decolor* Staudinger, 1888
Sabah (Eastern Malaysia) and the Philippines, including Palawan but excluding Panay, Cebu and Masbate (262). Usually rare and local, except in Mindoro, where it is common in season. Not known to be threatened. Five subspecies (262). Often confused with *G. (P.) euphrates*, as by D'Abrera (52).

128 *Graphium (Pathysa) antiphates* (Cramer, 1775)
North-eastern and southern India, Nepal, Sri Lanka, Burma, south-eastern China (including Hainan (Guangdong prov.)), Thailand, Laos, Vietnam, Kampuchea, Peninsular and Eastern Malaysia, Brunei and Indonesia (Sumatra, Mentawai Is, Nias, Bangka, Bali, Lesser Sunda Is (except Sumba), Sulawesi, Bunguran (Natuna Is) and Kalimantan) (52, 191, 262). Not threatened as a species but the subspecies in Sri Lanka is apparently very rare (259). Twelve subspecies. Common name: Five-bar Swordtail (45, 80, 251). Other refs: 4, 87, 131, 141, 143, 251, 287.

129 *Graphium (Pathysa) agetes* (Westwood, 1841)
Northern India (Assam and Sikkim), Burma, Thailand, Laos, Vietnam, south-eastern China (including Guangdong prov.), Kampuchea (?), Peninsular and Eastern Malaysia (Sarawak and Sabah) and Indonesia (Sumatra, Kalimantan (?)) (52, 262). Widespread, often common and not threatened. Uncommon in Peninsular Malaysia; found on open hilltops above 1000 m. Five subspecies (262). Common name: Four-bar Swordtail (80). Other refs: 45, 87, 131, 259.

130 *Graphium (Pathysa) stratiotes* (Grose–Smith, 1897)
Eastern Malaysia (Sabah and Sarawak), possibly also Brunei and Kalimantan (Indonesia) (52, 262). Not particularly common but not recognized as threatened. Other refs: 131, 259.

Species-group: *macareus* Godart

131 *Graphium (Pathysa) phidias* (Oberthür, 1896)
Vietnam, Laos (?). Recorded from only a small area and possibly Vulnerable but very little information available. Refs: 52, 262.

132 *Graphium (Pathysa) encelades* (Boisduval, 1836)
Indonesia (Sulawesi). Not uncommon and not recognized as being threatened. Refs: 52, 262.

133 *Graphium (Pathysa) idaeoides* (Hewitson, 1853)
Rare—refer to section 6, p. 215.

Philippines (Luzon, Samar, Leyte and Mindanao) (146). In view of the widespread habitat destruction in the Philippines, this species is a probable candidate for Vulnerable status in the future. A mimic of the sympatric but more widespread danaine *Idea leuconoe*. Much prized by Japanese and other collectors (274). Other refs: 52, 147, 148, 262.

134   *Graphium (Pathysa) delesserti* (Guérin, 1839)
Thailand, Peninsular and Eastern Malaysia, Philippines (Palawan), Brunei and Indonesia (Sumatra, Nias, Bangka, Bunguran, Java (probably extinct) and Kalimantan) (52, 262). Often common and no threats known although it appears to have become extinct in Java (45). If this species is extinct in Java it is likely to be due to loss of habitat; deforestation is very extensive on this densely populated island. The extremely scarce female resembles *Ideopsis gaura* (Danainae) (45). Four subspecies. Common name: Zebra (251), Malayan Zebra (45). Other refs: 87, 131.

135   *Graphium (Pathysa) xenocles* (Doubleday, 1842)(= *leucothoe* Westwood)
Northern India (Assam and Sikkim), Nepal, Bhutan, Burma, Thailand, Laos, Vietnam and China (Hainan (Guangdong prov.)) (52). Apparently common and not threatened. Four subspecies. Common name: Great Zebra (80), Greater Zebra (251). Other refs: 4, 259.

136   *Graphium (Pathysa) macareus* (Godart, 1819)
Northern India (Assam, Sikkim and Manipur), Nepal, Burma, Thailand, Laos, Vietnam, Kampuchea, China (Hainan (Guangdong Prov.)), Peninsular and Eastern Malaysia, Brunei, Philippines (Palawan) and Indonesia (Sumatra, Java, Bali and Kalimantan) (52, 262). Relatively common and not threatened. Eleven subspecies (52, 262). Common name: Lesser Zebra (80, 251). Other refs: 45, 87, 131, 259.

137   *Graphium (Pathysa) ramaceus* (Westwood, 1872)
Thailand, Peninsular and Eastern Malaysia, Brunei and Indonesia (Sumatra and Kalimantan) (52, 262). Uncommon (45), but not rare and not thought to be threatened. Three subspecies. Common name for ssp. *G. (P.) r. pendleburyi*: Pendlebury's Zebra (45). Other ref: 87.

138   *Graphium (Pathysa) megarus* (Westwood, 1841)
Northern India (Assam and Sikkim), Burma, Thailand, Laos, Vietnam, south-eastern China (including Hainan (Guangdong prov.)), Kampuchea, Peninsular and Eastern Malaysia, Brunei and Indonesia (Sumatra, Java, Bali and Kalimantan) (52, 262). Rather common and not considered to be threatened, although the nominate subspecies is protected by law in India (182). Eight subspecies. Common name: Spotted Zebra (80, 282). Other refs: 45, 87, 131, 259.

139   *Graphium (Pathysa) megaera* (Staudinger, 1888)
Indeterminate—refer to section 6, p. 217.
Philippines (Palawan). This species merits Indeterminate status because of its restricted range in a habitat that is being increasingly disrupted. Refs: 52, 262.

140   *Graphium (Pathysa) stratocles* (C. & R. Felder, 1862)
Philippines (Luzon, Mindoro, Marinduque, Bohol, Mindanao, Calamians and Palawan) (52, 262). Not uncommon and not known to be threatened. Three subspecies.

141 *Graphium (Pathysa) deucalion* (Boisduval, 1836)
Indonesia (Sulawesi, Moluccas (Halmahera, Bacan and Ternate) and Irian Jaya (Biak)(115)). Not known to be threatened. Two (262) or three (104) subspecies. Other refs: 48, 52.

142 *Graphium (Pathysa) thule* (Wallace, 1865)
Indonesia (Irian Jaya) and Papua New Guinea. Certainly not common, but not known to be threatened. Rarely collected but possibly overlooked as it mimics the common danaines *Ideopsis juventa* and *Tirumala hamata*. One subspecies with three forms (104). Other ref: 48.

Subgenus: *Arisbe* Hübner

Species-group: *angolanus* Goeze

143 *Graphium (Arisbe) endochus* (Boisduval, 1836)
Madagascar only. Apparently well distributed and not currently threatened, but it is a forest species and its status in a country which is being rapidly deforested should be carefully monitored. Refs: 33, 34, 50, 204, 206, 285.

144 *Graphium (Arisbe) angolanus* (Goeze, 1779) (= *pylades* F., 1793)
Southern and tropical Africa: distributed in woodland over most of the continent. Common, often abundant and not threatened. Three subspecies. Common names: Angola White Lady Swallowtail (282, 285), White Lady Swallowtail (34, 44, 214). Other refs: 20, 33, 50, 63, 97, 165. *See note after References.*

145 *Graphium (Arisbe) taboranus* (Oberthür, 1886)
Zaire, Tanzania, Angola, Zambia, Malawi and Namibia. Not uncommon and not known to be threatened. Two subspecies. Common name: Tabora Swallowtail (215, 285). Other refs: 33, 34, 50, 63, 97, 165. *See note after References.*

146 *Graphium (Arisbe) morania* (Angas, 1849)
South Africa (Cape Province, Natal and Transvaal), Namibia, Zimbabwe, southern Malawi and Mozambique. Not uncommon and not threatened. Common name: White Lady Swallowtail (285), Lesser White Lady Swallowtail (214). Other refs: 34, 50, 63, 115.

Species-group: *ridleyanus* White

147 *Graphium (Arisbe) ridleyanus* (White, 1843)
Lowland forest in West Africa: Sierra Leone, Liberia, Ivory Coast, Ghana, Togo, Benin, Nigeria, Cameroon, Central African Republic (?), Gabon, Congo, Zaire, Uganda, western Tanzania, Zambia and Angola. Rather widely distributed and not known to be threatened. Common names: Acraea Swallowtail (282, 285), Red Graphium (34), Ridley's Swallowtail (215). Other refs: 20, 33, 50, 115.

Species-group: *philonoe* Ward

148 *Graphium (Arisbe) philonoe* (Ward, 1873)
Forest and woodland in East Africa: Kenya, Tanzania, southern Sudan, northern Uganda, south-west Ethiopia, Mozambique and Malawi.

Common in much of its range and not threatened. Two subspecies.
Common names: White-dappled Swallowtail (285), Eastern Graphium
(34). Other refs: 33, 50, 97.

Species-group: *adamastor* Boisduval

149 *Graphium (Arisbe) almansor* (Honrath, 1884)
Cameroon, Central African Republic, Sudan, Ethiopia, Equatorial Guinea
(?), Gabon, Congo, Zaire, Uganda, western Kenya, Rwanda (?), Burundi
(?), north-west Tanzania, Angola and Zambia. Uncommon and local,
though widely distributed. Not threatened. Four subspecies. Refs:
33, 34, 50, 151, 285, 291.

150 *Graphium (Arisbe) poggianus* (Honrath, 1884)
Angola, Zaire and Zambia (115). Given as a form of *G. (A.) almansor* by
Berger (16) but accepted as a full species by Hancock (109), under the name
carchedonius Karsch. The name *poggianus* follows Hancock (115). Uncommon but no evidence of being threatened. *See note after References.*

151 *Graphium (Arisbe) adamastor* (Boisd. 1836) (= *carchedonius* Karsch)
Guinea, Sierra Leone, Liberia, Ivory Coast, Ghana, Togo, Benin and
Nigeria; also a subspecies in Central African Republic and northern Zaire
(? disjunct range). Uncommon and local but not threatened. Two subspecies. Refs: 20, 33, 34, 50, 285.

152 *Graphium (Arisbe) agamedes* (Westwood, 1842)
Lowland forest in Ghana, Togo, Benin, Nigeria, Cameroon, Central
African Republic, Equatorial Guinea (?), Gabon (?), northern Congo and
northern Zaire. Local and uncommon but not regarded as threatened.
Common name: Glassy Graphium (34). Other refs: 20, 33, 50, 285.

153 *Graphium (Arisbe) aurivilliusi* (Seeldrayers, 1896)
Insufficiently Known—refer to section 6, p. 219.
West Africa: Type specimen labelled "Congo" but probably present-day
Zaire (34). Known only from the type series and possibly not specifically
distinct (form of *agamedes* ? (250)), but recognized by Carcasson (34).
Other refs: 33, 50, 285.

154 *Graphium (Arisbe) olbrechtsi* (Berger, 1950)
Zaire. Uncommon and scarce. This recently-described species may deserve
Rare status when more information becomes available. Two subspecies.
Refs: 33, 34, 50, 115, 285. *See note after References.*

155 *Graphium (Arisbe) odin* (Strand, 1910)
Cameroon, Congo, Central African Republic and Zaire. Not common but
not threatened. Three (possibly only two) subspecies. Refs: 34, 50, 285. *See
note after References.*

156 *Graphium (Arisbe) auriger* (Butler, 1877)
West Africa: Gabon. Uncommon but not known to be threatened. More
information needed. Refs: 34, 50, 285. *See note after References.*

157 *Graphium (Arisbe) hachei* (Dewitz, 1881)
Lowland forest in Cameroon, Equatorial Guinea (?), Central African
Republic, Gabon, Congo, western Zaire and Angola. Uncommon, but not

threatened. Two or three subspecies. Common name: Milky Graphium (34). Other refs: 33, 50, 285.

158 *Graphium (Arisbe) weberi* (Holland, 1917)
Insufficiently Known—refer to section 6, p. 220.
Cameroon. Known only from the type specimen in Pittsburgh, U.S.A. (50). More information required. Common name: Weber's Swallowtail (285). Other refs: 33, 34. *See note after References.*

159 *Graphium (Arisbe) fulleri* (Grose–Smith, 1883)
West Africa: Cameroon, Gabon and the Congo. Possibly a form of *G. (A.) ucalegonides*, but regarded as a full species by Carcasson (34). Uncommon, but not known to be threatened. Common name: Fuller's Swallowtail (270). Other refs: 33, 50, 285. *See note after References.*

160 *Graphium (Arisbe) ucalegonides* (Staudinger, 1884)
Ghana, Togo, Benin, Nigeria, Cameroon, Central African Republic, Equatorial Guinea (?), Gabon, Congo, Zaire and Angola. Not uncommon and not threatened. Two subspecies. Refs: 34, 50, 285. *See note after References.*

161 *Graphium (Arisbe) ucalegon* (Hewitson, 1865)
Lowland forest in Nigeria, Cameroon, Central African Republic (?), Gabon, Congo, Zaire, Uganda, Rwanda (?), Burundi (?), Tanzania and Angola. Not uncommon and not considered to be threatened. Two subspecies. Common name: Creamy Graphium (34). Other refs: 20, 33, 50, 151, 285.

162 *Graphium (Arisbe) simoni* (Aurivillius, 1899)
Cameroon, Gabon, Congo and Zaire. Possibly a form of *G. (A.) ucalegon* (250), although regarded as a full species by Carcasson (34). Uncommon but not known to be threatened. Common name: Simon's Swallowtail (285). Other refs: 33, 50.

Species-group: *leonidas* Fabricius

163 *Graphium (Arisbe) cyrnus* (Boisduval, 1836)
Madagascar. Apparently common throughout the island. Not known to be threatened. Two subspecies (206), not recognized by D'Abrera (50). Other refs: 33, 34.

164 *Graphium (Arisbe) leonidas* (Fabricius, 1793)
Throughout tropical and southern Africa, common in forest, woodland and gardens but absent from dry and montane areas. Not threatened. Three subspecies, two on small islands and so very restricted in range. Larva now generally found on the introduced Custard Apple (*Annona reticulata* (Annonaceae)), but this is presumably not the original foodplant. Sexes similar. Common names: Veined Swallowtail (44, 285), Common Graphium (34). Other refs: 20, 33, 50, 63, 97, 214, 215.

165 *Graphium (Arisbe) pelopidas* (Oberthür, 1879)
Tanzania (Pemba Island and possibly Zanzibar). Recognized as a species by D'Abrera (50) but not by other authors, including Munroe (188), Carcasson (34), Smart (250) and Hancock (109, 113). Should possibly be regarded as a subspecies of *G. (A.) leonidas* as are the populations on Sao Tomé and

Principé in the Gulf of Guinea. Status of Zanzibar population particularly uncertain. Should the Pemba Island population be considered as a full species it would certainly be given Rare or even Vulnerable status. Pemba Island is largely covered with clove plantations but natural habitat may still be found in the Ngezi forest, north of Pemba (264).

166 *Graphium (Arisbe) levassori* (Oberthür, 1890)
Vulnerable—refer to section 6, p. 222.
Comoros (Grande Comoro) only (205). Range very circumscribed and shrinking. Other refs: 33, 34, 50, 206.

Species-group: *tynderaeus* Fabricius

167 *Graphium (Arisbe) tynderaeus* (Fabricius, 1793)
Lowland forest in West Africa: Sierra Leone, Liberia, Ivory Coast, Ghana, Togo, Benin, Nigeria, Cameroon, Central African Republic (?), Gabon (?), Congo and Zaire. Widely distributed, not uncommon and no threats recognized. Common name: Green-spotted Swallowtail (285). Other refs: 20, 34, 50.

168 *Graphium (Arisbe) latreillanus* (Godart, 1819)
Lowland forest in West Africa: Sierra Leone, Liberia, Ivory Coast, Ghana, Togo, Benin, Nigeria, Cameroon, Central African Republic (?), Gabon (?), Congo, Zaire, Uganda and Angola. Not threatened. Two subspecies. Common names: Coppery Swallowtail (285), Olive Graphium (34). Other refs: 20, 33, 50.

Subgenus: *Graphium* Scopoli

Species-group: *policenes* Cramer

169 *Graphium (Graphium) junodi* (Trimen, 1893)
Restricted to Mozambique. There has also been a single record from eastern Zimbabwe but this was probably a vagrant (115). Uncommon or scarce, but not known to be threatened. Common names: Mozambique Swallowtail (285), Junod's Swordtail (44), Scarce Swordtail (44). Other refs: 33, 34, 50, 214.

170 *Graphium (Graphium) nigrescens* (Eimer, 1889)
Cameroon, Gabon and Zaire. Species of uncertain status (50) but said to be rare (285). More information required. Common name: Dusky Swordtail (285). Other refs: 33, 34.

171 *Graphium (Graphium) polistratus* Grose–Smith, 1889
East Africa, largely coastal from southern Somalia southwards to Delagoa Bay, i.e. Somalia, Kenya, Tanzania, Malawi and Mozambique. Not particularly common but no threats known. *Graphium sisenna* (Mabille, 1890) was listed by Munroe (188) but recognised as a synonym of *G. (G.) polistratus* by Hancock (109). Common name: Dancing Swordtail (285), Dusky Swordtail (214). Other refs: 33, 34, 50, 63.

172 *Graphium (Graphium) policenes* (Cramer, 1775)
Tropical and southern Africa, widely distributed. Common in woodland and forest and not threatened. Most authors do not accept described subspecies to be valid (33, 50, 97). Larvae feed on *Artobotrys*

(Annonaceae). *G. (G.) boolae* was shown to be an aberration of *G. (G.) policenes* (41). Common names: Small-striped Swordtail (285), Small Striped Swordtail (214), Common Swordtail (34). Other refs: 20, 63.

Species-group: *porthaon* Hewitson

173 *Graphium (Graphium) porthaon* (Hewitson, 1865)
East Africa, fairly widely distributed in woodland in western Transvaal (South Africa), Mozambique, Botswana, Zimbabwe, Malawi, Zambia, Zaire, Tanzania and eastern Kenya. Apparently not uncommon and no threats known. Two subspecies of which *G. (G.) p. tanganyikae* is rare in Tanzania (34, 151). Common names: Cream-striped Swordtail (285), Cream Swordtail (34), Pale Spotted Swordtail (214). Other refs: 33, 50, 63, 97, 165.

Species-group: *illyris* Hewitson

174 *Graphium (Graphium) illyris* (Hewitson, 1873)
Lowland forest in Guinea, Sierra Leone, Liberia, Ivory Coast, Ghana, Togo, Benin, Nigeria, Cameroon, Gabon (?), Congo and Zaire (34). Not uncommon but not known to be threatened. Common names: Yellow-banded Swordtail (285), Cream-banded Swordtail (34). Other refs: 20, 33, 50, 103.

175 *Graphium (Graphium) gudenusi* (Rebel, 1911)
Eastern Zaire, south-western Uganda, Rwanda and Burundi, a rather restricted range in highland forests (263). Scarce and local, but not threatened at present. However, human population pressure on some parts of its range, e.g. the highlands of Rwanda, is increasing and the species needs further study and monitoring. Common name: Kigezi Swordtail (285). Other refs: 33, 34, 50.

176 *Graphium (Graphium) kirbyi* (Hewitson, 1872)
East Africa: Kenya and Tanzania. Local but sometimes common in coastal forests and not threatened. Common name: Kirby's Swordtail (34, 285). Other refs: 33, 50.

Species-group: *colonna* Ward

177 *Graphium (Graphium) colonna* (Ward, 1873)
East Africa, coastal woodlands and forests in South Africa (Natal, Transvaal (?)), Swaziland (?), Mozambique, Malawi, Tanzania, Kenya, Somalia and southern Ethiopia (34). Not known to be threatened. Common names: Mamba Swordtail (285), Black Swordtail (34). Other refs: 33, 50, 63, 214.

Species-group: *antheus* Cramer

178 *Graphium (Graphium) antheus* (Cramer, 1779)
Most of continental Africa. Common and not in danger. *G. (G.) mercutius* Grose–Smith and Kirby, is not listed by Munroe (188), D'Abrera (50), Carcasson (34) and most other authors and is only a form of *G. (G.) antheus* (115). Common name: Large Striped Swordtail (44, 285). Other refs: 20, 33, 63, 97, 165, 214, 215.

179 *Graphium (Graphium) evombar* (Boisduval, 1836)
Madagascar, distributed over the whole island. Common and not threatened. Refs: 33, 34, 50, 206.

Species-group: *eurypylus* Linnaeus

180 *Graphium (Graphium) arycles* (Boisduval, 1836)
Southern Burma, Thailand, Kampuchea (?), Peninsular and Eastern Malaysia, Philippines (Palawan), Brunei and Indonesia (Sumatra, Bangka, Java and Kalimantan) (52, 262). Scarce but not known to be threatened although the nominate subspecies is protected in India (182). Three (262) or four (52) subspecies. Common name: Spotted Jay (80). Other refs: 45, 87, 231.

181 *Graphium (Graphium) bathycles* (Zinken–Sommer, 1831)
Southern Thailand (?), Peninsular and Eastern Malaysia, Philippines (Palawan), Brunei and Indonesia (Sumatra, Java and Kalimantan) (52, 262). Not rare and not threatened. Two subspecies, a third, *chiron*, having recently been raised to the specific level (52, 262) (see below). Common name: Veined Jay (111). Other refs: 45, 80, 87, 131, 231, 259.

182 *Graphium (Graphium) chiron* (Wallace, 1865)
Nepal, northern India (Assam), Burma, southern China, Thailand, Laos, Vietnam, Kampuchea and Peninsular Malaysia (76). Formerly regarded as a subspecies of *G. (G.) bathycles* but raised to specific level by Saigusa *et al.* (237). There is doubt concerning the validity of the name *chiron*. Hancock (115) considers it to be a homonym requiring replacement. Both *chironides* Honrath, 1884 and *clanis* (Jordan, 1909) are available and Hancock suggests the latter as being more appropriate, *chironides* having been used as an infrasubspecific title. We do not presume to alter the nomenclature here, but adopt the most recent published judgement. Three subspecies. Common name: Veined Jay (80). Other refs: 52, 109, 259, 262.

183 *Graphium (Graphium) leechi* (Rothschild, 1895)
China. Little information available and not known to be threatened. Ref: 188.

184 *Graphium (Graphium) procles* (Grose–Smith, 1887)
Indeterminate—refer to section 6, p. 224.
Montane forest in Eastern Malaysia (Sabah). Considered to be a good species by D'Abrera (52), Tsukada and Nishiyama (262) and Robinson (231) (who notes its similarity to *G. doson*) Other ref: 131.

185 *Graphium (Graphium) meyeri* (Hopffer, 1874)
Indonesia (Sulawesi and Sula) (262). Not a well-known species, but not considered to be threatened. Two subspecies (52).

186 *Graphium (Graphium) eurypylus* (Linnaeus, 1758)
North-eastern India, Bangladesh, Burma, Thailand, Andamans, Nicobar Is, southern China (including Hainan (Guangdong prov.)), Vietnam, Laos, Kampuchea, South East Asia to Papua New Guinea and north-eastern Australia (52, 262). Collected for trade in Papua New Guinea (36). *G. (G.) e. macronius* is protected by law in India (182). Not threatened. About twenty subspecies (262). Common name: Great Jay (80), Pale Green Triangle (282). Other refs: 5, 14, 40, 45, 48, 87, 176, 259.

187 *Graphium (Graphium) doson* (C. & R. Felder, 1864)
Northern and southern India, Nepal, Sri Lanka, Bangladesh, Burma, Thailand, south-east China (including Hainan (Guangdong prov.)), Taiwan, Japan, Vietnam, Laos, Kampuchea, Malaysia, Brunei, Indonesia (Sumatra, Nias, Mentawai Is, Bangka, Java, Bali, Lombok, Sumbawa, Bawean and Kalimantan) and the Philippines (including Palawan) (52, 191, 262)). Widespread and often common, but scarce in Honshu, Japan, where it only occurs in the southern part (82). Fourteen subspecies (262). Common name: Common Jay (45, 80, 251). Other refs: 87, 141, 149, 246, 259.

188 *Graphium (Graphium) evemon* (Boisduval, 1836)
North-eastern India (Assam), Burma, Thailand, southern China, Laos, Vietnam, Kampuchea, Malaysia, Brunei and Indonesia (Sumatra, Simeulue, Nias, Mentawai Is, Java and Kalimantan) (52, 191, 262). Fairly common and not threatened but one subspecies, *G. (G.) e. albociliates* is protected by law in India (182). Seven subspecies (262). Common name: Lesser Jay (3, 80). Other refs: 45, 87, 131, 231, 259.

Species-group: *agamemnon* Linnaeus

189 *Graphium (Graphium) agamemnon* (Linnaeus, 1758)
Southern and northern India, Nepal, Sri Lanka, Andamans, Nicobar, Bangladesh, Burma, Thailand, Laos, Kampuchea, southern China (including Hainan (Guangdong prov.)), Taiwan, South East Asia to Papua New Guinea, Bougainville, Solomon Is and Australia (northern Queensland) (52, 262). Farmed and collected for trade in Papua New Guinea (36, 68). Range similar to that of *G. (G.) sarpedon* but not reaching Japan. Common and not threatened. Twenty or twenty-one subspecies. Common names: Tailed Jay (45, 80, 251), Tailed Green Jay, Green-spotted Triangle (282). Other refs: 5, 14, 40, 48, 87, 141, 176, 191, 220, 231, 246, 287.

190 *Graphium (Graphium) macfarlanei* (Butler, 1877)
Indonesia (Buru, Obi, Halmahera, Batjan, Seram (?) and Irian Jaya), Papua New Guinea, Bismarck Archipelago (including New Britain) and north-eastern Australia. Farmed and collected for trade in Papua New Guinea (36, 68). Not threatened. Three subspecies. Other refs: 14, 40, 48, 141, 176.

191 *Graphium (Graphium) meeki* (Rothschild and Jordan, 1901)
Rare—refer to section 6, p. 226.
Solomon Is (Santa Isabel (171) and Choiseul) and Papua New Guinea (Bougainville)(219). Despite D'Abrera's statement that it is probably extinct (48), examples are sent in from time to time to the Bulolo Agency (36). The species is little known and may be classed as Rare until more details are available. A second subspecies *G. (G.) meeki inexpectatum* was described from Bougainville in 1981 (181). Other ref: 142.

Species-group: *sarpedon* Linnaeus

192 *Graphium (Graphium) gelon* Boisduval, 1859
New Caledonia (including Loyalty Is). Restricted range, but stated to be 'not rare' (48) and not recognizably threatened. Other ref: 133.

193   *Graphium (Graphium) macleayanum* (Leach, 1814)
Australia including Tasmania, and Lord Howe and Norfolk Is. Recently
recorded from mainland New Guinea (48). Not threatened as a species.
Four subspecies. Common name: Macleay's Swallowtail or Swordtail
(282). Other refs: 14, 95, 141, 176, 289.

194   *Graphium (Graphium) weiskei* (Ribbe, 1900)
All of New Guinea (Indonesia (Irian Jaya) and Papua New Guinea) (48) in
suitable habitat. Very common in some areas although the foodplant is still
unknown. It is collected for trade but this is occasionally halted for periods
of up to one year (36). It is not known to be threatened. Common name:
Purple-spotted Swallowtail (282). Other ref: 120.

195   *Graphium (Graphium) stresemanni* Rothschild, 1916
Rare—refer to section 6, p. 228.
Known only from the island of Seram in the Moluccas (Indonesia), where it
is Rare (48). Not listed by Munroe (188) as a species but recognized by most
other authors. Other refs: 48, 196.

196   *Graphium (Graphium) codrus* (Cramer, 1776)
Phlippines (not Palawan, Calamian Is, or Sulu Is), Indonesia (Sulawesi,
Moluccas (including Sula, Buru, Ambon and Seram but not Obi) and Irian
Jaya (including Waigeo, Biak, Kai and Aru), Papua New Guinea, Bismarck
Archipelago (including New Britain and the Admiralty Is), Bougainville
and the Solomon Is (not Rennell) (262). It is collected for trade in Papua
New Guinea and *G. (G.) c. auratus* is farmed on Manus I. (36). Some local
populations may be threatened but the species is not. Fourteen subspecies.
Other refs: 48, 52, 68, 127, 141, 181, 218.

197   *Graphium (Graphium) empedovana* (Corbet, 1941)(= *empedocles* F.,
1787)
Peninsular and Eastern Malaysia, Philippines (Palawan), Indonesia (Sumatra, Bangka, Java and Kalimantan) and Brunei (262). Sometimes regarded
as a subspecies of the Papuan *G. (G.) codrus* (45, 52). Uncommon generally
and possibly Vulnerable in Malaysia (10). *G. (G.) empedocles* is regarded
by some as a senior synonym of *G. (G.) empedovana* (262). Not known to be
threatened over most of its range. Other ref: 87.

198   *Graphium (Graphium) cloanthus* Westwood, 1841
Disjunct range: northern India, Nepal (251), Bhutan, northern Burma,
southern and central China, Taiwan, northern Thailand and Indonesia
(Sumatra)(52, 262). Common and not threatened. Three subspecies includ-
ing *G. (G.) clymenus* Leech, a little known form from central and western
China, but not including *G. (G.) sumatranum* (see below) (52, 141, 259).
Common name: Glassy Bluebottle (80, 251).

199   *Graphium (Graphium) sumatranum* (Hagen, 1894)
Indonesia (Sumatra). Recognized as a subspecies of *G. (G.) cloanthus* by
most authors (52, 262) but given species rank by Hancock (109) on the
strength of its isolated distribution and morphological differences. No
known threats, but more information needed.

200   *Graphium (Graphium) monticolum* (Fruhstorfer, 1897)
Indonesia (southern Sulawesi). Regarded by some as a subspecies of *G.
(G.) sarpedon* (121, 262), but by others as a full species (52, 109). *Graphium*

sarpedon textrix Tsukada and Nishyama, 1980, belongs to this species. No threats are known, although more information is needed.

201 *Graphium (Graphium) milon* C. & R. Felder, 1864
Indonesia (Sulawesi (including Banggai) and Moluccas: Sula, Buru, Ambon and Seram but not Morotai) (262). Status revised by Murayama in 1978 (190) but not accepted as a good species by Hancock (109). Not known to be threatened. Six subspecies (262).

202 *Graphium (Graphium) sandawanum* Yamamoto, 1977
Vulnerable—refer to section 6, p. 231.
A recent discovery (1977), known only from the Philippines (Mindanao: Mt Apo) (52, 262). Other ref: 289.

203 *Graphium (Graphium) sarpedon* Linnaeus, 1758
Southern India, Sri Lanka, northern India, Nepal, Bangladesh (?), Burma, south-eastern and western China (including Hainan (Guangdong prov.)), Taiwan, South Korea (?), Japan, Thailand, Laos, Kampuchea, Malaysia, Philippines, Brunei, Indonesia (not Moluccas), Papua New Guinea (including New Britain), Bismarck Archipelago, Solomon Is, Vanuatu (New Hebrides) (102) and north-eastern Australia (52, 262). Very common and not threatened. Widely collected for trade. Sixteen recognized subspecies (262) not including *milon* and *monticolum* which are regarded as separate species (52, 109, 262) or *textrix* which is a synonym of *monticolum* (52). G. *(G.) protensor* Gistel, is regarded as a synonym of *G. (G.) sarpedon* (109). Common names: Blue Triangle (282), Common Bluebottle (45, 80, 251). Other refs: 4, 14, 36, 40, 48, 87, 131, 141, 149, 176, 207, 220, 259.

Species-group: *mendana* Godman and Salvin

204 *Graphium (Graphium) mendana* (Godman and Salvin, 1888)
Rare—refer to section 6, p. 234.
Papua New Guinea (Bougainville) and Solomon Is. A little-known species though specimens are occasionally sent to the Bulolo Insect Trading Agency from Bougainville (36). Four subspecies. Other refs: 48, 99, 181, 218.

Species-group: *wallacei* Hewitson

205 *Graphium (Graphium) wallacei* (Hewitson, 1859)
Indonesia (Moluccas (possibly not Buru or Seram), Aru and Irian Jaya), and Papua New Guinea where it is collected for trade (36). Not threatened. Two subspecies. Other ref: 48.

206 *Graphium (Graphium) hicetaon* (Mathew, 1886)
Solomon Is. Not rare (48) and not known to be threatened. Other refs: 219, 220.

207 *Graphium (Graphium) browni* (Godman and Salvin, 1879)
Bismarck Archipelago (including New Britain). Fairly restricted range, but not rare (48) and not known to be threatened.

Subfamily: PAPILIONINAE
Tribe: Troidini

Genus: *Battus* Scopoli

Species-group: *philenor* Linnaeus

208　*Battus philenor* (Linnaeus, 1771)
Southern Canada, U.S.A. and Mexico. Not uncommon and not known to be threatened. Two subspecies (55). Common names: Pipevine Swallowtail (154, 216), Blue Swallowtail (216). Other refs: 62, 137, 141, 234, 269.

209　*Battus zetides* Munroe, 1971
Vulnerable—refer to section 6, p. 236.
Known only from high elevations in Haiti and the Dominican Republic. "Apparently very rare" (230), but "locally common under the right conditions and at the right times of year" (243). Further information needed. Originally described as *zetes* (Westwood, 1847) (230, 234). Common name: Zetides Swallowtail (230). Other refs: 51, 55, 139.

210　*Battus devilliers* (Godart, 1823)
Cuba and the Bahamas: Andros I. (230, 234). Not known to be threatened in Cuba, but the population on Andros is rare and precarious (268). Status needs clarification and careful monitoring. Very high prices demanded by dealers for this species. Common name: Devilliers Swallowtail (230). Other refs: 51, 55.

Species-group: *polydamas* Linnaeus

211　*Battus polydamas* (Linnaeus, 1758)
Southern U.S.A., Mexico, Central America, South America, Antigua, Bahamas, Cuba, Dominica, Dominican Republic, Grenada, Guadeloupe, Haiti, Jamaica, Martinique, Puerto Rico, St. Kitts, St. Lucia, Trinidad and Tobago. Very common and not threatened; often associated with disturbed habitats, rarely entering the forest (61). Well differentiated local races (51). Thirteen or fourteen subspecies. Common names: Polydamas Swallowtail (154, 216, 230), Gold Rim (9, 216), Black Page (9, 282). Other refs: 21, 26, 55, 62, 137, 234, 269.

212　*Battus streckerianus* (Honrath, 1884)
Peru; uncommon, but no threats known. Refs: 51, 55, 234.

213　*Battus archidamas* (Boisduval, 1836)
Coastal areas of central Chile. Appears to be common and not threatened. A male labelled Mendoza, Argentina, was described in 1925 by Ehrmann as the type specimen of *Papilio lindeni* (51). Other refs: 55, 234.

214　*Battus polystictus* (Butler, 1874)
Brazil, Paraguay and Argentina. Fairly widely distributed, not uncommon and not thought to be threatened. Ranched in Brazil (24). Two subspecies. Other refs: 51, 55, 122, 234.

215　*Battus philetas* (Hewitson, 1869)
Northern Peru and Ecuador. Not uncommon and not recognized as threatened. Two subspecies. Refs: 51, 55, 234.

216　*Battus madyes* (Doubleday, 1846)
Peru, Bolivia and Argentina. Not uncommon and not known to be threatened. Up to six subspecies. Refs: 51, 55, 234.

217 *Battus eracon* (Godman and Salvin, 1897)
Western Mexico. Local and uncommon, but not known to be threatened. A possible candidate for listing because of its restricted range. Further study and reserarch are necessary. Refs: 51, 55, 234, 269.

218 *Battus belus* (Cramer, 1777)
Mexico, Central America, Trinidad, Colombia, Venezuela, Guyana, Surinam, French Guiana, Ecuador, Peru, Brazil and Bolivia. Apparently not threatened, but is encountered as solitary individuals in humid forest and is rarely abundant (at least in Costa Rica) (61). Five or six subspecies; *B. b. chalceus* is only known from the type collection (174). Common name: the Belus (9). Other refs: 51, 55, 234, 269.

219 *Battus crassus* (Cramer, 1777)
Costa Rica, Panama (?), Colombia, Venezuela, Guyana, Surinam, French Guiana (?), Ecuador, Peru, Bolivia, Brazil, northern Argentina and possibly Paraguay. Widespread and not threatened, but solitary and rarely collected (61). Other refs: 26, 51, 55, 234.

220 *Battus laodamas* (Felder, 1859)
Mexico, Central America and Colombia. Little known but not threatened. Extremely local, but fairly common in places (61). Four subspecies. Other refs: 51, 55, 57, 62, 234, 269.

221 *Battus lycidas* (Cramer, 1777)
Mexico, Guatemala, Honduras, Nicaragua, Costa Rica, Panama, Trinidad, Colombia, Venezuela, Ecuador, Surinam, Peru, Bolivia and Brazil. Little known but not threatened. Very rare in Costa Rica (61). Common name: Lycidas Swallowtail (9, 282). Other refs: 51, 55, 150, 234.

Genus: *Euryades* C. & R. Felder

222 *Euryades duponchelii* (Lucas, 1839)
Woodlands in southern Brazil, Paraguay, Argentina and Uruguay (51). Less common than *E. corethrus* but not known to be threatened. Very high prices demanded by dealers for this species. The two species belonging to the genus *Euryades* are closer to the Australasian *Cressida* than to other Neotropical *Aristolochia* feeders (51). Other ref: 55.

223 *Euryades corethrus* (Boisduval, 1836)
Pampas in southern Brazil, Paraguay, Argentina and Uruguay. Not uncommon but not known to be threatened. Very high prices demanded by dealers for this species. Refs: 51, 55, 141.

Genus: *Cressida* Swainson

224 *Cressida cressida* (F., 1775) (= *heliconides* Swainson, 1832)
Indonesia (Lesser Sunda Is: Timor, Leti, Moa, Babar, Wetar and Tanimbar Is (?)), eastern Papua New Guinea and Australia (north Northern Territory and north and eastern Queensland) (262). Generally common and not threatened. Five subspecies. Common name: Big Greasy Butterfly (282). Other refs: 14, 40, 48, 141, 176.

Genus: *Parides* Hübner

Species-group: *ascanius* Cramer

225   *Parides gundlachianus* (C. & R. Felder, 1864)
Restricted to forests in Cuba. Not exceptionally rare but requires monitoring. *Parides columbus* (Herrich–Schaeffer, 1862) is a homonym. Common name: Gundlach's Swallowtail (230). Other refs: 51, 55, 234.

226   *Parides alopius* (Godman and Salvin, 1890)
According to standard works this species is found in western Mexico, Guatemala, Honduras and Nicaragua (51, 55, 61, 62, 234), but recent unpublished work by Racheli shows it to be a Mexican endemic (221). The position should be clarified as soon as possible. It is certainly uncommon (269) and known from very few localities, but is not known to be threatened.

227   *Parides photinus* (Doubleday, 1844)
Mexico, Guatemala, Honduras, Nicaragua and Costa Rica. In many places abundant (269); generally common and not threatened. In Costa Rica moderately common in remnant patches of forest, 300–800 m (61). Life cycle known and described (61). *Parides dares* (Hewitson, 1869) from Nicaragua may be a hybrid, form (250), aberration (109), or synonym (268) of *Parides photinus*. Other refs: 51, 55, 62, 234.

228   *Parides montezuma* (Westwood, 1842)
Mexico, Guatemala, Honduras, Nicaragua and Costa Rica. Occurs from sea level to 700 m in dry forest habitats (61). Generally fairly common north of Costa Rica, but rare or locally common within Costa Rica (61). Not threatened. Other refs: 51, 55, 62, 234, 269.

229   *Parides phalaecus* (Hewitson, 1869)
Eastern Ecuador and possibly Peru. Very scarce and perhaps meriting Rare status, but more information required. Refs: 51, 55, 234.

230   *Parides agavus* (Drury, 1782)
Brazil, Paraguay and Argentina. Not uncommon and not thought to be threatened. Ranched in Brazil (24). Other refs: 51, 55, 234.

231   *Parides proneus* (Hübner, 1825)
South-eastern Brazil. Not uncommon and not known to be threatened. Refs: 26, 51, 55, 234.

232   *Parides ascanius* (Cramer, 1775) (= *orophobus* D'Almeida, 1942)
Vulnerable—refer to section 6, p. 240.
Only known from scattered points on the coast of Rio de Janeiro state, Brazil where it is declining rapidly due to habitat destruction. Protected by law in Brazil. Common names: Fluminense Swallowtail, Ascanius Swallowtail (282). Other refs: 51, 55, 234.

233   *Parides bunichus* (Hübner, 1821) (= *chamissonia* Eshscholtz, 1821)
Brazil and northern Argentina. Not uncommon and in no known danger. Two subspecies including *Parides bunichus diodorus* (Hopffer) which is treated by D'Almeida (55) as a separate species. Other refs: 26, 51, 122, 234.

234 *Parides diodorus* (Hopffer, 1866)
Brazil (Piaví, Goías and Minas Gerais states). Given as a subspecies of
bunichus (= *chamissonia*) by Rothschild & Jordan (234); not listed by
D'Abrera (51). Not known to be threatened.

Species-group: *aeneas* Linnaeus

235 *Parides perrhebus* (Boisduval, 1836)
South-eastern Brazil, northern Argentina, Paraguay and Uruguay. Generally common and not threatened. Two subspecies, but probably better
placed with *P. bunichus*. Refs: 51, 55, 122, 234.

236 *Parides hahneli* (Staudinger, 1882)
Rare—refer to section 6, p. 242.
Brazil. Although a few specimens have featured in dealers' lists in the last
few years, the species must be regarded as very scarce. Occasionally
ranched. Common name: Hahnel's Amazonian Swallowtail. Refs: 27, 51,
55, 234, 283.

237 *Parides mithras* (Grose–Smith, 1902) (= *triopas* Godart, 1819)
Guyana, Surinam, French Guiana and northern Brazil (Obidos). Not
uncommon but not known to be threatened. Further data needed in order to
assess conservation status. Two subspecies. Refs: 51, 55, 234.

238 *Parides chabrias* (Hewitson, 1852)
Peru and Ecuador (Upper Amazon). An uncommon species, probably
conspecific with *P. mithras*. Further data needed in order to assess
conservation status. Refs: 51, 55, 234.

239 *Parides quadratus* (Staudinger, 1890)
Brazil (Amazonas) and Peru (Upper Amazon). Apparently a rare species
(221). More information required. Two subspecies. Other refs: 51, 55, 234.

240 *Parides pizarro* (Staudinger, 1884)
Insufficiently Known—refer to section 6, p. 244.
Peru (Upper Amazon) and Brazil (Acre State). Again, a fairly scarce and
little known species. Two subspecies including *Parides pizarro kuhlmanni*
May (25, 156), with *Parides steinbachi* from Bolivia possibly a third (156).
Other refs: 51, 55, 234.

241 *Parides steinbachi* (Rothschild, 1905)
Insufficiently Known—refer to section 6, p. 246.
Eastern Bolivia. Again a very scarce and uncommon species. It has been
suggested that it may be a subspecies of *Parides pizarro* (25, 156). D'Abrera
regards *Parides pizarro kuhlmanni* from Peru and western Brazil as a
subspecies of *Parides steinbachi* (51). Other refs: 55, 139, 234.

242 *Parides coelus* (Boisduval, 1836)
Insufficiently Known—refer to section 6, p. 248.
Known only from French Guiana. Little is known about this scarce species,
but it may be under threat. More details needed. Refs: 51, 55, 139, 234.

243 *Parides tros* (Fabricius, 1793) (= *dardanus* Fabricius)
Brazil (Rio de Janeiro) and northern Argentina. Generally common (55)
and not threatened. Other refs: 26, 51, 54, 122, 234.

244 *Parides aeneas* (Linnaeus, 1758)
Colombia, Guyana, Surinam, French Guiana, Brazil, Ecuador, Peru and
Bolivia. Not uncommon and not known to be threatened. *Parides schuppi*
(Röber, 1927) appears to be a form of *Parides aeneas* (109). Six subspecies.
Other refs: 51, 55, 234.

245 *Parides klagesi* (Ehrmann, 1904)
Insufficently Known—refer to section 6, p. 250.
Venezuela only. Another scarce and little-known species; the male, taken
only in 1983, has not been described. More data are required. Refs:
51, 55, 139, 234.

246 *Parides orellana* (Hewitson, 1852)
Peru and Brazil (Upper Amazon). Not at all common, but not known to be
threatened. More information required. Refs: 51, 55, 234.

247 *Parides childrenae* (Gray, 1832)
Mexico, Guatemala, Honduras, Nicaragua, Costa Rica, Panama, Colombia and Ecuador. Extremely rare in Mexico (62). Moderately common in
primary rain forest (at least in Costa Rica), but intolerant of areas without
substantial forest cover (61). Not believed to be threatened, but poor
forest management might cause a decline. Three subspecies. Other refs:
51, 55, 234, 269.

248 *Parides sesostris* (Cramer, 1779)
Mexico, Central America (possibly excluding El Salvador), Trinidad and
South America (excluding Chile, Uruguay and possibly Paraguay); widely
distributed and not threatened. Four subspecies. Common name:
Southern Cattle Heart (9, 282). Other refs: 51, 55, 57, 61, 62, 122,
234, 269.

249 *Parides burchellanus* (Westwood, 1872)
Vulnerable—refer to section 6, p. 252.
Central Brazil. A little-known and very scarce species; more information is
needed. Refs: 26, 51, 55, 145, 234.

250 *Parides polyzelus* (C. & R. Felder, 1865)
Mexico, Guatemala, Belize and Honduras. Generally common and not
threatened. Two subspecies. Refs: 51, 55, 57, 62, 234, 269.

251 *Parides iphidamas* (Fabricius, 1793)
Mexico, Central America, Colombia, Venezuela, Ecuador and northern
Peru. Generally common and not threatened. Occurs from sea level to
1200 m in Costa Rica, tolerant of many habitats (61). Five or six subspecies. Other refs: 51, 55, 57, 62, 234, 269.

252 *Parides vertumnus* (Cramer, 1779)
Colombia, Venezuela, Guyana, Surinam, French Guiana, Ecuador, Brazil, Peru and Bolivia. Generally common and not threatened. Five to seven
subspecies. Refs: 51, 55, 234.

253 *Parides cutorina* (Staudinger, 1898)
Eastern Peru and Ecuador (Upper Amazon). Scarce and localized but not
known to be threatened. More information required. Refs: 51, 55, 234.

254 *Parides lycimenes* (Boisduval, 1870)
Mexico, Central America, Colombia, Venezuela and Ecuador. Fairly common and not threatened. Three subspecies. Refs: 55, 57, 61, 62, 234, 269.

255 *Parides phosphorus* (Bates, 1861)
Colombia, Venezuela (?), Guyana, Brazil, Ecuador and eastern Peru. Not uncommon but not known to be threatened. Status uncertain, further data required. Three subspecies. Refs: 51, 55, 234.

256 *Parides drucei* (Butler, 1874)
Ecuador, Peru, Brazil and Bolivia. Not uncommon and not threatened. May be a subspecies of *P. anchises*. Refs: 55, 234.

257 *Parides erlaces* (Gray, 1852)
Ecuador, Peru, Brazil, Bolivia and northern Argentina. Not uncommon nor threatened. Five subspecies. Refs: 51, 55, 122, 234.

258 *Parides nephalion* (Godart, 1819)
South-eastern Brazil, northern Argentina and Paraguay. Generally common and not threatened. May be a subspecies of *P. anchises*. Ranched in Brazil (24). Other refs: 26, 51, 55, 122, 234.

259 *Parides anchises* (Linnaeus, 1758)
Colombia, Venezuela, Ecuador, Peru, Trinidad, Guyana, Surinam, French Guiana, Brazil, eastern Bolivia, Paraguay and northern Argentina. Widely distributed and not uncommon. Ten subspecies. *Parides hedae* (Foetterle, 1902) from Brazil is not included as a species by Munroe (188) or Hancock (109) and considered to be possibly a form (234, 250) or subspecies (55) of *P. anchises*. *P. eversmanni* (Ehrmann, 1925) is an invalid name, the male of this supposed species was in fact *P. anchises* while the female was *P. eurimedes* (55, 109). Common name: the Cattle Heart (9). Other refs: 26, 122.

260 *Parides erithalion* (Boisduval, 1836)
Costa Rica, Panama, Colombia and northern Venezuela. Generally common and not threatened, sea level to 700 m (61). Four subspecies. Other refs: 55, 234.

Species-group: *lysander* Cramer

261 *Parides panthonus* (Cramer, 1780)
Brazil, Guyana, French Guiana and Surinam. Not common but not regarded as threatened. However, *P. p numa* has not been recorded since 1920 (23). Two subspecies. Other refs: 26, 51, 55, 234.

262 *Parides aglaope* (Gray, 1852)
Brazil, southern Peru and eastern Bolivia. Not common, but far from rare and not threatened. Refs: 26, 51, 55, 234.

263 *Parides castilhoi* (D'Almeida, 1967)
Brazil (Castilho, Rio Parana, São Paulo) (51). More information is required on this little known species.

264 *Parides lysander* (Cramer, 1775)
Colombia, Brazil, Guyana, Surinam, French Guiana, Ecuador and Peru. Often common (55) and not threatened. Two subspecies. Other ref: 51.

265 *Parides echemon* (Hübner, 1813) (= *echelus* Hübner, 1815)
Guyana, Surinam, French Guiana and Brazil (lower Amazon). Common
and not threatened. Two subspecies. Refs: 51, 55, 234.

266 *Parides zacynthus* (Fabricius, 1793)
Eastern Brazil. Fairly common and not threatened. *P. z. polymetus* is
ranched in Brazil (24). Two subspecies. Other refs: 51, 55, 234.

267 *Parides neophilus* (Geyer, 1837)
Colombia, Venezuela, Guyana, Surinam, French Guiana, Trinidad,
southern and western Brazil, Ecuador, Peru, Bolivia, Paraguay and
north-eastern Argentina. Common and not threatened. Six subspecies.
Common name: Spear-winged Cattle Heart (9, 282). Other refs:
234, 55, 51.

268 *Parides eurimedes* (Stoll, 1780) (= *arcas* Cramer, 1777)
Mexico, Guatemala, Belize (?), El Salvador, Honduras, Nicaragua, Costa
Rica, Panama, Colombia, Venezuela, Ecuador, Guyana, Surinam and
French Guiana. Generally common and not threatened. Adoption of the
name *eurimedes* follows Hancock (115), *arcas* being preoccupied. *Parides
eversmanni* (Ehrmann, 1925) is an invalid name, the male of this supposed
species was in fact *P. anchises* while the female was *P. eurimedes* (55, 109).
Five, possibly six subspecies. Other refs: 51, 57, 61, 62, 234, 269.

269 *Parides timias* (Gray, 1852)
Ecuador. Fairly common, despite the restricted range. Two subspecies,
possibly conspecific with *P. eurimedes*. Refs: 51, 55, 234.

Genus: *Atrophaneura* Reakirt
Subgenus: *Pharmacophagus* Haase

Species-group: *antenor* Drury

270 *Atrophaneura (Pharmacophagus) antenor* (Drury, 1773)
Madagascar. This species is the only Afrotropical representative of the
Troidini. Well distributed and apparently not uncommon but the status of
the species needs to be continually monitored. Collected commercially.
Refs: 33, 34, 50, 141, 204, 206.

Subgenus: *Atrophaneura* Reakirt

Species-group: *latreillei* Donovan

271 *Atrophaneura (Atrophaneura) daemonius* (Alpheraky, 1895)
China. Very little information available and none on current status. Further
research required. Ref: 109.

272 *Atrophaneura (Atrophaneura) plutonius* (Oberthür, 1907)
South-western China, Bhutan, Nepal and north-eastern India (Sikkim,
Manipur, Nagaland, Mizoram and Assam (?)) (52, 94). A rare species of
high elevations, not known to be threatened but requires monitoring. The
subspecies *pembertoni* (Pemberton's Chinese Windmill (80)) and *tytleri*
(Tytler's Chinese Windmill (80)) are protected by law in India (182). Three
subspecies. Common name: Chinese Windmill (94, 251), Pemberton's
Chinese Windmill (80), Tytler's Chinese Windmill (80). Other ref: 259.

273 *Atrophaneura (Atrophaneura) alcinous* (Klug, 1836)
Southern China, northern Vietnam (?), northern Laos (?), Taiwan, North Korea, South Korea, Japan (to northern Honshu) and U.S.S.R. (Primorskiy Kray). Apparently common and not in danger as a species but rare in the U.S.S.R. and listed in the U.S.S.R. Red Data Book, category Rare (8, 260). Seven subspecies. Common name: Chinese Windmill (80). Other refs: 52, 141, 149, 246, 247.

274 *Atrophaneura (Atrophaneura) latreillei* (Donovan, 1826)
Afghanistan (136), northern India (including Uttar Pradesh, Sikkim, Assam, Manipur, Nagaland and Meghalaya (?)), Nepal, Bhutan (?), northern Burma, southern China and northern Vietnam (?) (52). Not rare and not thought to be threatened although the subspecies, *A. (A.) l. kabrua* is protected by law in India (182). Common name: Rose Windmill (80, 251). Three subspecies. Other refs: 4, 141, 259.

275 *Atrophaneura (Atrophaneura) polla* (de Nicéville, 1897)
North-eastern India (Assam, Manipur, Chin Hills of Nagaland, Arunachal Pradesh), Burma (eastern Bhamo and Bernandmyo of the Shan states), northern Thailand (?), south-western China (?) and northern Laos (?) (52, 284). Recorded as being very rare by Evans (80) and Talbot (259). Protected by law in India (182). More information needed on distribution and status. Common name: De Nicéville's Windmill (80, 94). Other ref: 109.

276 *Atrophaneura (Atrophaneura) crassipes* (Oberthür, 1879)
North-eastern India (Manipur), Burma (southern Shan states), northern Thailand, northern Laos, northern Vietnam (Tonkin) (94) and possibly southern China. Described as very rare (259) and perhaps deserving some conservation status, but more information required. It is protected by law in India (182). Common name: Black Windmill (80, 94). Other ref: 52.

277 *Atrophaneura (Atrophaneura) adamsoni* (Grose–Smith, 1896)
Burma and Thailand. Locally common in the latter country (168) but said to be generally rare (259). Not known to be threatened, but further research needed. Common name: Adamson's Rose (80). Other ref: 52.

278 *Atrophaneura (Atrophaneura) nevilli* (Wood–Mason, 1896)
North-eastern India (Assam), Burma (Shan states) and western China (52, 94). Said to be very common in western China though very rare in India (259) where it is protected (182) but not known to be threatened. Common name: Nevill's Windmill (80, 94). Other ref: 5.

279 *Atrophaneura (Atrophaneura) laos* (Riley and Godfrey, 1921)
Northern Thailand and Laos. Said to be very rare (168), but status not known with any certainty. More data required for conservation assessment. Other ref: 52.

280 *Atrophaneura (Atrophaneura) mencius* (Felder, 1862)
Sichuan (Szechwan), central and south-eastern China (52, 207). Two subspecies, although D'Abrera notes that *A. (A.) m. rhadinus* is closer to *A. (A.) nevilli* and may be a distinct species (52). No information on threats; more research needed.

281 *Atrophaneura (Atrophaneura) impediens* (Rothschild, 1895)
China including Sichuan (Szechwan) (207) and Taiwan. No data on
conservation status; more research needed. Two subspecies (52).

282 *Atrophaneura (Atrophaneura) febanus* (Fruhstorfer, 1908)
Restricted to Taiwan. Apparently common, but status should be monitored
in view of the limited distribution. D'Abrera lists *febanus* as a subspecies of
impediens but this has not been demonstrated adequately (52). Not listed by
Hancock (109). Other refs: 141, 246.

283 *Atrophaneura (Atrophaneura) hedistus* (Jordan, 1928)
Southern China (Yunnan). Also little known, with no information available
on status. Regarded by D'Abrera (52) as a subspecies of *A. (A.) dasarada*.

284 *Atrophaneura (Atrophaneura) dasarada* (Moore, 1857)
Northern India, Nepal, Bhutan, Burma, south-eastern China (including
Hainan (Guangdong prov.)) (52). Not rare and not threatened. Five
subspecies. Common name: Great Windmill (80, 251). Other refs:
4, 141, 168, 259.

285 *Atrophaneura (Atrophaneura) polyeuctes* (Doubleday, 1842)
Northern India, Nepal, Bhutan (?), Burma, northern Thailand, Laos,
Vietnam, southern China (including Yunnan) and Taiwan (52). Up to five
subspecies. The species *lama* (Oberthür) described from western China is
generally regarded as a subspecies of *polyeuctes*. *A. (A.) philoxenus* Gray is
a synonym. Common name: Common Windmill (80, 251). Other refs:
4, 141, 168, 246, 259.

Species-group: *nox* Swainson

286 *Atrophaneura (Atrophaneura) semperi* (C. & R. Felder, 1861)
Philippines (not Cebu). Not common but not known to be threatened.
Seven subspecies (52). *A. (A.) erythrosoma* Reakirt is a synonym. Other
refs: 141, 147, 262.

287 *Atrophaneura (Atrophaneura) kuehni* (Honrath, 1886)
Indonesia (eastern and northern Sulawesi). Little known about status,
possibly very rare (121). It is certainly very rare in collections. More data are
required on its conservation status. Two subspecies (121). *A. (A.)
mesolamprus* Rothschild is believed to be a synonym (52). Other ref: 262.

288 *Atrophaneura (Atrophaneura) luchti* (Roepke, 1935)
Rare—refer to section 6, p. 254.
Indonesia (Java). Restricted to the mountains in the far east of Java. Little is
known about this species and the specific status has been brought in
question due to its similarity with *A. (A.) priapus* (118). Other refs: 52, 262.

289 *Atrophaneura (Atrophaneura) hageni* (Rogenhofer, 1889)
Indonesia: restricted to high plateau of Sumatra. Uncommon, but not
known to be threatened. Refs: 52, 141.

290 *Atrophaneura (Atrophaneura) priapus* (Boisduval, 1836)
Indonesia (Java and southern Sumatra) (262). The three 'species' *sycorax*,
priapus and *hageni* are often confused and may represent one, two or three
good species. Not known to be threatened, though status uncertain. More

information and close monitoring are needed. Two (52) or three (262) subspecies. Common name: White-head Batwing (3).

291 *Atrophaneura (Atrophaneura) sycorax* (Grose–Smith, 1885)
Southern Burma, Thailand, Peninsular Malaysia and Indonesia (Sumatra and western Java) (52, 262). Rather rare (45) but not known to be threatened. Three subspecies (52, 262). Common name: White-head Batwing (80). Other refs: 87, 168, 259 (as *A. (A.) priapus*).

292 *Atrophaneura (Atrophaneura) horishanus* (Matsumura, 1910)
Restricted to Taiwan. Apparently common but status should be kept under review. *A. (A.) sauteri* Heyne is a synonym. Refs: 52, 141, 246.

293 *Atrophaneura (Atrophaneura) aidoneus* (Doubleday, 1845)
Northern India (including Uttar Pradesh, Sikkim, Assam, Meghalaya (?), Manipur (?) and Nagaland (?)), Bhutan, Burma, northern Laos, northern Vietnam, southern China (including Hainan (Guangdong prov.)) (52). Not common but not regarded as threatened. Common name: Lesser Batwing (80). Other refs: 4, 259.

294 *Atrophaneura (Atrophaneura) varuna* (White, 1868)
North-eastern India (including Sikkim), eastern Nepal (?), Burma, Thailand, northern Laos, northern Vietnam and Peninsular Malaysia (52). Not rare across most of its range and not known to be threatened. In Peninsular Malaysia it is uncommon and flies at high elevations (45). Believed to be extinct in Singapore (45). Two subspecies. Common name: Common Batwing (3, 80). Other refs: 4, 87, 141, 168, 259.

295 *Atrophaneura (Atrophaneura) zaleucus* (Hewitson, 1865)
Burma, Thailand, northern Laos, northern Vietnam and possibly south-western China (52). Not generally common, but not known to be threatened. Sometimes regarded as a subspecies of *A. (A.) varuna* (259). Common name: Burmese Batwing (80). Other refs: 141, 168.

296 *Atrophaneura (Atrophaneura) nox* (Swainson, 1822–23)
Peninsular and Eastern Malaysia, southern Thailand (?), Brunei (?) and Indonesia (Sumatra, Nias, Mentawai Is, Java, Bali and Kalimantan) (52, 262). A widespread but often local species of well-wooded localities; extinct in Singapore (45). Nine subspecies. Common name: Malayan Batwing (45). Other refs: 87, 131, 141.

297 *Atrophaneura (Atrophaneura) dixoni* (Grose–Smith, 1901)
Indonesia (northern and central Sulawesi). Status uncertain, said to be uncommon to rare, presumably localized (121) More data are required. Other refs: 52, 262.

Subgenus: *Losaria*

Species-group: *coon* Fabricius

N.B. The next four species represent a distinctive facies in the genus *Atrophaneura* (called the *coon* group by Hancock (109)) and are here placed in the subgenus *Losaria*. Tsukada and Nishiyama treat *Losaria* as a full genus (262).

298 *Atrophaneura (Losaria) neptunus* (Guérin–Méneville, 1840)
Southern Burma, southern Thailand, Peninsular and Eastern Malaysia, Philippines (Palawan), Brunei and Western Indonesia (Sumatra, Nias,

Simeulue and Kalimantan). Less common than *A. (L.) coon* but widely distributed and not known to be threatened. Eight subspecies. Common name: Yellow-bodied Clubtail (80), Yellow–red Clubtail (3). Other refs: 45, 52, 87, 131, 168, 259, 262.

299   *Atrophaneura (Losaria) palu* (Martin, 1912)
Insufficiently Known—refer to section 6, p. 256.
Indonesia (western Sulawesi (Palu)). Taxonomic status no longer in doubt (114). Said to be a subspecies of *A. (L.) coon* by Tsukada and Nishiyama (262) but it is apparently distinct and isolated from other members of the *coon* group (114, 121). Believed to be restricted to the west coast where it is rare and localized (172).

300   *Atrophaneura (Losaria) coon* (Fabricius, 1793)
Northern India (Assam), southern Burma, Thailand, Laos, Vietnam, Kampuchea, China (Hainan (Guangdong prov.)), Peninsular Malaysia, Nicobar Is, Indonesia, (Sumatra, western Java and Bawean Is) (52, 262). The record from Assam (80) seems to be erroneous, as do some attributions of subspecies (52). *A. (L.) c. sambilanga* from the Nicobar Is is protected under Indian law (182) and is said to be very rare (5). *A. (L.) c. doubledayi* is extinct in Singapore (45). Eight subspecies excluding *palu*. Common name: Common Clubtail (45, 80, 282). Other refs: 4, 87, 107, 141, 168, 259.

301   *Atrophaneura (Losaria) rhodifer* (Butler, 1876)
Restricted to the Andaman Is. Possibly to be classified when its status is clearer but described as not rare (5, 80, 259) and so not listed as threatened. Common name: Andaman Clubtail (80). Other ref: 52.

Subgenus: *Pachliopta*

Species-group: *polydorus* Boisduval

N.B. The genus *Pachliopta* has been synonymized with *Atrophaneura* (107) but we retain *Pachliopta* at the subgeneric level as representing a distinct facies within the genus. Both genera have in the past been called *Polydorus*.

302   *Atrophaneura (Pachliopta) hector* (Linnaeus, 1758)
Southern and eastern India, Sri Lanka, Andamans and possibly the coast of western Burma (52). Generally common and not threatened but said to be very rare in the Andamans (5, 259). Protected by law in India (182). Common names: Common Rose (3, 282), Crimson Rose (80). Other refs: 4, 287.

303   *Atrophaneura (Pachliopta) jophon* (Gray, 1852)
Vulnerable—refer to section 6, p. 258.
Sri Lanka. Declining due to loss of habitat (52, 53). *A. (P.) pandiyana* from southern India is sometimes regarded as a subspecies of *A. (P.) jophon*. Common name: Ceylon Rose (80). Other refs: 259, 287.

304   *Atrophaneura (Pachliopta) pandiyana* (Moore, 1881)
Southern India. Uncommon but not considered to be threatened. (This species and *A. (P.) jophon* are sometimes regarded as being conspecific). Common name: Malabar Rose (80). Other refs: 4, 52.

305   *Atrophaneura (Pachliopta) oreon* (Doherty, 1891)
Indonesia: Lesser Sunda Is (Sumba, Alor, Flores and Solor) (52, 262).
Possibly threatened in view of the very narrow range, but insufficient
information available on status. Two subspecies (262). Other ref: 109.

306   *Atrophaneura (Pachliopta) liris* (Godart, 1819)
Indonesia: Lesser Sunda Is (Timor, Sawu, Wetar, Moa, Leti, Kisar, Babar,
Damar, Romang and Tanimbar) (262). Apparently common and not
threatened. Six subspecies. Other refs: 48, 130.

307   *Atrophaneura (Pachliopta) polyphontes* (Boisduval, 1836)
Indonesia (Sulawesi, Salayar, Talaud, Sangihe (?) and the Moluccas (Sula,
Ternate, Bacan, Halmahera and Morotai)) (52, 262). Five subspecies (262).
Other refs: 48, 141.

308   *Atrophaneura (Pachliopta) schadenbergi* (Semper, 1886)
Vulnerable—refer to section 6, p. 261.
Philippines: Luzon and the Babuyan Is. A locally distributed species with
seasonally limited occurrence in lowland and medium elevation wooded
areas. Threatened by habitat destruction. Two subspecies. Refs: 52, 177,
262.

309   *Atrophaneura (Pachliopta) mariae* (Semper, 1874)
Philippines (Luzon, Samar, Leyte, Bohol, Mindanao, Sibuyan and Polillo).
Neither rare nor threatened. Three subspecies. Refs: 52, 141, 147, 177, 262.

310   *Atrophaneura (Pachliopta) phegeus* (Hopffer, 1886)
Southern Philippines (Mindanao, Samar, Leyte and Cebu). Not rare and
not considered to be threatened. *A. (P.) leytensis* (Murayama) is believed
to be a synonym. Refs: 52, 141, 147, 177, 262.

311   *Atrophaneura (Pachliopta) phlegon* (C. & R. Felder, 1864)
Philippines (Mindoro, Panay, Guimaras, Marinduque and Mindanao
(262)). Not uncommon and not known to be threatened. Two subspecies
including Bryk's (1930) *strandi* from the Philippines (262). *A. (P.) phlegon*
is regarded as a subspecies of *annae* by D'Abrera (52) but *annae* is normally
regarded as a synonym of this species as is *sabinae* (Seyer) (107). Other refs:
52, 177.

312   *Atrophaneura (Pachliopta) atropos* (Staudinger, 1888)
Indeterminate—refer to section 6, p. 263.
Restricted to Philippines (Palawan). Not much is known about its status,
but it seems to be uncommon and is listed as Indeterminate until more
information is available. Refs: 52, 141, 262.

313   *Atrophaneura (Pachliopta) kotzebuea* (Eschscholtz, 1821)
Widely distributed in the Philippine Is south of Babuyan Is, but absent from
Palawan (262). Flies on edges of rain forest (262). *A. (P.) kotzebuea* is
recognized by Tsukada and Nishiyama (262), but not by Munroe (188) or
Hancock (109). Not specifically known to be threatened, although any
Philippine endemic dependent upon rain forest must be considered to be at
some risk. Five subspecies (262).

314   *Atrophaneura (Pachliopta) aristolochiae* (Fabricius, 1775)
Afghanistan (118), India, Nepal, Sri Lanka, Burma, Thailand, southern
and eastern China (including Hainan (Guangdong prov.)), Hong Kong,

Taiwan, Japan (south-western Okinawa only), Laos, Vietnam, Kampuchea, Andamans, Nicobar Is, Peninsular and Eastern Malaysia, Brunei, Philippines (Palawan and Leyte), Indonesia (Sumatra, Nias, Enggano, Bangka, Java, Bali, Kangean, Lombok, Sumbawa, Sumba, Flores, Tanahjampea and Kalimantan) (262). Generally common and not threatened. Up to twenty subspecies. Common name: Common Rose (45, 80, 251). Other refs: 4, 5, 52, 87, 131, 141, 143, 147, 168, 246, 259, 287.

315   *Atrophaneura (Pachliopta) polydorus* (Linnaeus, 1758)
Indonesia (Moluccas (not Morotai, but including Seram and Ambon), Tanimbar, Irian Jaya, Kai, Aru, Waigeu and Skouten Is), Papua New Guinea, Bismarck Archipelago (including New Britain), Bougainville, Solomons (including San Cristobal and Ulawa I.), Trobriand Is, D'Entrecasteaux Is, Louisiade Archipelago and Australia (northern Queensland)(262). Generally common and not threatened as a species, though some subspecies may be threatened (138). About four subspecies are farmed in Papua New Guinea (36). Up to thirty-one subspecies are known (262). Common name: Red-bodied Swallowtail (282). Other refs: 14, 40, 48, 68, 141, 176.

Genus: *Trogonoptera* Rippon

N.B. Both species of *Trogonoptera* are included on Appendix II of the Convention on International Trade in Endangered Species of Wild Fauna and Flora (CITES). The 87 countries that are party to CITES are obliged to invoke national legislation implementing the Convention. In line with this requirement import and export restrictions have been implemented by national legislation in the Federal Republic of Germany, the Netherlands and the United Kingdom, and by Regulation 3626/82 of the Commission of the European Community.

316   *Trogonoptera brookiana* (Wallace, 1855)
Peninsular and Eastern Malaysia, Indonesia, (Sumatra, Mentawai Is (284) and Kalimantan) and Brunei (52, 262). Locally common (especially males) and not thought to be threatened, despite the large numbers taken for the decorative trade. Protected by law in Malaysia and Indonesia but still collected commercially, possibly as many as 125 000 specimens are exported per year (10). In Malaysia the species is 'protected' but not 'totally protected'. This means that licences and permits may be issued for the taking, import, export or possession of specimens. A permit is also required to accompany the possession of a specimen. Five subspecies. Common name: Raja(h) Brooke's Birdwing (3, 45, 49, 282). Other refs: 87, 131, 141, 201.

317   *Trogonoptera trojana* (Honrath, 1886)
Philippines (Palawan, Balabac) (52, 262, 281). Less common and with a much more restricted range than *T. brookiana* but not presently regarded as being threatened. Nevertheless males are collected in great numbers, particularly in the rain forests of northern Palawan, and the forests themselves are under growing pressure from timber interests (56). In common with D'Abrera (49) most authors consider *T. trojana* to be a distinct species, but Igarashi (141) regards it as a subspecies of *T. brookiana*.

Genus: *Troides* Hübner

N.B. All species of *Troides* are included on Appendix II of the Convention on International Trade in Endangered Species of Wild Fauna and Flora (CITES). The 87 countries that are party to CITES are obliged to invoke national legislation implementing the Convention. In line with this requirement import and export restrictions have been implemented by national legislation in the Federal Republic of Germany, the Netherlands and the United Kingdom, and by Regulation 3626/82 of the Commission of the European Community.

Subgenus: *Ripponia* Haugum & Low

Species-group: *hypolitus* Cramer

318 *Troides (Ripponia) hypolitus* (Cramer, 1775)
Indonesia (Sulawesi (including Sangihe and Talaud) and the Moluccas: Sula, Morotai, Halmahera, Ternate, Buru, Seram, Saparua, Ambon (52, 262) and Obi (252)). Not rare and not regarded as being threatened, though protected in Indonesia. Four subspecies. Other refs: 48, 141, 188.

Subgenus: *Troides* Hübner

Species-group: *amphrysus* Cramer

319 *Troides (Troides) cuneifer* (Oberthür, 1879)
Southern Thailand, Peninsular Malaysia and Indonesia (Sumatra and Java) (52, 262). Not rare and not known to be threatened across most of it range, but considered to be Vulnerable and in need of some protection in Peninsular Malaysia (10). Three subspecies. Common name: Golden Birdwing (3). Other refs: 45, 49, 87, 141.

320 *Troides (Troides) amphrysus* (Cramer, 1782)
Southern Burma (Mergui Archipelago), southern Thailand, Peninsular and Eastern Malaysia, Brunei, Indonesia (Kalimantan, Sumatra, Simeulue (?), Nias, Batu, Mentawai Is, Enggano and Java) (49, 52, 191, 262). Not rare and not known to be threatened except in Malaysia where said to be Vulnerable (10). Protected in Indonesia. Six subspecies. Common names: Golden Birdwing (3), Malay Birdwing (80). Other refs: 45, 87, 131, 141, 168, 231.

321 *Troides (Troides) miranda* (Butler, 1869)
Brunei, Eastern Malaysia (Sabah and Sarawak), Labuan I. and Indonesia (northern Kalimantan and western Sumatra) (52, 262). Igarashi's map (141) erroneously shows most of the Malaysian peninsula, as well. Not rare and not known to be threatened, though protected in Indonesia. Two subspecies. Other refs: 49, 131, 231.

322 *Troides (Troides) andromache* (Staudinger, 1892)
Indeterminate—refer to section 6, p. 266.
Eastern Malaysia (Sabah and Sarawak) (49, 132, 231) but not recorded from Indonesia (Kalimantan) although it is protected there. Very localised. Two subspecies. Other refs: 11, 52, 131.

Species-group: *helena* Linnaeus

323 *Troides (Troides) magellanus* (C. & R. Felder, 1862)
Philippines (Batan Is, Babuyan Is, Luzon, Polillo, Marinduque, Cuyo I.,
Samar, Cebu, Leyte, Bohol and Mindanao) and Taiwan (Hung-t'ou Hsü Is
only) (262). Not uncommon and not thought to be threatened. Three
subspecies (262) Other refs: 49, 52, 141, 147, 246.

324 *Troides (Troides) prattorum* (Joicey and Talbot, 1922)
Indeterminate—refer to section 6, p. 269.
Indonesia (Buru) only; a very restricted range. The species is rare and little
known as Buru was previously almost impossible to visit due to its use as a
prison camp (49). It is now apparently accessible to collectors and
specimens are beginning to appear on dealers' lists. Other ref: 48.

325 *Troides (Troides) minos* (Cramer, 1779)
Western and south-western India (49, 52). Despite its restricted range the
species is not known to be threatened, but continuous monitoring is
recommended. Some authorities have regarded it as a subspecies of *T. (T.)
helena*. Other refs: 4, 15, 80, 259.

326 *Troides (Troides) aeacus* (C. & R. Felder, 1860)
Northern India, Nepal, Burma, China, Thailand, Laos (?), Vietnam (?),
Taiwan, Kampuchea, Peninsular Malaysia (49, 52) and Indonesia (Suma-
tra) (118). Generally common and not threatened though classified as
Vulnerable and possibly in need of protection in Peninsular Malaysia (10).
It is very uncommon in Sumatra (118). In 1974 the Entomological Suppliers
Association of Great Britain agreed not to trade in the endemic Taiwan
subspecies *T. a. kaguya* on the grounds that such trade could threaten the
well-being of the taxon (123). Three subspecies. Common name: Golden
Birdwing (80, 251). Other refs: 4, 45, 87, 141, 168, 207, 246, 259.

327 *Troides (Troides) plateni* Staudinger, 1888
Philippines (only known positively in Palawan (281), but possibly occurs in
Calamian Is) (262). A common species easily raised in captivity and not
thought to be at risk (56). Igarashi (141) and, by implication, Munroe (188),
Smart (250) and Hancock (109) regard it as a subspecies of *T. (T.)
rhadamantus*, but D'Abrera emphasises its specific distinction (49, 52).
Other ref: 119.

328 *Troides (Troides) rhadamantus* (Lucas, 1835)
Philippines (except Palawan, Dumaran, Bugsuk and Calamian Is) (49, 52,
262). Present in Luzon, Polillo, Mindoro, Marinduque, Romblon, Masbate, Samar, Panay, Guimaras, Negros, Cebu, Leyte, Panaon, Dinagat,
Bohol, Camiguin, Mindanao, Basilan, Jolo, Tawitawi (262). Not uncommon and not known to be threatened. Listing of this species under
Indonesian legislation refers to *T. (T.) dohertyi*, sometimes considered to
be a subspecies of *T. (T.) rhadamantus* (118, 262). Other refs: 141, 147, 281.

329 *Troides (Troides) dohertyi* Rippon, 1893
Vulnerable—refer to section 6, p. 271.
Indonesia (Talaud Is and Sangihe) (262); listed as threatened on account of
its very restricted range. However, the remarks on the specific distinctness
of *T. (T.) plateni* apply in exactly the same way to *T. (T.) dohertyi*.

Protected in Indonesia under the name *T. (T.) rhadamantus*, of which it has been considered to be a subspecies (118, 246). Common name: Talaud Black Birdwing. Other refs: 49, 52, 119, 141.

330   *Troides (Troides) helena* (Linnaeus, 1758)
From north-eastern India through Indo–China and southern China to Indonesia i.e. north-eastern India (including Orissa and Sikkim), Nepal, Bangladesh, Burma, southern China (including Hainan (Guangdong prov.)), Hong Kong, Thailand, Vietnam, Laos, Kampuchea (?), Andaman Is, Nicobar Is, Peninsular and Eastern Malaysia, Indonesia (Sumatra, Nias, Enggano, Java, Bawean, Kangean Is, Bali, Lombok, Sumbawa, Bunguran (Natuna), Sulawesi, Butung, Tukangbesi and Kalimantan) and Brunei (49, 52, 262). Widely distributed and often common, but classified as Vulnerable and possibly requiring protection in Peninsular Malaysia (10). Protected in Indonesia. Seventeen subspecies. Common name: Common Birdwing (45, 80, 251). Other refs: 4, 5, 14, 87, 131, 141, 143, 168, 231, 259.

331   *Troides (Troides) oblongomaculatus* (Goeze, 1779)
Indonesia (southern Sulawesi, Salayar (Kabia), Tanahjampea, Sula, Buru, Seram, Ambon, Seramlaut, Watubela, Banda Is, Salawati, Skouten Is (Yapen) and Irian Jaya) and Papua New Guinea (49, 52, 262). Generally common and certainly not threatened. It is easily reared and commonly sold on the butterfly market in Papua New Guinea (17). Either six (49) or nine (262) subspecies. The Sulawesi records may refer to hybrids (115). Other refs: 14, 36, 48, 217.

332   *Troides (Troides) darsius* (Gray, 1852)
Sri Lanka only. Not rare and not threatened. Regarded as a subspecies of *T. (T.) helena* by earlier authors (80, 259, 287) but now generally accepted as a full species. Other refs: 49, 52.

333   *Troides (Troides) criton* (C. & R. Felder, 1860)
Indonesia, confined to the Moluccas (Morotoi, Ternate, Halmahera (Jailolo Gilolo), Bacan and Obi) (49). May be confused with the hybrid *celebensis* from Sulawesi which, though *criton*-like, has nothing whatever to do with this species (118). Reasonably common (48) and not thought to be threatened, though protected in Indonesia. Possibly two subspecies.

334   *Troides (Troides) riedeli* (Kirsch, 1885)
Indonesia (Tanimbar Is). The species has a restricted range and might be considered for threatened status. However, it is said not to be rare (48) and it is protected by law in Indonesia. The species should be carefully monitored and extensive suitable habitat protected. Any serious habitat disruption would necessitate immediate listing as Vulnerable and appropriate conservation measures. Other refs: 49, 262.

335   *Troides (Troides) plato* Wallace, 1865
Indonesia (Timor). Apparently very rare (48) but locally common (118). More information is required on its status and monitoring is recommended. It is protected in Indonesia. Earlier regarded as a subspecies of *haliphron* and not listed as a good species by several authors (109, 141, 188). Reinstated as a species by Haugum and Low who had studied its genitalia (118, 119). Other refs: 49, 262.

336 *Troides (Troides) vandepolli* (Snellen, 1890)
Indonesia (north-west Java and western Sumatra) (49, 52, 262). Uncommon, but not known to be threatened though it is protected in Indonesia.
Two subspecies (262), which are possibly full species (49).

337 *Troides (Troides) haliphron* (Boisduval, 1836)
Southern Indonesia: Sulawesi, Salayar (Kabia), Tanahjampea, Kalao,
Bonerate, Sumbawa, Sumba, Komodo (?), Rinja (?), Flores, Solor Is,
Alor, Wetar, Romang, Damar, Leti, Moa, Babar and the Tanimbar Is (49,
262). Not rare and not known to be threatened but it is protected in Indonesia. Either nine (49) or eleven (262) subspecies. Other refs: 48, 52, 141.

338 *Troides (Troides) staudingeri* (Röber, 1888)
Confined to the Babar (Babber) group in Indonesia east of Timor (49).
Raised to a full species by Haugum and Low (118, 119) who had studied the
genitalia, but previously considered to be a subspecies of *haliphron* by other
authors. Protected in Indonesia under *T. (T.) haliphron*. Not known to be
threatened. Other ref: 109.

Genus: *Ornithoptera* Boisduval

N.B. All species of *Ornithoptera* are included on Appendix II of the Convention on International Trade in Endangered Species of Wild Fauna and
Flora (CITES). The 87 countries that are party to CITES are obliged to
invoke national legislation implementing the Convention. In line with this
requirement import and export restrictions have been implemented by
national legislation in the Federal Republic of Germany, the Netherlands
and the United Kingdom, and by Regulation 3626/82 of the Commission of
the European Community. Hancock (109) reduced *Ornithoptera* to a
subgenus of *Troides* but we choose to retain *Ornithoptera* as a full genus for
practical reasons of usage.

Species-group: *tithonus* de Haan

339 *Ornithoptera goliath* Oberthür, 1888
Indonesia (Seram, Waigeo and Irian Jaya), Papua New Guinea and Goodenough I. The species is protected in Indonesia; the Papua New Guinea
subspecies is also protected but has been recommended as a good candidate
for farming (36). Adults have been noted as scarce in nature (119), but a
recent report by M.J. Parsons (in litt.) records that *O. goliath* may also be
locally common and does not require protected status. In view of this
reassessment, *O. goliath* is not listed as threatened. Ten subspecies have
been described but this has been reduced to two (49) or five (119). Common
name: Goliath Birdwing (282). Other refs: 48, 183, 217.

340 *Ornithoptera tithonus* de Haan, 1840
Insufficiently Known—refer to section 6, p. 273.
Indonesia (Irian Jaya, Waigeo, Salawati and Mysol (Misoöl) Is). A relatively rare species (119) with a restricted range (117) but one subspecies has
been repeatedly collected in recent years (119). Protected in Indonesia.
Three subspecies, one as yet undescribed. Other refs: 48, 49.

341 *Ornithoptera rothschildi* Kenrick, 1911
Indeterminate—refer to section 6, p. 276.
Indonesia (endemic to the Arfak Mountains, Irian Jaya). Although the

species has an extremely limited distribution, it is 'not rare, localized, and occurs in some abundance at certain sites' (119). It is under heavy commercial collecting pressure (117), despite being protected in Indonesia. *Ornithoptera akakeae* Kobayashi and Koiwaya, 1978 from Irian Jaya (Arfak Mountains) is considered by Haugum and Low (119) to be a natural hybrid of *O. rothschildi* and *O. priamus poseidon*. Common name: Rothschild's Birdwing (49). Other refs: 48, 155.

342 *Ornithoptera chimaera* (Rothschild, 1904)
Indeterminate—refer to section 6, p. 278.
Papua New Guinea and Indonesia (Irian Jaya). A montane species, scarce but widely distributed in the east (119); protected in Papua New Guinea and Irian Jaya but recommended as a good candidate for farming in the former (36). Two subspecies are recognized in the most recent study. Common name: Chimaera Birdwing. Other refs: 48, 49, 117, 183.

343 *Ornithoptera paradisea* Staudinger, 1893
Indeterminate—refer to section 6, p. 281.
Papua New Guinea and Indonesia (Irian Jaya): widespread but localised in seven main areas (119). The species is protected in both Papua New Guinea and Irian Jaya. Six named subspecies, of which the nominate subspecies has been extinct for many years. Common names: Paradise Birdwing (49, 282), Tailed Birdwing (282). Other refs: 14, 48, 141, 183, 217.

344 *Ornithoptera meridionalis* (Rothschild, 1897)
Vulnerable—refer to section 6, p. 284.
Extreme south-eastern Papua New Guinea (where it is protected) and Indonesia (Irian Jaya). Very rare and localized but not uncommon where it occurs (119). It must be considered Vulnerable due to the danger from habitat destruction, sedentary behaviour of adults and the extremely low number of eggs laid by each female (118). Dealers' prices are exceptionally high (3). One or two subspecies. Other refs: 48, 49, 108, 141, 183.

Species-group: *priamus* Linnaeus

345 *Ornithoptera alexandrae* (Rothschild, 1907)
Endangered—refer to section 6, p. 288.
Restricted to a very small part of south-eastern Papua New Guinea where it is protected from collectors (183) but under threat from habitat destruction. Conservation of this species is a priority for any programme on swallowtails. Common name: Queen Alexandra's Birdwing (49). Other refs: 14, 48, 119, 141.

346 *Ornithoptera victoriae* (Gray, 1856)
Widely distributed throughout the Solomons Is and Bougainville (Papua New Guinea) (48, 49, 119, 141, 188, 219, 250) and not rare (171), but in demand by collectors so that Vulnerable status could be applied to some subspecies. May be threatened by agriculture and forestry in localized sites (171). Declining rapidly on Malaita due to intense deforestation (220). Protected on Bougainville (Papua New Guinea), but may be struck off the protected list in the near future to allow farming, since it occurs in large numbers and the larvae feed on the common and easily grown *Aristolochia tagala* (36). Seven subspecies (32). *O. allottei* Rothschild (Abbé Allotte's Birdwing) from the Solomon Archipelago (Bougainville and the south of

Malaita) is generally regarded as a hybrid of *O. victoriae* x *O. priamus urvillianus*; Racheli (219) gives references to the different views. The hybrid is protected on Bougainville (Papua New Guinea). Common name: Queen Victoria's Birdwing (49). Other ref: 183.

347 *Ornithoptera aesacus* (Ney, 1903)
Indeterminate—refer to section 6, p. 292.
Indonesia; known only from the the island of Obi in the Moluccas (48, 49, 119). Very rare and with a restricted distribution on this heavily logged island. When more information is available Vulnerable or Endangered status may prove appropriate.

348 *Ornithoptera croesus* Wallace, 1859
Vulnerable—refer to section 6, p. 294.
Indonesia: Moluccas (Halmahera, Ternate, Tidore, Bacan, Morotai and Sanana (Sula Is)). Not common, conservation status in some doubt, but possibly threatened by insecticide sprays (119) and certainly in demand by collectors. Two subspecies are distinct whilst there is some doubt about a third which has recently been described from Sanana (262). *O. c. lydius* is unusual in that the female mimics unpalatable species of Danainae. Other refs: 49, 141.

349 *Ornithoptera priamus* (Linnaeus, 1758)
Indonesia (Ambon, Seram, Waigeo, Misool, Kai Is, Aru Is, Tanimbar Is and Irian Jaya), Papua New Guinea and offshore islands, Bismarck Archipelago (including New Britain), Trobriand Is, D'Entrecasteaux Is, Woodlark I., Louisiade Archipelago, Bougainville and the Solomon Is (excluding San Cristobal) (119). Protected in Indonesia. Following Hancock (109), the Australian forms *euphorion* and *richmondia* (north-eastern coastal regions southwards to New South Wales) are here given full species status, but other authors may include these as subspecies of *priamus* (115). While many races are generally common, some of the eleven to fifteen subspecies merit special consideration. Some of the Papua New Guinea subspecies are farmed (36). *O. p. caelestis* (Rothschild), generally regarded as a subspecies of *O. priamus* (119, 141), from the Louisiade Archipelago suffers from a very restricted range, pressures on its habitat and overcollecting, although it is now being farmed (36, 83, 138). *O. p. miokensis* (Ribbe) from the Duke of York Is, *O. p. boisduvali* (Montrouzier) from Woodlark I. and *O. p. demophanes* (Frühstorfer) (possibly only a form of *O. p. poseidon* (Doubleday); the New Guinea Birdwing) from the Trobriand Is are all threatened by pressures on their habitat, a restricted range and to a much lesser extent overcollecting (138). *O. p. urvillianus* (Guérin–Méneville), D'Urville's Birdwing, from the Bismarck Archipelago (New Ireland), Bougainville and the Solomon Is (excluding San Cristobal and Rennell) (119, 141, 219, 220) is farmed in the first two areas (36). Common name: Priam's Birdwing (49). Other refs: 14, 40, 48, 68, 176, 217, 255.

350 *Ornithoptera euphorion* (Gray, 1853)
Cooktown to Mackay, northern Queensland, Australia. Listed as a subspecies of *priamus* by D'Abrera (49) and Haugum and Low (119), but accepted as a full species by Hancock (109). Protected in Queensland under the name *O. priamus*, the Cairns Birdwing (258). This species

should be monitored because of its restricted distribution in an area of extensive habitat destruction. Common name: Cairn's Birdwing (49).

351 *Ornithoptera richmondia* (Gray, 1852)
Eastern coastal region of Australia: southern Queensland and northern New South Wales, a restricted range. Specific identity doubtful and generally regarded as a subspecies of *O. priamus* (40, 119, 141, 176), although accepted as a full species by Hancock (109) and D'Abrera (49). Protected in Queensland (115). As for *O. euphorion*, monitoring of this species is highly recommended. Common name: Richmond Birdwing (49). Other ref: 48.

Subfamily: PAPILIONINAE
Tribe: Papilionini

Genus: *Papilio* Linnaeus

N.B. The genus *Papilio sensu lato* is retained throughout. The generic names proposed by Hancock (109) have been used at the subgeneric level.

Subgenus: *Pterourus* Scopoli

Species-group: *glaucus* Linnaeus

352 *Papilio (Pterourus) glaucus* L., 1758 (= *turnus* L., 1771)
Canada, eastern and south-eastern U.S.A. (including Alaska and Texas) and Mexico. Common and not generally threatened, but listed as a Special Species by Washington State Department of Game. Three subspecies. Common name: Tiger Swallowtail (154, 216). Other refs: 55, 62, 137, 141, 234, 269.

353 *Papilio (Pterourus) alexiares* Hopffer, 1886
Restricted to eastern Mexico, but not known to be threatened. Two subspecies. Refs: 51, 55, 62, 175, 234, 269.

354 *Papilio (Pterourus) multicaudatus* Kirby, 1884 (= *daunus* Bois., 1836)
South-western Canada (British Columbia and Alberta), western and mid-western U.S.A. (Washington, Montana, North Dakota, South Dakota, Oregon, Idaho, Wyoming, Nebraska, California, Nevada, Utah, Colorado, Kansas, Arizona, New Mexico and Texas), Mexico and Guatemala. Fairly widely distributed and not threatened. Common names: Two-tailed Tiger Swallowtail (216), Three-tailed Swallowtail (282). Other refs: 55, 62, 137, 169, 234, 269.

355 *Papilio (Pterourus) rutulus* Lucas, 1852
Western and mid-western U.S.A. (Washington, Idaho, Montana, South Dakota (Black Hills), Wyoming, Oregon, California, Nevada, Utah, Colorado, Arizona and New Mexico), south-western Canada (British Columbia) and northern Mexico (Baja California). Common and not threatened but rare east of the Rockies. Three subspecies (269) of doubtful validity (137). Common name: Western Tiger Swallowtail (216). Other refs: 55, 62, 141, 234.

356 *Papilio (Pterourus) eurymedon* Lucas, 1852
South-western Canada (British Columbia), western U.S.A. (Washington, Idaho, Montana, Oregon, Wyoming (?), California, Nevada, Utah,

Colorado, Arizona and New Mexico) and northern Mexico (Baja California). According to D'Almeida (55) it is also found further south in Mexico, Belize and possibly Guatemala but this is questionable. Not uncommon and not known to be threatened. Common name: Pale Tiger Swallowtail (216). Other refs: 62, 137, 234, 269.

Species-group: *troilus* Linnaeus

357   *Papilio (Pterourus) pilumnus* Boisduval, 1836
Mexico and Guatemala; occasional records from adjacent U.S.A. (Texas and southern Arizona), Honduras and El Salvador. Not threatened. Refs: 51, 55, 62, 137, 169, 216, 234, 269.

358   *Papilio (Pterourus) palamedes* Drury, 1773
South-eastern U.S.A. (Pennsylvania, Maryland, West Virginia, North Carolina, South Carolina, Georgia, Florida, Tennessee (?), Alabama, Mississippi, Missouri, Arkansas, Louisiana, Texas and Nebraska (?)) and Mexico. Not uncommon and not known to be threatened. Two subspecies. Common name: Palamedes Swallowtail (154, 216, 230). Other refs: 55, 137, 169, 269.

359   *Papilio (Pterourus) troilus* Linnaeus, 1758
Eastern Canada and eastern and south-eastern U.S.A. (to Texas). Generally common and not threatened. Two subspecies. Common names: Spicebush Swallowtail (154, 216, 230), Green-clouded Swallowtail (216). Other refs: 55, 137, 141, 234, 269.

360   *Papilio (Pterourus) ascolius* C. & R. Felder, 1864
Nicaragua (?), Costa Rica, Panama, Colombia and western Ecuador, a subspecies is also found in French Guiana and possibly Guyana and Surinam (55). Unusual in being an ithomiine mimic. Rare in Costa Rica (61). Scarcer than *P. (P.) zagreus* but not thought to be threatened. More data required. Five subspecies. Other ref: 234.

361   *Papilio (Pterourus) neyi* Niepelt, 1909
Ecuador. A little-known species, probably only a form of *P. (P.) zagreus* (250). Further information is needed on both the taxonomic and conservation status of this species. Other refs: 51, 55.

Species-group: *zagreus* Doubleday

362   *Papilio (Pterourus) zagreus* Doubleday, 1847
Colombia, Venezuela, Ecuador, Peru and Bolivia. Not uncommon and not known to be threatened. Two subspecies. *P. (P.) chrysoxanthus* Fruhstorfer, 1915 is a synonym. Refs: 51, 55, 169, 234.

363   *Papilio (Pterourus) bachus* C. & R. Felder, 1865
Colombia, Ecuador, Peru and Bolivia. Not uncommon and not known to be threatened. Two subspecies. Refs: 51, 55, 234.

Species-group: *scamander* Boisduval

364   *Papilio (Pterourus) hellanichus* Hewitson, 1868
Uruguay, Brazil and Argentina. Not common, but not recognized as threatened. Refs: 51, 55, 122, 234.

365   *Papilio (Pterourus) scamander* Boisduval, 1836
Argentina, Brazil and possibly Paraguay. Not uncommon and not known to be threatened. Ranched in Brazil (24). Three (possibly four) subspecies. Considered synonymous with *P. (P.) grayi* Boisduval, 1836, *P. (P.) eurymander* Rothschild and Jordan, 1906 and *P. (P.) joergenseni* Röber, 1925 (51). Other refs: 26, 55, 122, 234.

366   *Papilio (Pterourus) xanthopleura* Godman and Salvin, 1868
Peru (Upper Amazon) and Ecuador. It may also occur in eastern Bolivia. Uncommon and could possibly be designated Rare, but more information needed. Refs: 51, 55, 234.

367   *Papilio (Pterourus) birchalli* Hewitson, 1863
Costa Rica, Panama and Colombia. Widely distributed in Costa Rica, but seldom collected (61). Uncommon generally, though possibly less so than *P. (P.) xanthopleura*. Not known to be threatened. Two subspecies. Other refs: 51, 55, 234.

Species-group: *homerus* Fabricius

368   *Papilio (Pterourus) cleotas* Gray, 1832
Costa Rica, Panama, Colombia, Venezuela, Brazil and northern Argentina (51, 55, 234). Possibly also in Paraguay and Bolivia. Rare in Costa Rica (61). Ranched in Brazil (24). Not known to be threatened.

369   *Papilio (Pterourus) menatius* Hübner, 1819
Panama, Colombia, Guyana, Surinam, French Guiana, Brazil, Ecuador, Peru and Bolivia. Generally common and not known to be threatened. Up to ten subspecies. *P. (P.) judicael* Oberthür, 1888 is considered a hybrid between *P. (P.) menatius* and *P. (P.) warscewiczi* by Hancock (109) but this has been questioned (221). *P. (P.) aristeus* Cramer, 1781 is a homonym. Other refs: 51, 55, 234.

370   *Papilio (Pterourus) phaeton* (Lucas, 1857)
Costa Rica, Panama, Colombia, Venezuela. Usually regarded as a subspecies of *cleotas*, but accepted by Hancock as a good species (109). Little information available, but not known to be threatened. Three subspecies. Other ref: 51.

371   *Papilio (Pterourus) victorinus* Doubleday, 1844
Mexico and Central America (excluding Panama). Rare in Costa Rica. Flies at canopy level (61). Not believed to be threatened. Three subspecies. *P. (P.) diazi* Racheli and Sbordoni, 1975 was recently discovered in Mexico (Morelos) (225). However according to Hancock (109, 112) and Tyler (268) it is of hybrid origin, probably between *P. (P.) victorinus* and *P. (P.) garamas*. Other refs: 51, 55, 62, 234, 269.

372   *Papilio (Pterourus) garamas* Geyer, 1829
Mexico. Not uncommon but not regarded as threatened. Up to five subspecies have been described from Central America but most of these probably belong to *P. (P.) abderus* (115, 234). Subspecies *syedra* is placed in *garamas* by DeVries (61), but this may be better placed under *abderus*. For the present purpose we retain *garamas* as a Mexican endemic, but a thorough taxonomic clarification is required. Other refs: 51, 62, 175, 261, 269.

373   *Papilio (Pterourus) abderus* Hopffer, 1856
Mexico and Central America. Often regarded as a subspecies of *P. (P.)
garamas* (see above), this taxon requires further study. Not known to be
threatened. Four subspecies. Refs: 55, 62, 234, 269.

374   *Papilio (Pterourus) homerus* Fabricius, 1793
Endangered—refer to section 6, p. 297.
Jamaica. This much-prized species was formerly classified as Vulnerable
but has now been raised to Endangered. Its mountain habitats seem to be
threatened and specimens are still in demand (130). Its status should be
closely monitored. Common name: Homerus Swallowtail (21, 230). Other
refs: 51, 55, 234, 267, 283.

375   *Papilio (Pterourus) warscewiczi* Hopffer, 1886
Ecuador, Peru and Bolivia. Not uncommon and not threatened. Three
subspecies. Refs: 51, 55, 234.

376   *Papilio (Pterourus) cacicus* Lucas, 1852
Colombia, Ecuador, Venezuela and Peru. Not common but not known to
be threatened. Up to four subspecies. Refs: 51, 55, 234.

377   *Papilio (Pterourus) euterpinus* Godman and Salvin, 1868
Colombia, Ecuador and northern Peru. Scarce, possibly deserving Rare
status, but not known to be threatened. More information needed. Two
subspecies. Refs: 51, 55, 234.

Subgenus: *Heraclides* Hübner

Species-group: *thoas* Linnaeus

378   *Papilio (Heraclides) esperanza* Beutelspacher, 1975
Vulnerable—refer to section 6, p. 299.
Mexico (Oaxaca state). Very rare and with a restricted range. Little
information available as its locality is kept secret. Refs: 18, 51, 62, 137, 175,
268, 276.

379   *Papilio (Heraclides) andraemon* Hübner, 1823
Cuba, Bahamas, Cayman Is and sporadically Florida (270). Introduced into
Jamaica where it is a pest of *Citrus*. Otherwise not common but not
threatened as a species. Three subspecies. Common names: Bahaman
Swallowtail (230), Andraemon Swallowtail (21). Other refs: 51, 55, 169,
234, 269.

380   *Papilio (Heraclides) machaonides* Esper, 1796
Confined to Hispaniola (Haiti and Dominican Republic) and the Cayman Is
(Grand Cayman) (51). A report of its occurrence in Puerto Rico (230)
seems doubtful (92). A common butterfly throughout Hispaniola (243)
despite reports to the contrary (230), occurring from sea-level to high
elevation. Common name: Machaonides Swallowtail (230). Other refs: 55,
234.

381   *Papilio (Heraclides) thersites* Fabricius, 1775
Confined to Jamaica, where it is 'local and rather rare but widespread' (230)
but is not threatened (266). Its status should be carefully monitored.
Common names: False Androgeus Swallowtail (230), Thersites Swallowtail
(21). Other refs: 55, 234.

382   *Papilio (Heraclides) astyalus* Godart, 1819
U.S.A. (southern Texas), Mexico, Central America and South America (except for Chile). At least the nominate subspecies is common and abundant. Encountered as solitary individuals along forest edges (61). Ranched in Brazil (24). Five subspecies (22). *P. (H.) lycophron* Hübner is a synonym. Common name: the Lycophron (9). Other refs: 26, 51, 55, 62, 137, 234, 269.

383   *Papilio (Heraclides) androgeus* Cramer, 1775
Mexico, Central America, Cuba, Hispaniola, Trinidad, Tobago, St. Lucia and South America (except Chile and Uruguay). Generally common and not threatened. Tolerant of open areas and secondary growth (61). Three subspecies. Common names: the Androgeus (9), Androgeus Swallowtail (230), Queen Page (9, 282). Other refs: 26, 51, 55, 62, 234, 269.

384   *Papilio (Heraclides) ornythion* Boisduval, 1836
U.S.A. (southern Texas), Mexico and Guatemala. Not uncommon, but not known to be threatened. Common name: Ornythion Swallowtail (154). Other refs: 51, 55, 57, 62, 137, 234, 269.

385   *Papilio (Heraclides) aristodemus* Esper, 1794
U.S.A. (Southern Florida), Cuba, Cayman Is, Hispaniola, Bahamas and possibly Puerto Rico. Not common anywhere, though probably not threatened as a species. Five subspecies (including two recently described from the Bahamas) (38). Common name: Dusky Swallowtail (230). Other refs: 51, 55, 216, 234, 269.

Papilio (Heraclides) aristodemus ponceanus Schaus, 1911
Endangered—refer to section 6, p. 301.
The Floridan subspecies, *ponceanus* Schaus, is an Endangered taxon and is protected under the U.S. Endangered Species Act (47, 137, 283). It has recently been relisted as Endangered (270). Common name: Schaus' Swallowtail (154, 216).

386   *Papilio (Heraclides) caiguanabus* Poey, 1851
Indeterminate—refer to section 6, p. 305.
Restricted to Cuba, where it is rare (230): more information on its status and prospects is required. Common name: Poey's Black Swallowtail (230). Other refs: 51, 55, 234.

387   *Papilio (Heraclides) aristor* Godart, 1819
Indeterminate—refer to section 6, p. 307.
Restricted to Haiti and the Dominican Republic (Hispaniola) where it is said to be very rare (230), but may be locally common in xeric areas (243). More data on status are required as there is little information available. Common name: Scarce Haitian Swallowtail (230). Other refs: 51, 55, 234.

388   *Papilio (Heraclides) thoas* Linnaeus, 1771
U.S.A. (Texas, Colorado and Kansas), Mexico, Central America, Cuba, Jamaica, Antigua, Trinidad and South America (possibly not Chile). Often common and not threatened. Ranched in Brazil (24). Seven, or eight subspecies. Common names: Thoas Swallowtail (154, 216, 230), King Page (9, 282), Citrus or Orange Swallowtail (282). Other refs: 21, 26, 51, 55, 61, 62, 137, 150, 175, 234, 269.

389 *Papilio (Heraclides) cinyras* Ménétriés, 1857
Peru and eastern Ecuador to Brazil (central Amazonas) and Bolivia to Brazil (Mato Grosso) (51). Often regarded as a subspecies of *thoas*, but with a very distinct appearance. Little information, but not believed to be threatened.

390 *Papilio (Heraclides) homothoas* Rothschild and Jordan, 1906
Trinidad, Colombia and Venezuela. Although the range is fairly restricted, the species is not threatened. Common name: Small King Page (9). Other refs: 51, 55, 234.

391 *Papilio (Heraclides) cresphontes* Cramer, 1777
Southern Canada, U.S.A. (east of the Rockies and including the southern states), Mexico, Central America, Colombia (?), Cuba, possibly Bermuda and Bahamas. Widely distributed and not threatened. Common names: Giant Swallowtail (60, 154, 216, 230), Orange Dog (216). Other refs: 51, 55, 57, 61, 62, 137, 234, 269.

392 *Papilio (Heraclides) paeon* Boisduval, 1836
Costa Rica, El Salvador, Colombia, Venezuela, Ecuador, Peru and Bolivia. Not uncommon and no threats recognized. Three, possibly four, subspecies. Refs: 51, 55, 57, 234.

Species-group: *torquatus* Cramer

393 *Papilio (Heraclides) garleppi* Staudinger, 1892
Insufficiently Known—refer to section 6, p. 309.
Guyana, Surinam, French Guiana, Bolivia, Brazil and Peru. Not known to be in danger but reported to be a very rare species (268). Other refs: 51, 55, 234.

394 *Papilio (Heraclides) torquatus* Cramer, 1777
Mexico, Central America, Trinidad and South America (except Chile and Uruguay). Generally common and not threatened. Six subspecies including *P. (H.) t. tolus* Godman and Salvin, 1890 from Mexico, possibly Belize and Guatemala. *P. (H.) tolus* is generally regarded as a subspecies of *P. (H.) torquatus* (51, 234) and by only a few authorities as a good species (175). *P. (H.) tasso* (Staudinger, 1884) is an aberration of *P. (H.) torquatus* (109). Common name: the Torquatus (9). Other refs: 26, 55, 61, 62, 169, 269.

395 *Papilio (Heraclides) hectorides* Esper, 1794
South-eastern Brazil, Paraguay, northern Argentina and Uruguay (51). Not uncommon and not considered to be threatened. Ranched in Brazil (24). Other refs: 55, 122, 169, 234.

396 *Papilio (Heraclides) lamarchei* Staudinger, 1892
Brazil (near the Argentine border), Bolivia and northern Argentina. Uncommon but not known to be threatened. Refs: 51, 55, 122, 234.

397 *Papilio (Heraclides) himeros* Hopffer, 1866
Vulnerable—refer to section 6, p. 311.
Brazil and according to D'Almeida (55) also from Corrientes in northern Argentina. Two subspecies, both of which are exceedingly rarely encountered and declining (23). Other refs: 26, 51, 122, 139, 234.

Species-group: *anchisiades* Esper

398   *Papilio (Heraclides) hyppason* Cramer, 1775
Venezuela (?), Guyana, Surinam, French Guiana, Brazil, Peru and Bolivia.
Not uncommon and apparently not threatened. Refs: 51, 55, 169, 234.

399   *Papilio (Heraclides) pelaus* Fabricius, 1775
Cuba, Jamaica, Hispaniola and Puerto Rico. Apparently not uncommon
but widely distributed and not threatened. Three subspecies. Common
names: Prickly Ash Swallowtail (230), Pelaus Swallowtail (21). Other refs:
51, 55, 234.

400   *Papilio (Heraclides) oxynias* (Geyer, 1827)
Cuba. Widespread in that country, according to Riley (230), but status
requiring clarification. Common name: Cuban Black Swallowtail (230).
Other refs: 51, 55, 169, 234.

401   *Papilio (Heraclides) epenetus* Hewitson, 1861
Known only from western Ecuador. Certainly uncommon but vulnerability
not known and more information needed. Larvae feed on *Citrus*. Refs: 51,
55, 169, 234.

402   *Papilio (Heraclides) erostratus* Westwood, 1847
Eastern Mexico (not western as given by D'Almeida (55)), Guatemala and
Belize. Not common, but not known to be threatened. Other refs: 51, 62,
234, 269.

403   *Papilio (Heraclides) erostratinus* Vazquez, 1947
Mexico. Described fairly recently but known from several localities and not
thought to be threatened. Tyler tentatively suggests it might be a northern
subspecies of *P. (H.) erostratus* (269). Other refs: 51, 55, 62.

404   *Papilio (Heraclides) pharnaces* Doubleday, 1846
Southern Mexico and possibly Guatemala (51). Not uncommon and,
despite the narrow range, apparently not threatened. Other refs: 55, 62,
234, 269.

405   *Papilio (Heraclides) chiansiades* Westwood, 1872
Ecuador and Peru (Upper Amazon). Generally common and not threatened. Refs: 51, 55, 169, 234.

406   *Papilio (Heraclides) dospassosi* Rütimeyer, 1969
South-east Colombia, Putumayo Valley (236). Listed by Hancock (109) but
not by D'Abrera (51). No further information on taxonomic status or
threats. This species is apparently only known from the type specimen.
Possibly only a subspecies of *Eurytides trapeza* (221). It could be a candidate
for conservation concern if shown to be a good species.

407   *Papilio (Heraclides) anchisiades* Esper, 1788
Southern U.S.A. (Texas), Mexico, Central America, South America
(excluding Chile and Uruguay) and Trinidad. Widespread and generally
very abundant; not threatened. Tolerant of a very wide range of habitats.
Ranched in Brazil (24). Three or four subspecies (all apparently common).
Common names: Ruby-spotted Swallowtail (216), Orange Dog (9, 282).
Other refs: 26, 51, 55, 57, 62, 141, 154, 169, 234, 269.

408 *Papilio (Heraclides) maroni* Moreau, 1923
Insufficiently Known—refer to section 6, p. 313.
Known only from French Guiana (male and female) (186). but very little
information available. Smart (250) notes its similarity to the common
species *P. (H.) isidorus*. Other refs: 51, 55, 139, 187.

409 *Papilio (Heraclides) rogeri* Boisduval, 1836
Mexico (Yucatan), Belize and possibly Guatemala (51), a restricted range.
The species seems to be fairly common, at least in Yucatan, and not
threatened. Other refs: 55, 62, 234, 269.

410 *Papilio (Heraclides) isidorus* Doubleday, 1846
Costa Rica (61), Panama, Colombia, Ecuador, Bolivia and Peru. Generally
common and not threatened. Apparently very rare, montane at its northern
limits in Costa Rica, not collected in recent years (61). Four subspecies.
Other refs: 51, 55, 234.

411 *Papilio (Heraclides) rhodostictus* Butler & Druce, 1874
Panama, Costa Rica, Colombia and western Ecuador. Appently not
uncommon, at least in the southern part of its range, but rarely recorded in
Costa Rica (61). Not known to be threatened. Three subspecies. Other refs:
51, 55, 234.

Subgenus: *Eleppone* Hancock

Species-group: *anactus* Macleay

412 *Papilio (Eleppone) anactus* Macleay, 1826
Eastern Australia and possibly New Caledonia (vagrants?) (133). Common
in at least parts of its range and not threatened. Common name: Dingy
Swallowtail (282). Other refs: 40, 48, 106, 141, 176.

Subgenus: *Chilasa* Moore

Species-group: *elwesi* Leech

413 *Papilio (Chilasa) elwesi* Leech, 1889
Eastern and central China. Scarce and not well known, but not thought to
be threatened. The subspecies *P. (C.) e. maraho* (see below) has been
raised to a full species. Refs: 52, 141.

414 *Papilio (Chilasa) maraho* (Shiraki and Sonan, 1934)
Vulnerable—refer to section 6, p. 315.
Taiwan. Sometimes regarded as a subspecies of *P. (C.) elwesi* (52), but
Igarashi (141), Hancock (109) and Haugum (118) believe that it is quite
distinct. It certainly merits conservation concern due to habitat destruction
within its restricted distribution and the high prices paid for it by collectors.
It has been suggested that this species is rarely collected because of its
montane distribution (118), but Taiwan is such a small and crowded island
that all populations of the butterfly are considered to be at risk. Other ref:
246.

Species-group: *clytia* Linnaeus

415 *Papilio (Chilasa) clytia* Linnaeus, 1758
India, Sri Lanka, Nepal, Bhutan (?), Bangladesh, Burma, Thailand,
southern China (including Hainan (Guangdong prov.)), Hong Kong,

Vietnam, Laos, Kampuchea, Peninsular Malaysia, Philippines, Andamans and Indonesia (Flores, Alor, Timor and Moa) (52, 262). Generally common and not threatened, but the nominate subspecies is protected by law in India (182). Up to seven subspecies. Common name: Common Mime (45, 80, 251). Other refs: 4, 5, 48, 87, 141, 143, 259, 287.

416 *Papilio (Chilasa) paradoxa* Zinken–Sommer, 1831
Northern India (including Assam), Bangladesh, Burma, southern China (?), Thailand, Vietnam, Laos, Kampuchea, Peninsular and Eastern Malaysia, Philippines (Palawan), Brunei and Indonesia (Sumatra, Bangka, Nias, Batu and Kalimantan) (52, 262). Never common (269) but not known to be threatened. However, *P. (C.) paradoxa telearchus* is protected by law in India (182). A number of forms, mimicking different Danainae. Seven subspecies. Common name: Great Blue Mime (3, 45, 80, 282). Other refs: 4, 87, 131.

Species-group: *veiovis* Hewitson

417 *Papilio (Chilasa) veiovis* Hewitson, 1853
Indonesia (Sulawesi). Not uncommon, but not known to be threatened. Refs: 52, 262.

418 *Papilio (Chilasa) osmana* Jumalon, 1967
Vulnerable—refer to section 6, p. 317.
Philippines (southern Leyte and northern Mindanao). Scarce everywhere in its restricted range (262). Other refs: 52, 148.

419 *Papilio (Chilasa) carolinensis* Jumalon, 1967
Vulnerable—refer to section 6, p. 320.
Philippines (north-eastern Mindanao). Restricted range, claimed by Tsukada and Nishiyama (262) to be the rarest Philippine swallowtail. Other refs: 52, 148.

Species-group: *agestor* Gray

420 *Papilio (Chilasa) agestor* (Gray, 1832)
Northern India, Nepal, Bhutan (?), Bangladesh (?), Burma, Thailand, central and southern China, Taiwan, Vietnam, Laos, Kampuchea (?) and Peninsular Malaysia (52). Not threatened across most of its range but considered to be Vulnerable in Peninsular Malaysia (10). Six subspecies. Common name: Tawny Mime (80, 251). Other refs: 45, 87, 141, 246, 259.

421 *Papilio (Chilasa) epycides* (Hewitson, 1862)
Northern India (Assam, Sikkim and Manipur), Bhutan, Burma, Thailand, southern China, Taiwan, Vietnam (?), Laos (?) and Peninsular Malaysia (?). No known threats to the species, although the nominate subspecies is protected by law in India (182). Six subspecies. Common name: Lesser Mime (80). Other refs: 4, 45, 87, 141, 246, 259.

422 *Papilio (Chilasa) slateri* (Hewitson, 1857)
Northern India (Sikkim and Assam), Bhutan (?), Burma, Thailand, southern China (?), northern Vietnam, Laos, Peninsular and Eastern Malaysia, Brunei and Indonesia (Kalimantan (?) and northern Sumatra) (52, 262). Not known to be threatened as a species but some subspecies may be rare and the nominate subspecies is protected by law in India (182). Four

(262) to six (52) subspecies. Common names: Blue-striped Mime (80), Brown Mime (3). Other refs: 4, 45, 87, 131, 141.

Species-group: *laglaizei* Depuiset

423 *Papilio (Chilasa) laglaizei* Depuiset, 1877
Indonesia (Irian Jaya, Waigeo and Aru) and Papua New Guinea. Local but not threatened. Refs: 14, 48, 141, 256.

424 *Papilio (Chilasa) toboroi* Ribbe, 1907
Rare—refer to section 6, p. 322.
Papua New Guinea (Bougainville (203)) and Solomons (Santa Isabel and Malaita). Said to be common (48), but the price of specimens is relatively high (142). It is rarely collected for trade in Papua New Guinea (36). More information is needed. Two subspecies. Other refs: 132, 141, 218, 219, 228, 256.

425 *Papilio (Chilasa) moerneri* Aurivillius, 1919
Vulnerable—refer to section 6, p. 324.
Restricted to the Bismarck Archipelago (New Ireland and possibly New Britain and New Hanover). Known from very few specimens. It was regarded as extinct by D'Abrera (48) but this is not certain as large areas of potential habitat remain unexplored (83). Two subspecies have been described. Other refs: 6, 217, 235.

Subgenus: *Papilio* Linnaeus

Species-group: *machaon* Linnaeus

426 *Papilio (Papilio) alexanor* Esper, 1799
Southern France, southern Italy, Sicily, Yugoslavia, Albania, Greece, Turkey, Syria, Lebanon, Israel, Jordan, Iraq, Iran, western Pakistan, Afghanistan (including Nangrahar province) and U.S.S.R. (Turkmenistan). Extremely rare in Pakistan (Baluchistan) (259). Uncommon in the western part of its range and listed as Vulnerable in Europe (123), but apparently not threatened as a species. Declining in the U.S.S.R. due to degradation of alpine meadows and included in the U.S.S.R. Red Data Book, category Vulnerable (8, 260). Extinct in Yugoslavia, Endangered in the Balkan Peninsula and Italy (123), protected by law in Greece. Six subspecies. Common names: Southern Swallowtail (125, 282), Tiger Swallowtail (158), Baluchi Yellow Swallowtail (80). Other refs: 141, 169.

427 *Papilio (Papilio) hospiton* Guenée, 1839
Endangered—refer to section 6, p. 326.
Endemic to Corsica and Sardinia; status not thoroughly known, but giving rise to concern and listed as Endangered (123). Trade is restricted by law in France and the Netherlands. Common name: Corsican Swallowtail (125). Other ref: 202.

428 *Papilio (Papilio) machaon* Linnaeus, 1758
Entire Palearctic region through the U.S.S.R. to China (164, 207) and Japan (including Nepal, Bhutan and Taiwan) as far south as some oases in Saudi Arabia and Oman and the high mountains of Yemen (161); Pakistan, northern India, northern Burma, Canada and Alaska. Often common and not threatened as a species. Thirty seven subspecies (141). *P. (P.) m.*

saharae flies in Yemen and parts of the Sahara (161). Larsen (161) lists it as a full species, noting that its larva is very different to that of *machaon* in other parts of Arabia. *Papilio (P.) machaon* is protected by law in six of the Austrian provinces, Czechoslovakia, Hungary, the U.K. (subspecies *britannicus*) and India (subspecies *verityi*) (182). It is listed as Vulnerable in South Korea (152), the U.S.S.R. Red Data Book (8) and the Austrian Red Data Book (93). Common names: Swallowtail (3, 125, 282), Old World Swallowtail (154, 216), Common Yellow Swallowtail (80, 251). Other refs: 52, 55, 137, 149,, 169, 202, 246, 259, 269.

429 *Papilio (Papilio) hippocrates* C. & R. Felder, 1864
Japan. Generally regarded as a subspecies of *P. (P.) machaon* (141, 149, 279), but accepted as a full species by Hancock (109).

430 *Papilio (Papilio) zelicaon* Lucas, 1852
Mexico, western U.S.A. including Alaska, and Canada (Yukon Territory (?), British Columbia, Alberta, Saskatchewan and Manitoba). Common and not threatened. *P. (P.) gothica* Remington from Canada (British Columbia and Alberta), western U.S.A. and Mexico is widely regarded as a biological race or subspecies of *P. (P.) zelicaon* or as the yellow form of *P. (P.) nitra* (109). Not threatened. Common name: Anise Swallowtail (216). Other refs: 55, 62, 137, 141, 169, 226, 234, 244, 269.

431 *Papilio (Papilio) indra* Reakirt, 1866
Western U.S.A. (Washington, Oregon, California, Nevada, Utah, Colorado and possibly Arizona). Uncommon but not generally threatened. Listed as a Special Species by Washington State Department of Game. Seven subspecies. Common names: Indra Swallowtail (216), Short-tailed Black Swallowtail (I75). Other refs: 62, 77, 78, 79, 137, 169, 234, 269.

432 *Papilio (Papilio) polyxenes* Fabricius, 1775 (= *ajax* L., 1758)
Southern Canada, U.S.A., Cuba, Mexico, Central America, Colombia, Venezuela, Ecuador and northern Peru. Widely distributed, often common and not threatened. Populations may fluctuate widely in numbers (61). Four subspecies. Common names: Eastern Black Swallowtail (216), American Black Swallowtail (230), Parsnip Swallowtail (154). Other refs: 55, 62, 137, 141, 169, 234, 269.

433 *Papilio (Papilio) brevicauda* Saunders, 1869
Eastern Canada (southern Labrador, Newfoundland, Quebec, New Brunswick, Nova Scotia), mainly maritime. It has been described as a subspecies of *P. (P.) polyxenes* (234) or of *P. (P.) ajax* (55) the latter a synonym of the former. Not particularly common but not known to be threatened. Three subspecies (not well defined) (137). Common names: Short-tailed Swallowtail (154, 216), Maritime Swallowtail (216). Other refs: 84, 85, 269.

434 *Papilio (Papilio) oregonia* Edwards, 1876
North-western U.S.A. (eastern Washington, Idaho, western Montana, North Dakota (?) and eastern Oregon) and south-western Canada (British Columbia and Alberta (?)). Possibly a form of *P. (P.) bairdi* (55, 234, 250). Not common but not recognized as being widely threatened. Listed as a Special Species by Washington State Department of Game. Two subspecies. Common name: Oregon Swallowtail (216). Other refs: 65, 137, 141, 213, 269.

435 *Papilio (Papilio) kahli* F. & R. Chermock, 1937
Canada (Manitoba and Saskatchewan). Specific status doubtful (250, 269).
Regarded as a subspecies of *P. (P.) nitra* by D'Almeida (55). Restricted
range, but not recognized as threatened. Other refs: 137, 216.

436 *Papilio (Papilio) nitra* Edwards, 1883
Canada (Alberta) and eastern side of Rocky Mountains in U.S.A.
(Montana, Wyoming and Colorado). Another taxon of doubtful specific
status, probably a form or subspecies of *P. (P.) zelicaon*. Occurs early in the
year in high mountains and so rarely taken, but not believed to be
threatened. Refs: 55, 137, 216, 234, 269.

437 *Papilio (Papilio) coloro* Wright, 1905 (=*rudkini* Chermock, 1977)
U.S.A. (southern Nevada, south-eastern Utah, southern California and
western Arizona). Specific status doubtful (250, 269), possibly a form of *P.
(P.) bairdi* (55), *P. (P.) polyxenes* (118) or *P. (P.) zelicaon*. Restricted
range, but not recognized as threatened. Common name: Desert Swallowtail (216). Other refs: 86, 137.

438 *Papilio (Papilio) bairdi* Edwards, 1869
Mid-western U.S.A. (Montana (?), Idaho, Wyoming, South Dakota,
Nebraska, Utah, Colorado, Arizona and New Mexico). Not known to be
threatened. Common names: Western Black Swallowtail (216), Baird's
Swallowtail (216). Other refs: 55, 62, 137, 141, 234, 269.

439 *Papilio (Papilio) joanae* Heitzman, 1974
U.S.A. (Missouri only). Specific status not certain, possibly a form of *P.
(P.) polyxenes* (250); it is not included in Howe (137). The range is
restricted, but the butterfly is not regarded as a threatened species. Other
refs: 216, 269.

Subgenus: *Princeps* Hübner

Species-group: *xuthus* Linnaeus

440 *Papilio (Princeps) xuthus* Linnaeus, 1760
Southern China, northern Burma, Japan, Taiwan, Guam, Ogasawara-
shoto (Bonin I.), Wake (?), Midway Is (?) and Hawaii. Common and not
threatened. Two subspecies. Common name: Chinese Yellow Swallowtail
(80). Other refs: 52, 92, 141, 149, 207.

441 *Papilio (Princeps) benguetanus* Joicey and Talbot, 1923
Vulnerable—see section 6, p. 329.
Philippines (Luzon only). Occasionally regarded as a subspecies of *P. (P.)
xuthus*, but very different in appearance. Refs: 52, 141, 148, 262.

Species-group: *demolion* Cramer

442 *Papilio (Princeps) euchenor* Guérin–Méneville, 1829
Indonesia (Irian Jaya), Papua New Guinea, Bismarck Archipelago (includ-
ing New Britain), D'Entrecasteaux Is, Trobriand Is (?) and the Louisiade
Archipelago (?). Not uncommon and not threatened as a species though
some subspecies may be threatened (138). Collected for trade in Papua New
Guinea. This species was separated as the type of a monobasic genus,
Sugurua, by K. Okano in 1983 (198) but this has not been generally
accepted. Thirteen subspecies. Other refs: 14, 48, 52, 141, 217, 262.

443 *Papilio (Princeps) gigon* C. & R. Felder, 1864
Indonesia (Sulawesi, Banggai Is (?), Sula, Talaud and Sangihe). Not uncommon and not known to be threatened. Three subspecies. Refs: 52, 141, 262.

444 *Papilio (Princeps) demolion* Cramer, 1779
Burma, Thailand, Peninsular and Eastern Malaysia, Brunei, Philippines (Palawan) and Indonesia (Sumatra, Mentawai Is, Nias, Bangka, Java, Lombok, Sumbawa and Kalimantan) (52, 191, 262). Fairly common and not threatened. Three subspecies. Common names: Banded Swallowtail (3, 45), Burmese Banded Swallowtail (80). Other ref: 141.

445 *Papilio (Princeps) liomedon* Moore, 1874
Southern India. Uncommon, but protected by law (182) and not known to be threatened. It has been considered to be only a subspecies of *P. (P.) demolion* (80), but is apparently distinct (144). Common name: Malabar Banded Swallowtail (80). Other refs: 4, 52.

Species-group: *protenor* Cramer

446 *Papilio (Princeps) protenor* Cramer, 1775
Northern India, (including Kashmir and Jammu, Sikkim and Assam), Bangladesh, Burma, southern China (including Hainan (Guangdong prov.)), northern Vietnam (?), northern Laos (?), Taiwan, North Korea, South Korea and Japan. Common and not threatened. Five subspecies. Common name: the Spangle (80, 251). Other refs: 4, 52, 141, 149, 153, 207, 246, 259.

447 *Papilio (Princeps) demetrius* Cramer, 1782
Japan, North Korea, South Korea and China. Not uncommon except in Korea where it is considered to be Extinct or Endangered (152, 153). Losses in Korea suggest that monitoring throughout its range might be advisable. Sometimes considered to be only a subspecies of *P. (P.) protenor* (141). Other refs: 149, 247.

448 *Papilio (Princeps) macilentus* Janson, 1877
Japan, North Korea, South Korea where it is Vulnerable (152) and eastern and central China. Common and not threatened. Refs: 141, 149, 247.

Species-group: *bootes* Westwood

449 *Papilio (Princeps) bootes* Westwood, 1842
North-eastern India (Uttar Pradesh, Sikkim, Assam, Meghalaya, Manipur and Nagaland), northern Burma, northern Laos and south-western China, a restricted range. Not common but not regarded as threatened. It is protected by law in India (182). Three subspecies (not including *P. (P.) janaka*—see below). Common name: Tailed Redbreast (80). Other refs: 4, 52, 141, 259.

450 *Papilio (Princeps) janaka* Moore, 1857
Northern India (including Sikkim) and Nepal. Often regarded as only a subspecies of *P. (P.) bootes* (see above) (80, 141). Not known to be threatened. Common name: Tailed Redbreast (80, 251). Other refs: 4, 52.

Species-group: *memnon* Linnaeus

451   *Papilio (Princeps) lampsacus* Boisduval, 1836
Indonesia (western Java), only known from the localities of Mt Gede and
Mt Mas. Apparently rather scarce although said to be abundant on Mt Gede
by Tsukada and Nishiyama (262) who say "It is regrettable that collecting is
not permitted on Mt Gede now as it rises within the National Park area".
Further information and monitoring of this restricted species are needed.
Signs of habitat disruption would necessitate immediate listing as threat-
ened. Other refs: 45, 52.

452   *Papilio (Princeps) forbesi* Grose–Smith, 1883
Indonesia (northern Sumatra). Despite the restricted range, abundant in
some localities, not uncommon and not known to be threatened. Female
mimics *Atrophaneura (Atrophaneura) hageni* (262). Other refs: 45, 52.

453   *Papilio (Princeps) acheron* Grose–Smith, 1887
Rare—refer to section 6, p. 331.
Northern Borneo: Sabah, Sarawak (Eastern Malaysia), Brunei and
possibly Kalimantan (Indonesia). Conservation concern is the result of its
restricted known range and recent evidence of deforestation on Mt
Kinabalu. The taxonomic status of this species in relation to the two
preceding needs clarification (45). Other refs: 11, 52, 131, 231, 262.

454   *Papilio (Princeps) oenomaus* Godart, 1879
Indonesia (Lesser Sunda Is: Timor, Wetar, Moa, Leti, Kisar and Romang)
(262). Common (48) and presumably not threatened. Two subspecies.

455   *Papilio (Princeps) ascalaphus* Boisduval, 1836
Indonesia (Sulawesi, Salayar and Sula Is). Restricted range, but not
uncommon and not known to be threatened. Two subspecies. Refs: 141,
262.

456   *Papilio (Princeps) rumanzovia* Eschscholtz, 1821
Philippines (except Palawan), Batu Is and Indonesia (Talaud and Sangihe).
Also known from stray butterflies in southern Taiwan but not established
(52, 262). Not uncommon and not thought to be threatened. Other ref: 141.

457   *Papilio (Princeps) deiphobus* Linnaeus, 1758
Indonesia (Moluccas). Not common but not thought to be threatened. At
least two subspecies. Ref: 109.

458   *Papilio (Princeps) alcmenor* C. & R. Felder, 1864
North-eastern India, Nepal, Bhutan, Burma and China. Generally
common and not threatened. This species was formerly known as *P. (P.)
rhetenor* Westwood, 1841. Common name: Redbreast (80, 251). Other refs:
4, 52, 207, 259.

459   *Papilio (Princeps) thaiwanus* Rothschild, 1898
Taiwan only. Common, despite its restricted range and not therefore listed,
though its status should clearly be monitored. Refs: 52, 141, 246.

460   *Papilio (Princeps) polymnestor* Cramer, 1775
Southern India and the east coast, and Sri Lanka. Not uncommon and not
thought to be threatened. At least two subspecies. Common name: Blue
Mormon (80). Other refs: 4, 52, 141, 259, 287.

461 *Papilio (Princeps) lowi* Druce, 1873
Philippines (Palawan and Balabac) (52, 262). D'Abrera claims that it is also found in northern Borneo (52). Despite restricted range apparently not uncommon and not threatened. Males are seen widely but the females are more secretive; the young can be raised on *Citrus* (56).

462 *Papilio (Princeps) memnon* Linnaeus, 1758
North-eastern India (Sikkim, Assam and Nagaland), Nepal, Bhutan (?), Bangladesh, Nicobar Is, Andaman Is (stragglers only, 5), Burma, western, southern and eastern China (including Hainan), Taiwan, southern Japan, Ryukyu Is, Thailand, Laos, Vietnam, Kampuchea, Eastern and Peninsular Malaysia, Brunei and Indonesia (Sumatra, Mentawai Is, Nias, Batu, Simeulue, Bangka, Java, Kalimantan and the Lesser Sunda Is (except Timor, Wetar, Babar and Tanimbar)) (52, 191, 262). Common (especially since the cultivation of *Citrus* (45)) and not threatened. Up to 13 subspecies, highly mimetic and polymorphic in the female. Common name: Great Mormon (45, 80, 251). Other refs: 87, 131, 141, 149, 207, 231, 246, 259.

463 *Papilio (Princeps) mayo* Atkinson, 1873
Endemic to the Andamans. A little known butterfly for which more information is needed. Not rare according to Evans (80), though Tsukada and Nishiyama (262) suggest that it may be uncommon. Arora and Nandi state that males are common, females rare (5), and it is protected by Indian law (182). The female mimics *Atrophaneura (Losaria) rhodifer* Atkinson. Common name: Andaman Mormon (80). Other refs: 52, 259.

Species-group: *helenus* Linnaeus

464 *Papilio (Princeps) noblei* de Nicéville, 1889
Burma, Thailand, Laos and Vietnam (52). Said to be rare (80, 259), but current status not known. Two subspecies. Common name: Noble's Helen (80). Other refs: 144, 168.

465 *Papilio (Princeps) antonio* Hewitson, 1875
Philippines (Mindanao, Leyte and possibly Negros) (52, 262). Not known to be threatened but possibly could be classed as Rare. Poorly known and locally distributed but not always rare in its locality when found (262). Other ref: 144.

466 *Papilio (Princeps) biseriatus* Rothschild, 1895
Indonesia (restricted to Timor). Usually regarded as a subspecies of *P. (P.) helenus*, but recognized as a full species by Hancock (109). Further data may reveal threats to this very restricted taxon.

467 *Papilio (Princeps) iswara* White, 1842
Southern Burma, southern Thailand, Eastern and Peninsular Malaysia, Brunei and Indonesia (Sumatra, Bunguran, Bangka and Kalimantan) (52, 262). Not uncommon and not threatened. Two subspecies. Common names: Great Helen (80), Large Helen (3). Other refs: 45, 87, 259.

468 *Papilio (Princeps) hystaspes* (C. & R. Felder, 1862)
Philippines from Luzon south to Mindanao, excluding Palawan. Sometimes regarded as a subspecies of *helenus* (52) but recognized as a full species by Hancock (109) and Tsukada and Nishiyama (262). Widespread and not known to be threatened.

469   *Papilio (Princeps) sataspes* C. & R. Felder, 1864
Indonesia (Sulawesi, Banggai and Sula). There is no evidence that it is
threatened. Three subspecies (262). Other ref: 109.

470   *Papilio (Princeps) helenus* Linnaeus, 1758
Southern and north-eastern India, Nepal, Sri Lanka, Burma, Thailand,
Bangladesh, Bhutan, southern China (including Hainan (Guangdong
prov.)), Vietnam, Laos, Taiwan, southern Japan, South Korea, Ryukyu Is,
Kampuchea, Peninsular and Eastern Malaysia, Brunei, Philippines and
Indonesia (Sumatra, Java, Bangka, Kalimantan and the Lesser Sunda Is
(not Tanimbar)) (52, 262). Common and not threatened. Up to thirteen
subspecies. Common name: Red Helen (45, 80, 251). Other refs: 4, 87, 131,
141, 149, 246, 259.

471   *Papilio (Princeps) iswaroides* Fruhstorfer, 1897
Peninsular Malaysia, Indonesia (northern Sumatra) and possibly the
extreme south of Thailand. It has been suggested that it may also occur in
northern Borneo (262), but there has been no record so far. Two
subspecies. Other refs: 45, 52.

Species-group: *polytes* Linnaeus

472   *Papilio (Princeps) jordani* Fruhstorfer, 1902
Rare—refer to section 6, p. 333.
Indonesia (endemic to northern Sulawesi) where it is reported to be
extremely rare (262). The female mimics *Idea blanchardii* (Danainae).
Other refs: 52, 121, 144.

473   *Papilio (Princeps) polytes* Linnaeus, 1758
India, Nepal, Sri Lanka, Burma, Thailand, southern and western China
(including Hainan (Guangdong prov.)), Taiwan, Ryukyu Is (Japan), Guam
(221), Vietnam, Laos, Kampuchea, Nicobar Is, Andamans, Eastern and
Peninsular Malaysia, Brunei and Indonesia (except Moluccas and Irian
Jaya) (52, 262). Common and not threatened. Here we follow Hiura and
Alagar (127) and Tsukada and Nishiyama (262) who regard the subspecies
from the Philippines and Moluccas as a separate species, *P. (P.) alphenor*
Cramer (see below). Seventeen subspecies (23 if *P. (P.) alphenor* is
included)(262). *P. (P.) walkeri* Janson from southern India is probably only
a form of *P. (P.) polytes* (4, 80, 259). It is very uncommon and would be
classed as Rare if a good species. *P. (P.) sakontala* Hewitson, 1864 from
northern India (including Sikkim and Assam) is another taxon thought by
some authors to be a form of *P. (P.) polytes* rather than a good species (4,
80). It also is uncommon. Common name: Common Mormon (45, 80).
Other refs: 5, 48, 87, 131, 141, 143, 144, 149, 207, 246, 287.

474   *Papilio (Princeps) alphenor* (Cramer, 1776)
Indonesia (Moluccas, Sangihe and Talaud Is) and the Philippines (including
Palawan). Earlier regarded as a subspecies of *P. (P.) polytes*, but separated
as a full species in the revisionary work by Hiura and Alagar (127).
Widespread and not known to be threatened. Other ref: 262.

475   *Papilio (Princeps) ambrax* Boisduval, 1832
Papua New Guinea (including Woodlark I. (184), and Fergusson I. in the
Entrecasteaux Is), Indonesia (Irian Jaya and Aru) and north-eastern

Australia. Collected for trade in Papua New Guinea (36). Not rare and not threatened. Five subspecies. Other refs: 14, 40, 48, 141, 176.

476 *Papilio (Princeps) phestus* Guérin–Méneville, 1830
Papua New Guinea (Bismarck Archipelago (including New Britain and Admiralty Is) and Bougainville) and the Solomons (except Malaita, Guadalcanal and San Cristobal) (219). Common and not known to be threatened. Four subspecies. Other refs: 48, 68.

Species-group: *nephelus* Boisduval

477 *Papilio (Princeps) diophantus* Grose–Smith, 1882
Indonesia (northern Sumatra). Uncommon. Not thought to be threatened, though more information needed. The female is very scarce in collections. Refs: 45, 52, 144, 262.

478 *Papilio (Princeps) nephelus* Boisduval, 1836
Nepal, north-eastern India (including Assam and Sikkim), Bhutan (?), Burma, Thailand, Vietnam, Laos, southern China (including Hainan (Guangdong prov.), Taiwan, Kampuchea, Eastern and Peninsular Malaysia, Brunei and Indonesia (Sumatra, Mentawai Is, Nias, Batu, Java and Kalimantan) (52, 191, 262). Common and not threatened. Eleven subspecies including *chaon* Westwood (115). *P. (P.) nubilus* Staudinger from Sumatra and northern Borneo was placed as a possible aberration of *P. (P.) nephelus*, or a hybrid, by Hancock (109) and is regarded as a hybrid between *P. (P.) nephelus* and *P. (P.) polytes* by Tsukada and Nishiyama (262). More information required. Common names: Yellow Helen (80, as *P. chaon*), Black and White Helen (3, 45), Banded Helen (3). Other refs: 87, 131, 141, 231, 246.

479 *Papilio (Princeps) castor* Westwood, 1842
North-eastern India (Assam, Sikkim), Bhutan, Thailand, southern China, Taiwan, Vietnam, Laos and possibly Kampuchea (52). Common and not threatened. Deceptively similar to members of the subgenus *Chilasa* (52). Seven subspecies. Common name: Common Raven (80). Other refs: 4, 141, 246.

480 *Papilio (Princeps) mahadeva* Moore, 1878
Burma, Thailand, Laos (?), Vietnam (?), Kampuchea (?) and northern Peninsular Malaysia. Definitely uncommon but probably not meriting conservation status, though more information is needed. May be Vulnerable and in need of protection in Peninsular Malaysia (10). Deceptively similar to members of the subgenus *Chilasa* (52). Possibly two subspecies. Common name: Burmese Raven (45, 80). Other ref: 87.

481 *Papilio (Princeps) dravidarum* Wood–Mason, 1880
Southern India. Uncommon, but not known to be threatened. Deceptively similar to members of the subgenus *Chilasa* (52). Common name: Malabar Raven (80, 282). Other ref: 4.

Species-group: *fuscus* Goeze

482 *Papilio (Princeps) hipponous* C. & R. Felder, 1862
Indonesia (Sangihe and Talaud Is and possibly northern Sulawesi as vagrants) Clarification of both the taxonomic and conservation status of this

species is urgently needed. Up to six subspecies. Other refs: 52, 110, 118, 121, 262.

483 *Papilio (Princeps) pitmani* Elwes & de Nicéville, 1886
Burma, Thailand, Laos, Vietnam, China (including Hainan in Guangdong prov.) and the Philippines. This distribution is correct according to recent information (249). Presumably the distributions in (52) and (262) are erroneous. The range in the Philippines differs according to authors (52, 262). A report of its occurrence in Assam is very doubtful (259). Often regarded as a subspecies of *hipponous* (52, 262), but given as a full species by Hancock (109, 110). Not known to be threatened. Common name: Pitman's Helen (80). Other refs: 118, 121.

484 *Papilio (Princeps) fuscus* Goeze, 1779
Andamans (5), Nicobar Is, Peninsular and Eastern Malaysia, Brunei, Indonesia (except Sumatra (?), Java and the Lesser Sunda Is), Papua New Guinea, Bismarck Archipelago (except Admiralty Is), Bougainville, Solomon Is (except Santa Cruz), Vanuatu (= New Hebrides) (Torres Is), Trobriand Is (?), D'Entrecasteaux Is (?), Louisiade Archipelago (?) and Australia (Queensland and New South Wales) (262). Common generally and not threatened although the subspecies *P. (P.) f. andamanicus* (Andaman Helen (80))from the Andamans and Nicobar Is is protected by Indian law (182). Two subspecies are farmed in Papua New Guinea (36). Up to twenty-two subspecies. It has been surmised that *P. (P.) heringi* is a rare natural hybrid, *P. (P.) fuscus* x *P. (P.) tydeus* (111, 254) but this has been disputed (224). Common name: Blue Helen (3). Other refs: 4, 14, 40, 45, 48, 52, 87, 110, 131, 141, 176, 219, 220.

485 *Papilio (Princeps) canopus* Westwood, 1842
Indonesia (Lesser Sunda Is, including Tanimbar), Vanuatu (New Hebrides) (242) and Australia (Northern Territory and northern Queensland). *P. (P.) c. hypsicles* has recently been revised to a full species by Hancock (110) who notes its close relationship to *P. (P.) woodfordi* and not to *P. (P.) canopus*, as was previously thought. Common and not known to be threatened. Twelve subspecies if *P. (P.) c. hypsicles* is included (262). Other refs: 40, 48, 176.

486 *Papilio (Princeps) albinus* Wallace, 1865
Papua New Guinea and Indonesia (Irian Jaya and Moluccas). Not rare, except in the Moluccas, and not considered to be threatened. Collected for trade in Papua New Guinea (36). Two subspecies. Other refs: 48, 118.

487 *Papilio (Princeps) hypsicles* Hewitson, 1868
Vanuatu (New Hebrides). Status recently revived to a full species by Hancock (110) who noted it to be close to *P. (P.) woodfordi* and not to *P. (P.) canopus* which it superfically resembles and of which it was thought to be a subspecies. Restricted range but not known to be threatened. Other ref: 262.

488 *Papilio (Princeps) woodfordi* Godman and Salvin, 1888
Solomons, (except San Cristobal and Santa Cruz) and Bougainville where it is farmed (36). Not rare (48) and not known to be threatened. Four subspecies including *P. (P.) w. ptolychus* Godman and Salvin, 1888 (219). Other ref: 220.

489   *Papilio (Princeps) heringi* Niepelt, 1924
      Indonesia (endemic to Halmahera, Moluccas). A little-known species that
      was accepted as a full species and placed in the *aegeus* group by Munroe
      (188), but regarded as a rare natural hybrid of *P. (P.) fuscus* and *P. (P.)
      tydeus* by Hancock (109, 111). In a recent reassessment *P. (P.) heringi* is
      placed as a full species near the *fuscus* group (224). It has rarely been
      collected and the rapid acceleration of deforestation on Halmahera is a
      matter of concern. More information is required on this species.

      Species-group: *amynthor* Boisduval

490   *Papilio (Princeps) amynthor* Boisduval, 1859
      New Caledonia and Norfolk I. Apparently not uncommon, feeding on
      cultivated *Citrus* (133) and not threatened. Formerly known as *P. (P.)
      ilioneus* Donovan. Other refs: 48, 95, 111.

491   *Papilio (Princeps) schmeltzii* Herrich–Schaffer, 1869
      Fiji. Despite its restricted range, the species is apparently common (48) and
      not threatened. Other ref: 111.

492   *Papilio (Princeps) godeffroyi* Semper, 1866
      Western Samoa. Could possibly be classed as Rare, but more information
      on status required. Refs: 111, 144.

      Species-group: *gambrisius* Cramer

493   *Papilio (Princeps) inopinatus* Butler, 1883
      Indonesia (Lesser Sunda Is: Romang, Babar, Damar and Tanimbar). Not
      rare and not threatened. Specific status slightly doubtful (48) Two subspe-
      cies. Other refs: 111, 262.

494   *Papilio (Princeps) bridgei* Mathew, 1886
      Throughout the Solomons (except for Santa Cruz) and Bougainville (219)
      where it is collected and farmed for trade (36). Common (48) and not
      threatened. Seven subspecies including *P. (P.) b. erskinei* Mathew, 1886
      (219) and the recently described *P. (P.) b. michae* from Malaita (220).
      Other refs: 111, 144.

495   *Papilio (Princeps) weymeri* Niepelt, 1914
      Rare—refer to section 6, p. 335.
      Bismarck Archipelago (Admiralty Is). A rare species according to
      D'Abrera (48) but its rain forest habitat is still fairly extensive (217). The
      rutaceous foodplant is common on Manus I. (217). Sometimes available in
      trade as ex pupae (36). *P. (P.) cartereti* Oberthür, 1914 is a synonym, but the
      order of priority of the two names has been difficult to establish, both being
      published in the first six months of 1914. Ebner (68), assisted by R.I.
      Vane–Wright, decided on *cartereti* as the senior name, but *weymeri* has
      been adopted by Hancock (109). Other ref: 111.

496   *Papilio (Princeps) gambrisius* Cramer, 1777
      Indonesia (Seram, Ambon and Buru). A good species according to
      D'Abrera (48), but specific status slightly in doubt (144). More information
      needed before conservation status can be properly assessed. Two subspe-
      cies. Other ref: 111.

497 *Papilio (Princeps) tydeus* C. & R. Felder, 1860
Indonesia (Moluccas: Bacan, Ternate, Halmahera, Morotai and Obi). Said to be common (48) and presumably not threatened. It has been surmised that *P. (P.) heringi* Niepelt is a rare natural hybrid *P. (P.) tydeus* x *P. (P.) fuscus* (111, 254), but this has been disputed (224). Two subspecies.

498 *Papilio (Princeps) aegeus* Donovan, 1805
Eastern Australia (including Lord Howe I. (95)), Indonesia (Irian Jaya), Papua New Guinea, Bismarck Archipelago, D'Entrecasteaux Is, Louisiade Archipelago (?) and Trobriand Is (?). Common and not threatened. Farmed and some collected for trade in Papua New Guinea (36). Seven (111) or eleven subspecies. *P. (P.) oberon* Grose–Smith, 1897 from the Solomons (Santa Cruz) is now thought to be a local form of *P. (P.) aegeus* rather than a good species (111, 241). Common name: Orchard Butterfly (282). Other refs: 14, 40, 48, 141, 176, 188, 219.

Species-group: *cynorta* Fabricius

499 *Papilio (Princeps) cynorta* Fabricius, 1793
Lowland forest in Sierra Leone, Guinea (?), Liberia, Ivory Coast, Ghana, Togo, Benin, Nigeria, Cameroon, Central African Republic (?), Gabon (?), Congo, Zaire, western Kenya, Uganda, Angola and an isolated population in Ethiopia. Generally common and not threatened. Two subspecies. Common name: Common White-banded Papilio (34). Other refs: 20, 33, 50, 285.

500 *Papilio (Princeps) plagiatus* Aurivillius, 1898
Forests in Nigeria, Cameroon, Central African Republic, southern Sudan, Congo and northern Zaire. Rather uncommon but not thought to be threatened. Refs: 20, 34, 50, 285.

501 *Papilio (Princeps) zoroastres* Druce,1878
Cameroon, Equatorial Guinea (Bioko, formerly known as Fernando Póo I.), Gabon (?), Congo, Zaire, south-eastern Sudan, Uganda, western Kenya, Rwanda, Burundi (?), north-western Tanzania, Zambia and Angola. Generally common and not threatened. Five subspecies. Refs: 20, 33, 34, 50, 214, 285.

502 *Papilio (Princeps) echerioides* Trimen, 1868
Forests in East Africa: southern Ethiopia, Kenya, Tanzania, Malawi, Zambia, Mozambique, Zimbabwe and South Africa (Natal and Transvaal). Not uncommon and not threatened. Six subspecies. Common names: White-banded Swallowtail (44, 285), Southern White-banded Papilio (34). Other refs: 33, 50, 63, 89, 97, 165, 214.

503 *Papilio (Princeps) jacksoni* Sharpe, 1891
Kenya, Uganda, Zaire, eastern Congo (?), Burundi (?), Rwanda, western Tanzania, Zambia and Malawi. Common in some places and not threatened as a species, but one subspecies is rare in Tanzania (151). Five subspecies. Common name: Jackson's Swallowtail (285). Other refs: 33, 34, 46, 50, 97.

504 *Papilio (Princeps) fuelleborni* Karsch, 1900
Tanzania and Malawi. Restricted range, but not known to be threatened. Common name: Fuelleborn's Swallowtail (285). Other refs: 33, 34, 50.

505   *Papilio (Princeps) sjoestedti* Aurivillius,1908
Rare—refer to section 6, p. 337.
Known from montane forest in only three localities, all in northern Tanzania: Mt Meru, Ngorongoro Crater and, as a different subspecies, Mt Kilimanjaro, where it is abundant. Mt Meru and Mt Kilimanjaro are both well protected national parks indicating that this species, although very narrowly distributed, is not threatened at the moment. Common name: Kilimanjaro Swallowtail (285). Other refs: 33, 34, 43, 50, 151.

Species-group: *rex* Oberthür

506   *Papilio (Princeps) rex* Oberthür, 1886
Highland forest in Nigeria, Cameroon, Sudan, Ethiopia, eastern Congo, Zaire, Uganda, western Kenya, Rwanda, Burundi and Tanzania. Not uncommon (though difficult to capture) and not threatened. It has been suggested that this species mimics the danaid *Danaus formosa* Godman (50). Seven subspecies. Common names: Regal Swallowtail (285), King Papilio (34). Other refs: 33, 105.

Species-group: *nireus* Linnaeus

507   *Papilio (Princeps) epiphorbas* Boisduval, 1833
Madagascar. Well distributed over the whole island (and Comoro Is, according to D'Abrera (50)) and not threatened. Other refs: 34, 204, 206.

508   *Papilio (Princeps) manlius* Fabricius, 1798
Indeterminate—refer to section 6, p. 339.
Mauritius. Despite the restricted range, widespread loss of vegetation and reports of nearing extinction (165), this species is currently quite common and widespread due to its ability to utilize cultivated *Citrus* (13). On the other hand there is concern that it may be unable to compete with the introduced and spreading *P. (P.) demodocus* (12). Other refs: 34, 50, 206.

509   *Papilio (Princeps) phorbanta* Linnaeus, 1771
Vulnerable—refer to section 6, p. 342.
Endemic to Réunion. Two specimens of a dwarf form, subspecies *nana* Oberthür, also included in D'Abrera (50), and allegedly confined to the Seychelles possibly represent wind-blown vagrants from Réunion and were not established on these islands (118, 167, 206). An alternative theory suggests that the taxon was artificially introduced to the Seychelles but did not become established (116). The restricted range and rarity of *P. (P.) phorbanta* is indicative of Vulnerable status. Protected since 1979. Common name: Papillon La Pature. Other ref: 34.

510   *Papilio (Princeps) charopus* Westwood, 1843
Highland forests in Cameroon, Equatorial Guinea (Bioko, formerly known as Fernando Póo I.), Zaire, Rwanda, Burundi and Uganda. The type locality of Ashanti, Ghana is apparently erroneous as the type almost certainly came from Cameroon (116). Fairly common and not threatened. Two subspecies (115, 116). Common names: Westwood's Swallowtail (285), Blue-banded Swallowtail (34). Other refs: 20, 33, 50.

511   *Papilio (Princeps) hornimani* Distant, 1879
Highland forest in East Africa: northern Tanzania and south-eastern Kenya. Not uncommon and not thought to be threatened although a

subspecies is reported to be rare in Tanzania (151). Common name: Horniman's Swallowtail (285). Other refs: 33, 34, 50.

512 *Papilio (Princeps) mackinnoni* Sharpe, 1891
Highland forest in Sudan, Kenya, Uganda, Zaire, Rwanda, Burundi, Zambia, Malawi, Tanzania and Angola. Common in places, (e.g. Kenya) and not threatened as a species, but one subspecies is rare in Tanzania. Four named subspecies (116). Common name: Mackinnon's Swallowtail (34, 285). Other refs: 33, 50.

513 *Papilio (Princeps) sosia* Rothschild and Jordan, 1903
Senegal, Gambia, Sierra Leone, Guinea, Liberia, Ivory Coast, Ghana, Togo, Benin, Nigeria, Cameroon, Gabon, Congo, Central African Republic, Zaire, Uganda and northern Angola. Less common than *P. (P.) nireus* but fairly frequent and not threatened. Two subspecies, of which *P. (P.) s. debilis* Storace appears to be of hybrid origin (116). Common name: Straight-banded Swallowtail (285). Other refs: 20, 33, 34, 50.

514 *Papilio (Princeps) aethiopsis* (Hancock, 1983)
Adopted by Hancock (109) as a replacement name for *aethiops* Rothschild and Jordan, 1905. Ethiopia and north-western Somalia. A restricted range and the butterfly is 'local and generally uncommon' (285). It is not known to be threatened though more information is desirable. Common name: Abyssinian Blue-banded Swallowtail (285). Other refs: 34, 50, 116.

515 *Papilio (Princeps) nireus* Linnaeus, 1758
Forests throughout southern and tropical Africa. Very common and not threatened. Three subspecies (four including *P. (P.) n. aristophontes*) (116). Common names: Green-banded Swallowtail (44), Narrow Blue-banded Swallowtail (34, 285). Other refs: 20, 33, 50, 63, 97, 156, 165, 214, 215.

516 *Papilio (Princeps) aristophontes* Oberthür, 1897
Indeterminate—refer to section 6, p. 345.
Comoro Is. Taxonomic status as a good species recently revived (50), but still open to doubt. Regarded as a subspecies of *P. (P.) nireus* by most authors including Carcasson (34) and Hancock (116). Information on conservation status required. Other refs: 206, 279.

517 *Papilio (Princeps) oribazus* Boisduval, 1836
Madagascar. Well distributed, except in the west, and apparently quite common and not threatened. Refs: 50, 204, 206.

518 *Papilio (Princeps) thuraui* Karsch, 1900
East Africa: highland forest in southern Tanzania, Malawi and northern Zambia (115). Considered to be scarce and very local (43, 50, 285), perhaps deserving conservation status and certainly requiring monitoring. It has been reported to be rare in Tanzania (43, 151) and possibly threatened by deforestation if this proceeds unchecked, although it is quite firmly established at present (42). *P. (P.) t. ufipa* is apparently Vulnerable in Tanzania where it is restricted to the Ufipa plateau (151). Five subspecies according to Hancock (116) including *P. (P.) t. occidua* previously placed under *P. (P.) desmondi*, and *P. (P.) t. cyclopsis* and *P. (P.) t. ufipa*

previously placed as subspecies of *P. (P.) bromius*. Common name: Blue-spotted Black Swallowtail (285). Other refs: 34, 97.

519 *Papilio (Princeps) desmondi* Van Someren, 1939
East Africa: Kenya, Tanzania, Malawi, and Zambia. Fairly common and not threatened. The taxonomy and nomenclature are confused. *P. (P.) magdae* Gifford, 1961 was a replacement name for *P. (P.) brontes* Godman, 1885 and, therefore, has the same type locality, Mt Kilimanjaro (115). *P. (P.) d. magdae* is reported by Kielland to be rare in Tanzania (151). Four subspecies including *P. (P.) d. teita. P. (P.) d. occidua* Storace, described from Zaire, and also occuring in northern Zambia and Malawi, does not belong here but appears to be a subspecies of *P. (P.) thuraui* (115, 116). Common name: Godman's Swallowtail (285). Other refs: 33, 34, 50, 96, 97.

Papilio (Princeps) desmondi teita Van Someren, 1960
Endangered—refer to section 6, p. 347.
South-eastern Kenya (Taita Hills) only. A recently-described taxon (271), treated as a subspecies of *P. (P.) desmondi* by D'Abrera (50) and Carcasson (34). Common name: Taita Blue-banded Swallowtail. Other refs: 39, 116, 272.

520 *Papilio (Princeps) interjecta* Van Someren, 1960
East Africa: Uganda and western Kenya. Recently described (271) and accepted as a good species by D'Abrera (50), Carcasson (34) and Hancock (109, 116). Conservation status not known; further research required.

521 *Papilio (Princeps) bromius* Doubleday, 1845
Forests in Guinea, Sierra Leone, Liberia, Ivory Coast, Ghana, Togo, Benin, Nigeria, Cameroon, Central African Republic, Sudan, Ethiopia (?), Equatorial Guinea (?), Sao Tomé, Gabon, Congo, Zaire, Uganda, Kenya, Rwanda, Burundi, Tanzania and Angola. Common and not threatened as a species. *P. (P.) b. cyclopsis* from northern Malawi and Zambia, and *P. (P.) b. ufipa* from south-western Tanzania do not belong here but appear to be subspecies of *P. (P.) thuraui* (116). Four subspecies (116). Common name: Broad Blue-banded Swallowtail (34, 285). Other refs: 20, 33, 50.

522 *Papilio (Princeps) chrapkowskii* Suffert, 1904
Highland forest in Kenya and eastern and central Uganda. Often regarded as a subspecies of *bromius*. D'Abrera follows this course, although the name is missing from his index (50).

Species-group: *zalmoxis* Hewitson

523 *Papilio (Princeps) zalmoxis* Hewitson, 1864
Lowland forest in Liberia, Ivory Coast, Ghana, Togo (?), Benin (?), Nigeria, Cameroon, Central African Republic, Equatorial Guinea (?), Gabon, Congo and Zaire. Males commonly collected (though females very scarce) and species not known to be threatened. Often placed in the genus *Iterus*. Common names: Giant Blue Swallowtail (285), Great Blue Papilio (34). Other refs: 20, 50, 105.

Species-group: *zenobia* Fabricius

524 *Papilio (Princeps) gallienus* Distant, 1879
Lowland forest in Nigeria, Cameroon, Gabon, Equatorial Guinea (?), Congo and Zaire. Uncommon (285), but not markedly so and not

threatened. Common name: Large White-banded Papilio (34). Other ref: 20.

525 *Papilio (Princeps) mechowi* Dewitz, 1881
Cameroon, Central African Republic, southern Sudan, Gabon (?), Congo, Zaire, Uganda and Angola. Common in places and not threatened. Two subspecies. Common name: Mechow's Swallowtail (285). Other refs: 33, 34, 50.

526 *Papilio (Princeps) zenobius* Godart, 1819
Lowland forest in Sierra Leone, Guinea (?), Liberia, Ivory Coast, Ghana, Togo, Benin, Nigeria, Cameroon, Central African Republic, southern Sudan, Uganda (?), Equatorial Guinea (Bioko, formerly known as Fernando Póo I.), Gabon (?), Congo, Zaire and Angola. Carcasson (34) and D'Abrera (50) recognize *cypraeofila* but do not cite *zenobius*. In view of the possibility of confusion between *zenobius* and *zenobia*, this seems a good idea. Generally common and not threatened. *P. (P.) cypraeofila* Butler is a synonym. Three subspecies. Other refs: 20, 285.

527 *Papilio (Princeps) mechowianus* Dewitz, 1885
Angola, Zaire, Central African Republic and the Congo according to Carcasson (33, 34) but Williams (285) and Lewis (169) state that it also occurs in West Africa from Liberia (i.e. also Ivory Coast, Ghana, Togo, Benin, Nigeria and Cameroon). Fairly common and not threatened. Other ref: 50.

528 *Papilio (Princeps) andronicus* Ward, 1871
Cameroon and possibly adjacent areas. Uncommon but not known to be threatened. Refs: 50, 133.

529 *Papilio (Princeps) zenobia* Fabricius, 1775
Sierra Leone, Guinea (?), Liberia, Ivory Coast, Ghana, Togo, Nigeria, Cameroon, Central African Republic, Congo, Zaire and western Uganda. Generally common and not known to be threatened. Common name: Zenobia Papilio (34). Other refs: 20, 33, 50, 285.

530 *Papilio (Princeps) maesseni* Berger, 1974
Ghana (recorded only from Likpe in the northern Volta Region (222)) and Togo (115). Not known to D'Abrera (50) or Carcasson (34), nor listed in the *Zoological Record* but listed without comment by Hancock (109) who believes it to be a valid species (115). Further information is needed on the distribution and conservation of this obscure butterfly.

Species-group: *antimachus* Drury

531 *Papilio (Princeps) antimachus* Drury, 1782
Rare—refer to section 6, p. 351.
Lowland forest in southern Guinea, Sierra Leone (200), Liberia (98), Ivory Coast, Ghana, Togo (?), Benin (?), Nigeria (159), Cameroon, Central African Republic, Equatorial Guinea (including Bioko, formerly known as Fernando Póo I.), Gabon, Congo, Zaire, Uganda, Rwanda and Angola (58). Males not uncommonly collected, females very scarce. Often placed in the genus *Druryia*. Common names: African Giant Swallowtail (285), Giant Papilio (34). Other refs: 20, 33, 50, 66, 67, 81, 89, 105, 170, 257.

Species-group: *menestheus* Drury

532 *Papilio (Princeps) ophidicephalus* Oberthür, 1878
Eastern Kenya, Tanzania, south-eastern Zaire, Malawi, Zambia, Mozambique, Zimbabwe and South Africa (Natal, Transvaal and Cape Province). Not uncommon and not threatened. Ten subspecies, several of them with restricted ranges (274). Common names: Emperor Swallowtail (44, 249, 285), Snake's Head Swallowtail (249). Other refs: 33, 34, 50, 63, 97, 165, 214, 215.

533 *Papilio (Princeps) menestheus* Drury, 1773
Senegal, Guinea (?), Sierra Leone, Liberia (?), Ivory Coast, Ghana, Togo, Benin, Nigeria and Cameroon. Common and not threatened. Common name: Drury's Emperor Swallowtail (285). Other refs: 20, 33, 34, 50.

534 *Papilio (Princeps) lormieri* Distant, 1874
Forests up to 2000 m in Nigeria, Cameroon, Central African Republic, south-western Sudan, Equatorial Guinea (?), Gabon (?), Congo, Zaire, Uganda, western Kenya and Angola. Not uncommon and not threatened. Replaced by *P. (P.) menestheus* west of the Niger and by *P. (P.) ophidicephalus* from the Kenya coast to South Africa. Sexes similar, three subspecies. Common names: Western Emperor Swallowtail (285), Emperor Swallowtail (34). Other refs: 33, 50.

Species-group: *demoleus* Linnaeus

535 *Papilio (Princeps) morondavana* Grose–Smith, 1891
Vulnerable—refer to section 6, p. 354.
Madagascar only. The rarest of the Malagasy endemics, though stated to be less rare than supposed by Paulian and Viette (206). Threatened by loss of habitat (280) and possibly its popularity with commercial collectors (205). Common name: Madagascan Emperor Swallowtail. Other refs: 34, 50, 81.

536 *Papilio (Princeps) grosesmithi* Rothschild, 1926
Rare—refer to section 6, p. 356.
Madagascar only, particularly the west. Probably less rare than *P. (P.) morondavana*, but commercial collecting and habitat destruction need to be monitored. Refs: 34, 50, 206.

537 *Papilio (Princeps) erithonioides* Grose–Smith, 1891
Madagascar only, mainly in the west of the island. Not uncommon and apparently not threatened. Refs: 34, 50.

538 *Papilio (Princeps) demodocus* Esper, 1798
Tropical and southern Africa, Saudi Arabia, Yemen, South Yemen, Oman; introduced to Madagascar, Mauritius and Réunion. Abundant in open habitats and not threatened. A minor pest on *Citrus*. Aggressive towards other butterflies flying in its vicinity (50). Sexes alike, two subspecies. Common names: Citrus Swallowtail (34, 44, 249, 285), Christmas Butterfly (44, 249), Orange Dog (34), African Lime Butterfly (161). Other refs: 20, 33, 63, 97, 204, 214, 215.

539 *Papilio (Princeps) demoleus* Linnaeus, 1758
Oman, United Arab Emirates, Saudi Arabia, Kuwait, Bahrain, Qatar, western and possibly also eastern Afghanistan, Sri Lanka, India, Nepal,

Vietnam, Laos, Andamans (5), Burma, Thailand, Kampuchea, southern China (including Hainan (Guangdong prov.)), Taiwan, Japan (rare strays), Peninsular Malaysia, Singapore, Philippines, Indonesia (Sumatra, Sula, Talaud, Flores, Alor and Sumba), Papua New Guinea, Australia (including Lord Howe I. (95)), apparently Hawaii and possibly other Pacific Ocean islands. Another aggressive and very common species. Six subspecies. Common names: Lime or Lemon Butterfly (45, 80, 251), Chequered Swallowtail (282). Other refs: 4, 33, 40, 48, 52, 87, 118, 141, 148, 149, 165, 176, 246, 262.

Species-group: *leucotaenia* Rothschild

540 *Papilio (Princeps) leucotaenia* Rothschild, 1908
Vulnerable—refer to section 6, p. 358.
Montane forest in East Africa: south-western Uganda, Rwanda, Burundi and north-eastern Zaire. A rare species threatened by deforestation. Common name: Cream-banded Swallowtail (285). Other refs: 33, 34, 50.

Species-group: *delalandei* Godart

541 *Papilio (Princeps) delalandei* Godart, 1824
Madagascar. Well distributed in the forests of Madagascar, especially in the east; not known to be threatened. Refs: 34, 50, 206.

542 *Papilio (Princeps) mangoura* Hewitson, 1875
Rare—refer to section 6, p. 362.
Madagascar. Distributed in the eastern rain forests and usually regarded as rare (263). However, Paulian & Viette (206) treat it as less rare than is usually thought. At present it may not be threatened, but with deforestation proceeding very quickly, its status needs to be carefully monitored. Local catchers decoy male *P. (P.) mangoura* with female *P. (P.) delalandei*, which has similar yellow-barred wings (263). Other refs: 34, 50, 195.

Species-group: *phorcas* Cramer

543 *Papilio (Princeps) constantinus* Ward, 1871
Woodland and forested rivers in Ethiopia, Kenya, Somalia, Tanzania, Congo, south-east Zaire, Malawi, Mozambique, Zambia, Zimbabwe and South Africa (Natal and northern Transvaal). Not uncommon and not threatened. Three subspecies. Common name: Constantine's Swallowtail (34, 44, 285). Other refs: 33, 50, 63, 97, 214.

544 *Papilio (Princeps) phorcas* Cramer, 1775
Forests throughout tropical Africa excluding Ethiopia. Very common in places and not threatened. Six subspecies. Common names: Green-patch Swallowtail (285), Green Swallowtail (34). Other refs: 20, 33, 50, 97, 215.

545 *Papilio (Princeps) dardanus* Brown, 1776
Wooded and forested areas throughout tropical and southern Africa. This species is a famous mimetic and polymorphic swallowtail. Not threatened as a species but three subspecies are Rare or Vulnerable in Tanzania (151). There are twelve subspecies including *P. (P.) d. meriones* from Madagascar and *P. (P.) d. humbloti* restricted to the Comoros (263). *P. (P.) d. flavicornis* is restricted to Mt Kulal and *P. (P.) d. ochracea* to Mt Marsabit and Mt Nyiru; all in northern Kenya (50). *P. (P.) nandina* Rothschild from

Kenya was recently demonstrated to be a rare natural hybrid between *P. (P.) phorcas* and *P. (P.) dardanus* (37). Common name: Mocker Swallowtail (34, 44, 249, 285). Other refs: 20, 33, 63, 97, 165, 206, 214, 215.

Species-group: *hesperus* Westwood

546 *Papilio (Princeps) euphranor* Trimen, 1868
Restricted to highland forest in South Africa (Cape Province, Natal and Transvaal). Uncommon (34); but no threats known. Has been proposed for inclusion in a Red Data Book for South Africa (192), although no such work has been started. Common names: Forest Kite, Bush Kite (285), Natal Swallowtail (34). Other refs: 33, 50, 63, 214.

547 *Papilio (Princeps) pelodurus* Butler,1895
Highland forest (800–1900m) in East Africa: Malawi, Zambia (115) and eastern Tanzania. Restricted range, limited to the distribution of its foodplant, *Cryptocarya liebertiana* (Lauraceae), in evergreen forests. It is apparently not uncommon, although it will be under threat in Tanzania if deforestation proceeds unchecked (42, 43). Two subspecies. Common name: Eastern Black and Yellow Swallowtail (285). Other refs: 33, 34, 50.

548 *Papilio (Princeps) hesperus* Westwood, 1843
Lowland forest in Ivory Coast, Ghana, Togo, Benin, Nigeria, Cameroon, Central African Republic, southern Sudan, Equatorial Guinea (Bioko, formerly known as Fernando Póo I.), Congo, Zaire, Uganda, Rwanda (?), Burundi (?), western Tanzania and north-western Zambia. Replaced by *P. (P.) pelodorus* in Malawi and Tanzania and by *P. (P.) horribilis* from the Ivory Coast to Sierra Leone (34). Not uncommon, though difficult to capture and not threatened. Three subspecies. Common names: Black and Yellow Swallowtail (285), Hesperus Swallowtail (34). Other refs: 33, 50.

549 *Papilio (Princeps) horribilis* Butler, 1874
West Africa; Sierra Leone, Liberia, Ivory Coast and Ghana. Given as a full species by D'Abrera (50) and Carcasson (34), but not mentioned by other authors (188, 250, 285); Gifford (97) treats it as synonymous with *P. (P.) pelodurus*. Not uncommon and probably not threatened. Other ref: 41.

550 *Papilio (Princeps) nobilis* Rogenhoffer, 1891
Highland forest in southern Sudan, Kenya, Uganda, Burundi, Rwanda, Tanzania and Kivu Province, Zaire (34, 285). Not uncommon and not recognized as being threatened. Three subspecies. Common name: Noble Swallowtail (34, 285). Other refs: 33, 50.

Species-group: *palinurus* Fabricius

551 *Papilio (Princeps) crino* Fabricius, 1792
Southern and western coastal India and Sri Lanka (52). Not threatened. Common name: Common Banded Peacock (80). Other refs: 4, 259, 287.

552 *Papilio (Princeps) blumei* Boisduval, 1836
Indonesia (Sulawesi). Much in demand for the decorative trade but prices have fallen recently and it is not known to be threatened. Monitoring of status is required. Two, possibly four subspecies (262). Other refs: 52, 141.

553 *Papilio (Princeps) buddha* Westwood, 1872
Southern India. Locally common and not rare (118). Protected by law

(182), but not known to be threatened. Common name: Malabar Banded Peacock (80). Other refs: 4, 52, 259.

554 *Papilio (Princeps) palinurus* Fabricius, 1787
Burma, western Thailand, Peninsular and Eastern Malaysia, Brunei and Indonesia (Sumatra, Nias and Kalimantan) (52, 262). Generally common, not threatened except in Malaysia where possibly Vulnerable and in need of protection (10). Two subspecies (262). The five subspecies listed by D'Abrera (52) include *P. (P.) daedalus* subspecies (see below) and a synonym. Common names: Banded Peacock (3, 45), Burmese Banded Peacock (80), Moss Peacock (3). Other refs: 87, 131, 141, 168, 259.

555 *Papilio (Princeps) daedalus* C. & R. Felder, 1861
Philippines, including Palawan but excluding Sulu Archipelago (52, 262). Formerly regarded as the eastern subspecies of *P. (P.) palinurus* (52), which is distributed further to the west, but raised to full species rank by Hiura and Alagar (127). Abundant and not threatened. Two subspecies.

Species-group: *paris* Linnaeus

556 *Papilio (Princeps) chikae* Igarashi, 1965
Endangered—refer to section 6, p. 364.
Philippines: Luzon; only known from the Baguio and Bontoc areas in the north of the island (262). The recent discovery of this species has created a high demand, particularly from Japanese collectors. This, along with its limited distribution, danger of habitat destruction and apparent ease of capture all indicate Endangered status. Other refs: 2, 52, 53, 140, 148.

557 *Papilio (Princeps) maackii* Ménétriés, 1859
Japan, southern China, North Korea, South Korea, Taiwan and extreme south-eastern U.S.S.R. (Sikhote–Alin, etc. and Kuril Is); a similar, but slightly more northerly, distribution to *P. (P.) bianor*. Two subspecies. Generally common but rapidly declining in the U.S.S.R. and listed in the U.S.S.R. Red Data Book (8, 260). Other refs: 52, 141, 149, 169.

558 *Papilio (Princeps) bianor* Cramer, 1776
North-eastern Burma, Sichuan (Szechwan) and southern China, Japan, South Korea, North Korea, northern Laos and Vietnam. Generally common and not threatened. Seven subspecies; *okinawaensis* is now rare on Okinawa (2). Common name: Chinese Peacock (80). Other refs: 141, 143, 149, 246, 259.

559 *Papilio (Princeps) syfanius* Oberthür, 1886
Western China (Yunnan and Sichuan). Usually considered to be a subspecies of *bianor* but treated as a full species by Hancock (109). Not known to be threatened.

560 *Papilio (Princeps) polyctor* Boisduval, 1836
Eastern Afghanistan, Pakistan, northern India (including Sikkim and Assam), Nepal, Bhutan (?), Burma, Thailand, northern Vietnam and Laos (?). Generally common and not threatened. Six subspecies, including the recently described *P. (P.) p. pinratanai* (223). Common name: Common Peacock (80, 251). Other refs: 4, 52, 136, 141, 168, 259.

561 *Papilio (Princeps) dialis* (Leech,, 1893)
Southern China (including Hainan (Guangdong prov.)), north-eastern

Burma, northern Vietnam, Laos and Taiwan. Generally common and not threatened. Five subspecies. Common name: Black-crested Spangle (94). Other refs: 52, 141, 246, 259.

562 *Papilio (Princeps) elephenor* Doubleday, 1845
North-eastern India (Jorehat, Cachir and Sadiya in Assam, Khasia in Meghalaya, Nagaland and Manipur), Burma and northern Thailand. Although status given variously as rare or very rare, it seems marginal for threatened status. It is protected by law in India (182). More information from the eastern part of its range is desirable. Two subspecies. Common names: Yellow-crested Spangle (80, 94), Black-crested Spangle (*P. (P.) e. schanus*) (80). Other refs: 4, 52, 168, 259.

563 *Papilio (Princeps) paris* Linnaeus, 1758
Southern and north-western India, Bangladesh, Bhutan (?), Burma, northern Thailand, Laos, Vietnam, Sichuan (Szechwan) and southern China (including Hainan (Guangdong prov.)), Peninsular Malaysia (76), Taiwan and Indonesia (Sumatra and Java) (52, 262). Common in many areas and not threatened. Ten subspecies. Common name: Paris Peacock (*P. (P.) p. paris*) (80, 282), Tamil Peacock (*P. (P.) p. tamilana*) (80). Other refs: 4, 45, 141, 143, 168, 207, 246, 259.

564 *Papilio (Princeps) karna* C. & R. Felder, 1864
Philippines (Palawan), Eastern Malaysia, Brunei and Indonesia (Kalimantan, Java and Sumatra). Generally uncommon but not known to be threatened. On Palawan females remain near the foodplant, said to be a species of large forest tree; males congregate at drinking spots; numbers stable (56). More information required from other areas. Possibly three or more subspecies. Other refs: 45, 52, 131, 148, 262.

565 *Papilio (Princeps) arcturus* Westwood, 1842
North-western India, Nepal, Bhutan, Bangladesh (?), Burma, western and central China and Thailand. Not uncommon and not threatened. Two subspecies. Common name: Blue Peacock (80, 282). Other refs: 4, 52, 141, 207, 259.

566 *Papilio (Princeps) hoppo* Matsumara, 1908
Taiwan. Not uncommon and not thought to be threatened, despite its restricted range. Requires monitoring because of the butterfly trade in Taiwan. Refs: 52, 141, 246.

567 *Papilio (Princeps) krishna* Moore, 1857
North-eastern India (including Manipur, Sikkim and Assam), Nepal and western China (52). A narrow range. Apparently uncommon and commanding high prices in the trade, but not thought to be threatened. Two subspecies. Common name: Krishna Peacock (80, 251). Other refs: 141, 259.

Species-group: *peranthus* Fabricius

568 *Papilio (Princeps) neumoegeni* Honrath, 1890
Vulnerable—refer to section 6, p. 367.
Known only from Sumba in the Lesser Sunda Is in Indonesia, whence some numbers have recently been distributed by the trade; very restricted range, current status uncertain. Refs: 52, 134, 262.

569 *Papilio (Princeps) peranthus* Fabricius, 1787
Indonesia (Java, Bali, Lombok, Bawean, Kangean, Sumbawa, Flores, Adonara, Solor, Pantar, Alor, Tanahjampea, Kalao, Bonerate, Salayar and Sulawesi) (52, 262). The Sulawesi form is probably specifically distinct (118) but further assessment is needed to confirm this. Not uncommon but not thought to be threatened. Six to eight subspecies. Other ref: 141.

570 *Papilio (Princeps) pericles* Wallace, 1865
Indonesia (Lesser Sunda Is (Timor, Wetar, Moa, Leti, Babar, Damar, Romang and Tanimbar)) (262). Not rare (48) or threatened, but not common.

571 *Papilio (Princeps) lorquinianus* C. & R. Felder, 1865
Indonesia (Moluccas (Morotai, Ternate, Halmahera, Bacan and Seram) and western Irian Jaya). Scarce throughout its range but not known to be threatened. Five subspecies. Refs: 48, 141.

Species-group: *ulysses* Linnaeus

572 *Papilio (Princeps) ulysses* Linnaeus, 1758
Indonesia (Moluccas and Irian Jaya), Papua New Guinea, Bismarck Archipelago (including New Britain), Trobriand Is, D'Entrecasteaux Is (?), Louisiade Archipelago (?), north-eastern Australia and the Solomons (except Malaita, San Cristobal and Santa Cruz). Much prized for its beauty, but not threatened as a species. Collected for trade in Papua New Guinea (36, 68). Protected by law in Queensland, Australia (258). Sixteen subspecies. Common names: Ulysses Butterfly (282), Blue Mountain Butterfly (282), Mountain Emperor. Other refs: 14, 40, 48, 141, 142, 176, 217, 219.

573 *Papilio (Princeps) montrouzieri* Boisduval, 1859
New Caledonia (including the Loyalty Is). Despite this restricted range, it is not rare (48) but neither is it ubiquitous (133). Its status should be closely monitored. Other ref: 219.

The following species were listed by Munroe (188) but not by Hancock (109):

1. *Parnassius pythia* Rothschild, 1932
2. *Parnassius rothschildianus* Bryk, 1932
3. *Parnassius stenosemus* Honrath, 1890
4. *Eurytides chibcha* (Fassl, 1912)
5. *Eurytides hipparchus* (Staudinger, 1884)
6. *Graphium sisenna* (Mabille, 1890)
7. *Papilio peleides* Esper, 1793

The three parnassians were listed as subspecies of other *Parnassius* species in a revision by Bryk (31), the *Eurytides* species were regarded as "aberrations" by Hancock (109), *Graphium sisenna* from Mozambique is synonymous with *Graphium (Graphium) polistratus*, and *Papilio peleides* is listed in the British Museum (Natural History) records as possibly an artefact of South American origin.

References

1. Ackery, P.R. (1975). A guide to the genera and species of Parnassiinae (Lepidoptera: Papilionidae). *Bulletin of the British Museum (Natural History). Entomology* 31: 71–105.
2. Ae, S.A (1983). *In litt.*, 18 March.

3. Anonymous (1980–1984). Various dealers' lists.
4. Antram, C.B. (1924). *Butterflies of India*. Thacker, Spink & Co., Calcutta and Simla. 226 pp.
5. Arora, G.S. and Nandi, D.N. (1980). On the butterfly fauna of Andaman and Nicobar Islands (India). 1. Papilionidae. *Records of the Zoological Survey of India* 77: 141–151.
6. Aurivillius, C. (1919). Eine neue Papilio–Art. *Entomologisk Tidsskrift* 1919: 177–8.
7. Bain, J.R. and Humphrey, S.R. (1982). *A Profile of the Endangered Species of Thailand. Vol. 1. Through Birds*. Report No. 4, Office of Ecological Services, Florida State Museum, Gainsville. 344 pp.
8. Bannikov, A.G. and Sokolov, V.I. (eds) (1984). *The Red Data Book of the USSR. Rare and Threatened Species of Animals and Plants*. Lesnaya Promiishlyennost Press, Moscow. (In Russian).
9. Barcant, M. (1970). *Butterflies of Trinidad and Tobago*. Collins, London. 314 pp.
10. Barlow, H.S. (1983). Butterfly protection in Peninsular Malaysia. Manuscript *in litt.*, 4pp.
11. Barlow, H.S. (1983). *In litt.*, 29 June.
12. Barnes, M.J.C. (1983). *In litt.*, 5 July 1983.
13. Barnes, M.J.C. (1983). *In litt.*, 9 May.
14. Barrett, C. and Burns, A.N. (1951). *Butterflies of Australia and New Guinea* N.H. Seward Ltd., Melbourne. 187 pp.
15. Bell, T.R. (1911). The common butterflies of the plains of India. *Journal of the Bombay Natural History Society* 1911: 1115–1136.
16. Berger, L.A. (1950). Catalogues raisonnées de la faune entomologique du Congo Belge. Lépidoptères–Rhopalocères, I, Fam. Papilionidae. *Annales du Musée Royale du Congo Belge Ser 4to (C: Zool.)* 8: 1–104.
17. Berger, L.A. (1974). Notes sur quelques Papilionidae du Musée Royal de l'Afrique Centrale. *Lambillionea* 72–73: 69–76.
18. Beutelspacher, C.R. (1975). Una especie nueva de *Papilio* L. (Papilionidae) *Revista Sociedad Mexicana de Lepidopterologiá* 1(1): 3–6.
19. Blanchard, F. (1871). Remarques sur la faune de la principauté thibétaine du Moupin. *Comptes Rendus Hebdomadaires des Séances de l'Academie de Science, Paris* 72: 807–813.
20. Boorman, J. and Roche, P. (1957). *The Nigerian Butterflies. Part I: Papilionidae*. Ibadan University Press, Ibadan. 7pp.
21. Brown, F.M. and Heinemann, B. (1972). *Jamaica and its Butterflies*. Classey, London. 478 pp.
22. Brown, J.W. and Faulkner, D.K. (1984). Distributional records of certain Rhopalocera in Baja California, Mexico, with the description of a new subspecies of *Papilio* (*Heraclides*) *astyalus* (Godart) (Lepidoptera: Papilionidae). *Bulletin of the Allyn Museum* 83: 1–9.
23. Brown, K.S., Jr. (1983). *In litt.*, 6 April.
24. Brown, K.S., Jr. (1983). *In litt.*, 19 September.
25. Brown, K.S., Jr. (1983). *In litt.*, 23 October.
26. Brown, K.S., Jr. and Mielke, O.H.H. (1967). Lepidoptera of the central Brazil plateau. 1. Preliminary list of Rhopalocera (continued): Lycaenidae, Pieridae, Papilionidae, Hesperiidae. *Journal of the Lepidopterists' Society* 21: 145–168.
27. Brown, K.S., Damman, A.J. and Feeny, P. (1981). Troidine swallowtails (Lepidoptera: Papilionidae) in southeastern Brazil: natural history and food plant relationships. *Journal of Research on the Lepidoptera* 19(4) (1980): 199–226.
28. Bryk, F. (1923). *Lepidopterorum Catalogus Part 27: Baroniidae, Teinopalpidae, Parnassiidae*. Junk, Berlin. 247 pp.
29. Bryk, F. (1929, 1930). *Lepidopterorum Catalogus Parts 35,37,39: Papilionidae*. Junk, Berlin. 676 pp.
30. Bryk, F. (1934). Lepidoptera. Baroniidae, Teinopalpidae, Parnassiidae. Part 1. *Tierreich* 64: 131 pp.
31. Bryk, F. (1935). Lepidoptera, Parnassiidae. Part 2. (Subfamily Parnassiinae. *Tierreich* 65: 51 + 790 pp., 698 figs.
32. Calderara, P. (1984). A new subspecies of *Ornithoptera victoriae* Gray (Papilionidae) from Choiseul, Solomon Islands. *Proceedings and Transactions of the British Entomological and Natural History Society* 17: 31–35.

33. Carcasson, R.H. (1960). The swallowtail butterflies of East Africa (Lepidoptera, Papilionidae). *Journal of the East African Natural History Society* Special Supplement 6: 33 pp + 11 pl. Reprinted by E.W. Classey, Faringdon, 1975.

34. Carcasson, R.H. (1981). *Collins Handguide to the Butterflies of Africa*. Collins, London. Hardback edition, 188 pp.

35. Chou, I. (1946). A list of butterflies collected from W. Szechnan and E. Sikong. *Insecta Sinensium* 1: 15–52.

36. Clark, P.B. (1983). *In litt.*, 27 July.

37. Clarke, C. (1980). *Papilio nandina*, a probable hybrid between *Papilio dardanus* and *Papilio phorcas*. *Systematic Entomology* 5: 49–67.

38. Clench, H.K. (1978). *Papilio aristodemus* (Papilionidae) in the Bahamas. *Journal of the Lepidopterists' Society* 32: 273–276.

39. Clifton, M.P. (1982). *In litt.*, 13 September.

40. Common, I.F.B. and Waterhouse, D.F. (1972). *Butterflies of Australia*. Angus and Robertson, Sydney. 498 pp.

41. Condamin, M. and Roy, R. (1963). Lepidoptera Papilionidae in (ch. 19) La réserve naturelle intégrale du Mont Nimba, fasc. 5. *Memoires de l'Institut Francais Afrique Noire* 66: 415–422.

42. Congdon, T.C.E. (1983). *In litt.*, 7 March.

43. Congdon, T.C.E. (1984). *In litt.*, 24 June.

44. Cooper, R. (1973). *Butterflies of Rhodesia*. Longman Rhodesia, Salisbury. 138 pp.

45. Corbet, A.S. and Pendlebury, H.M. (1978). *The Butterflies of the Malay Pensinsula* (Third ed. revised by J.N. Eliot). Malayan Nature Society, Kuala Lumpur. 578 pp.

46. Cottrell, C.B. (1963). Two new subspecies of *Papilio jacksoni*. E. Sharpe (Lepidoptera: Papilionidae) from Tanganyika and the northern Rhodesia–Nyasaland border. *Proceedings of the Royal Entomological Society of London* (B) 32: 125–128.

47. Covell, C.V. and Rawson, G.W. (1973). Project ponceanus: a report on first efforts to survey and preserve the Schaus swallowtail (Papilionidae) in southern Florida. *Journal of the Lepidopterists' Society* 27: 206–210.

48. D'Abrera, B. (1971). *Butterflies of the Australian Region*. Lansdowne Press, Melbourne. 415 pp.

49. D'Abrera, B. (1975). *Birdwing Butterflies of the World*. Lansdowne Press, Melbourne. 260 pp.

50. D'Abrera, B. (1980). *Butterflies of the Afrotropical Region*. Lansdowne Editions, Melbourne. xx + 593 pp.

51. D'Abrera, B. (1981). *Butterflies of the Neotropical Region. Part I. Papilionidae and Pieridae*. Lansdowne Editions, Melbourne. xvi + 172 pp.

52. D'Abrera, B. (1982). *Butterflies of the Oriental Region. Part 1. Papilionidae and Pieridae*. Hill House, Victoria, Australia. xxxi + 244 pp.

53. D'Abrera, B. (1983). *In litt.*, 12 March.

54. D'Almeida, R.F. (1924). Les papilionides de Rio de Janeiro. Description de deux chenilles. *Annales de la Société Entomologique de France* 93: 23–30.

55. D'Almeida, R.F. (1965). *Catalogo dos Papilionidae Americanos*. Sociedade Brasileira de Entomologia. São Paulo, Brasil.

56. Dacasin, G.A. (1984). *In litt.*, 25 April.

57. Davis, F.L. (1928). *Notes on the Butterflies of British Honduras*. Old Royalty Book Publishers, London. 101 pp.

58. De Carvalho, J.P. (1983). *In litt.*, 4 April.

59. De Viedma, M.G. and Gómez–Bustillo, M.R. (1976). *Libro Rojo de los Lepidópteros Ibéricos*. Publicaciones del Ministerio de Agricultura Secretaría General Técnica, Madrid. 120 pp.

60. DeVries, P.J. (1983). *Papilio cresphontes* (Lechera, Papilio Grande, Giant Swallowtail). In: *Costa Rican Natural History*. Ed. D.H. Janzen. University of Chicago Press, Chicago and London. 816 pp.

61. DeVries, P.J. (in press). *The Butterflies of Costa Rica and Their Natural History*. Princeton University Press.

62. Díaz Francés, A. and Maza E., J. de la (1978). Guía ilustrada de las mariposas Mexicanas.

Sociedad Mexicana de Lepidopterologiá A.C., Publicaciones Especiales 3, 15 pp.

63. Dickson, C.G.C. and Kroon, D.M. (eds) (1978). *Pennington's Butterflies of Southern Africa.* Ad Donker, Johannesburg and London. 670 pp.

64. Dollinger, P. (1982). *In litt.*, 17 June.

65. Dornfeld, E.J. (1980). *The Butterflies of Oregon.* Timber Press, Oregon. 276 pp.

66. Drury, D. (1782). *Illustrations of Natural History.* Vol. 3. London. 76 pp + 50 plates.

67. Drury, D. (1837). *Illustrations of Exotic Entomology.* Vol. III. New ed, J.O. Westwood.

68. Ebner, J.A. (1971). Some notes on the Papilionidae of Manus Island, New Guinea. *Journal of the Lepidopterists' Society* 25: 73–80.

69. Ehrlich, P.R. (1958). The comparative morphology, phylogeny and higher classification of the butterflies. (Lepidoptera: Papilionidae). *University of Kansas Science Bulletin* 39: 305–370.

70. Eisner, C. (1966). Parnassiidae—Typen in der Sammlung J.C. Eisner. *Zoologische Verhandelingen, Leiden* 81: 1–190.

71. Eisner, C. (1968). Parnassiana Nova XLIII Nachträgliche betrachtungen zu der Revision der Subfamilie Parnassiinae (forsetzung 16). *Zoologische Mededelingen* 43: 9–17, 2 pls.

72. Eisner, C. (1969). Parnassiana Nova XLIV. Nachträgliche betrachtunge zu der Revision der Subfamilie Parnassiinae (Fortesetzung 17). *Zoologische Mededelingen* 43: 173–176.

73. Eisner, C. (1976). Parnassiana Nova XLIX. Die Arten und Unterarten der Parnassiidae (Lepidoptera) (zweiter Teil). *Zoologische Verhandelingen, Leiden* 146: 97–266.

74. Eisner, C. and Naumann, C.M. (1980). Parnassiana Nova LVII. Beitrag zur Ökologie und Taxonomie der Afghanischen Parnassiidae (Lepidoptera). *Zoologische Verhandelingen* 178: 35 pp + 9 pls.

75. Eisner, T. *et al.* (1970). Defense Mechanisms of Arthropods XXVII. Osmeterial secretions of papilionid caterpillars (*Baronia, Papilio, Eurytides*). *Annals of the Entomological Society of America* 63: 914–915.

76. Eliot, J.N. (1982). On three swallowtail butterflies from Peninsular Malaysia. *Malaysian Nature Journal* 35: 179–182.

77. Emmel, J.F. and Emmel, T.C. (1964). The life history of *Papilio indra minori. Journal of the Lepidopterists' Society* 18: 65–73.

78. Emmel, J.F. and Emmel, T.C. (1968). The population biology and life history of *Papilio indra martini. Journal of the Lepidopterists' Society* 22: 46–52.

79. Emmel, T.C. and Emmel, J.F. (1967). The biology of *Papilio indra kaibabensis* in the Grand Canyon. *Journal of the Lepidopterists' Society* 21: 41–49.

80. Evans, W.H. (1932). *The Identification of Indian Butterflies.* Bombay Natural History Society. 2nd ed., revised. 454 pp.

81. FAO/UNEP (1981). *Tropical Forest Resources Assessment Project. Forest Resources of Tropical Africa. Part 1: Regional Synthesis.* FAO, Rome, 108 pp.

82. Feeny, P. (1984). Pers. comm., 1 August.

83. Fenner, T.L. (1983). *In litt.*, 15 March.

84. Ferguson, D. (1950). Collecting a little-known *Papilio. Lepidopterist's News*, 4: 11–12.

85. Ferguson, D.C. (1954). The Lepidoptera of Nova Scotia. *Proceedings of the Nova Scotian Institute of Science* 23: 161–375.

86. Ferris, C.D. and Emmel, J.F. (1982). Discussion of *Papilio coloro* W.G. Wright (= *Papilio rudkini* F. and R. Chermock) and *Papilio polyxenes* Fabricius. *Bulletin of the Allyn Museum* 76: 13 pp.

87. Fleming, W.A. (1975). *Butterflies of West Malaysia and Singapore.* 2 vols. Classey, Faringdon. Vol. 1: x + 64 pp., Vol. 2: x + 92 pp.

88. Ford, E.B. (1944). Studies on the chemistry of pigments in the Lepidoptera, with reference to their bearing on systematics. 4. The classification of the Papilionidae. *Transactions of the Royal Entomological Society of London* 94: 201–223.

89. Fox, R.M., Lindsey, Jr., A.W., Clench, H.K. and Miller, L.D. (1965). The butterflies of Liberia. *Memoirs of the American Entomological Society* No. 19, 438 pp.

90. Freina, J. de (1980). Eine neue Unterart von *Parnassius nordmanni* Nordmann aus Kleinasien (Lepidoptera, Papilionidae). *Nachrichtenblatt Bayerischer Entomologische* 29: 50–62.

91. Gabriel, A.G. (1942). A new species of *Bhutanitis* (Lep. Papilionidae). *The Entomologist*. 75 (952):1

92. Garrison, R.W. (1983). *In litt.*, 7 April.

93. Gepp, J. (1983). *Rote Listen Gefährdeter Tiere Österreichs*. Bundesministeriums für Gesundheit und Umweltschutz, Wien. 242 pp.

94. Ghosh, S.K. and Mandal, D.K. (1981). Species of butterflies (Lepidoptera: Rhopalocera) considered threatened in India. Unpublished manuscript, 12 pp.

95. Gibbs, G.W. (1980). *New Zealand Butterflies: Identification and Natural History*. Collins, London. 207 pp.

96. Gifford, D. (1961). Notes on two Nyasaland *Papilios* (Lepidoptera: Papilionidae). *Entomologist* 94: 287–289.

97. Gifford, D. (1965). *A List of the Butterflies of Malawi*. Society of Malawi, Blantyre. 151 + vi pp.

98. Godfray, H.C.J. (1983). *In litt.*, 6 April.

99. Godman, F.D. and Salvin, O. (1888). New species of butterflies collected by Mr. C.M. Woodford in the Solomon Islands. *The Annals and Magazine of Natural History*. (6)1: 209–214.

100. Gómez Bustillo, M.R. and Fernández–Rubio, F. (1974). *Mariposas de la Península Ibérica (tomo 1)*. Servicio de Publicaciones del Ministerio de Agricultura, Madrid. 198 pp.

101. Gómez Bustillo, M.R. and Fernández–Rubio, F. (1974). *Mariposas de la Península Ibérica (tomo 2)*. Servicio de Publicaciones del Ministerio de Agricultura, Madrid. 258 pp.

102. Gross, G.F. (1975). The land invertebrates of the New Hebrides and their relationships. *Philosophical Transactions of the Royal Society. Series B* 272: 391–421.

103. Guilbot, R. and Plantrou, J. (1978). Note sur *Graphium illyris* (Hewitson) et révision systématique de l'espèce. *Bulletin de la Société Entomologique de France* 83: 68–73.

104. Hancock, D.L. (1979). Systematic notes on *Graphium felixi* (Joicey and Noakes) (Lepidoptera: Papilionidae) *Australian Entomological Magazine* 7: 11–12.

105. Hancock, D.L. (1979). Systematic notes on three species of African Papilionidae (Lepidoptera). *Arnoldia Rhodesia* 8(33): 1–6.

106. Hancock, D.L. (1979). The systematic position of *Papilio anactus* Macleay (Lepidoptera: Papilionidae) *Australian Entomological Magazine* 6: 49–53.

107. Hancock, D.L. (1980). The status of the genera *Atrophaneura* Reakirt and *Pachliopta* Reakirt (Lepidoptera: Papilionidae). *Australian Entomological Magazine* 7: 27–32.

108. Hancock, D.L. (1982). A note on the status of *Ornithoptera meridionalis tarunggarensis* (Joicey and Talbot) (Lepidoptera: Papilionidae). *Australian Entomological Magazine* 8: 93–95.

109. Hancock, D.L. (1983). Classification of the Papilionidae (Lepidoptera): a phylogenetic approach. *Smithersia* 2: 1–48.

110. Hancock, D.L. (1983). Phylogeny and relationships of the *Papilio fuscus* group of swallowtails (Lepidoptera: Papilionidae). *Australian Entomological Magazine* 9: 63–70.

111. Hancock, D.L. (1983). *Princeps aegeus* (Donovan) and its allies (Lepidoptera: Papilionidae): systematics, phylogeny and biogeography. *Australian Journal of Zoology* 31: 771–797.

112. Hancock, D.L. (1983). *In litt.*, 3 March.

113. Hancock, D.L. (1983). *In litt.*, 30 May.

114. Hancock, D.L. (1984). A note on *Atrophaneura palu* (Martin) 1912. *Papilio International* 1(3): 71–72.

115. Hancock, D.L. (1984). *In litt.*, 25 June.

116. Hancock, D.L. (1984). The *Princeps nireus* group of swallowtails (Lepidoptera: Papilionidae) systematics, phylogeny and biogeography. *Arnoldia Zimbabwe* 9(12): 181–215.

117. Haugum, J. (1983). *In litt.*, 2 June.

118. Haugum, J. (1984). *In litt.*, 18 June.

119. Haugum, J. and Low, A.M. (1978–1983). *A Monograph of the Birdwing Butterflies*. Vol. 1 (1): 1–84; (2): 85–192; (3): 193–308. Vol. 2 (1): 1–104; (2): 105–240. Scandinavian Science Press, Klampenborg.

120. Haugum, J. and Samson, C. (1980). Notes on *Graphium weiskei*. *Lepidoptera Group of 1968 Supplement* 8, 12 pp., 2 maps.
121. Haugum, J., Ebner, J. and Racheli, T. (1980). The Papilionidae of Celebes (Sulawesi). *Lepidoptera Group of '68 Supplement* 9: 21 pp, 1 map, 2 pl.
122. Hayward, K.J. (1967). *Genera et Species Animalium Argentinorum. 4. Insecta. Lepidoptera (Rhopalocera). Familiae Papilionidarum et Satyridarum*. Guillermo Kraft Lbla, S.G.; Bonariae General. Eds J.A.H. Rossi and M. Lillo. 447 pp.
123. Heath, J. (1981). *Threatened Rhopalocera (Butterflies) in Europe*. Nature and Environment Series (Council of Europe) No. 23, 157 pp.
124. Hidaka, T. (1983). *In litt.*, 29 March.
125. Higgins, L.G. and Riley, N.D. (1980). *A Field Guide to the Butterflies of Britain and Europe*. 4th ed., revised. Collins, London. 384 pp.
126. Hiura, I. (1980). A phylogeny of the genera of Parnassiinae based on analysis of wing pattern, with description of a new genus. *Bulletin of the Osaka Museum of Natural History* 33: 71–95.
127. Hiura, I. and Alagar, R.E. (1971). Studies on the Philippine butterflies chiefly collected by the co-operative survey by the Osaka Museum of Natural History and the National Museum of the Philippines. *Bulletin of the Osaka Museum of Natural History* 24: 29–44, 4 pls.
128. Hoffman, C.C. (1922). Restos de una antigua fauna del Norte entre los lepidópteros mexicanos. *Revista Mexicana de Biologia* 3: 1–23.
129. Hoffman, C.C. (1936). Relaciones zoogeográficas de los lepidópteros mexicanos. *Anales del Instituto de Biologiá, Universidad Nacional Autónoma de México* 7: 47–58.
130. Hoffmann, C.C. (1933). La fauna de Lepidopteros del distuito de Socnusco (Chiapas). *Anales del Instituto de Biologia, Universidad Nacional Autonoma de México* 4: 207–225.
131. Holloway, J.D. (1978). Butterflies and Moths. In *Kinabalu Summit of Borneo*. Sabah Society Monograph, 25–278.
132. Holloway, J.D. (1983). *In litt.*, 28 February.
133. Holloway, J.D. and Peters, J.V. (1976). The butterflies of New Caledonia and the Loyalty Islands. *Journal of Natural History* 10: 273–318.
134. Honrath, E.G. (1890). Diagnosen von zwei neuen Rhopaloceren. *Entomologische Nachrichten*. 16: 127.
135. Hope, F.W. (1843). On some rare and beautiful insects from Silhet, chiefly in the collection of Frederick John Parry, Esq., F.L.S. etc. *Transactions of the Linnean Society of London* 19: 131–136 + 2 plates.
136. Howarth, T.G. and Povolny, D. (1973). Beiträge zur Kenntnis der Fauna Afghanistans. *Rhop. Lep. I. Cas. Morav. Mus. (Acta Musei Moraviae)* 58: 131–158.
137. Howe, W.H. (1975). *The Butterflies of North America*. Doubleday and Co., Inc. 633 pp.
138. Hutton, A.F. (1983). *In litt.*, 28 March.
139. IUCN (1982). *IUCN Directory of Neotropical Protected Areas*. Tycooly International, Dublin. 436 pp.
140. Igarashi, S. (1965). *Papilio chikae*, an unrecorded Papilionid butterfly from Luzon island, the Philippines. *Tyo To Ga (Transactions of the Lepidopterists' Society of Japan)* 16: 41–49.
141. Igarashi, S. (1979). *Papilionidae and Their Early Stages*. Vol. 1:219 pp., Vol. 2:102 pp. of plates. Kodansha, Tokyo. (In Japanese).
142. Insect Farming and Trading Agency. (1979). Special Butterfly Price List 1979, 1980. Division of Wildlife, Bulolo, Morobe Province, Papua New Guinea.
143. Johnston, G. and Johnston, B. (1980). *This is Hong Kong: Butterflies*. Crown copyright, Hong Kong. 224 pp.
144. Jordan, K., in Seitz, A. (1908). *The Macrolepidoptera of the World. 9: The Rhopalocera of the Indo–Australian faunal region*. Papilionidae 11–118.
145. Jordan, K., in Seitz, A. (1907). *The Macrolepidoptera of the World. 5: The Macrolepidoptera of the American faunistic region*. Papilionidae 1–45.
146. Jumalon, J.N. (1964). Haunt and habits of *Graphium idaeoides* (Hewitson) (Papilionidae: Lepidoptera). *Philippine Journal of Science* 93(2): 207–217.
147. Jumalon, J.N. (1967). Two new papilionids. *Philippine Scientist* 1(4): 114–118.

148. Jumalon, J.N. (1969). Notes on the new range of some Asiatic papilionids in the Philippines. *The Philippine Entomologist* 1(3): 251–257.

149. Kawazoe, A. and Wakabayashi, M. (1976). *Colored Illustrations of the Butterflies of Japan*. Hoikusha, Osaka, Japan. 422 pp.

150. Kaye, W.J. (1919). A geographical table to show the distribution of the American papilios. *Novitates Zoologicae* 26: 320–355.

151. Kielland, J. (1983). *In litt.*, 6 March.

152. Kim, Hon Kyu (1982). Status survey of Alpine butterfly fauna, South Korea. In *WWF Yearbook 1982*. World Wildlife Fund, Gland. 492 pp.

153. Kim, Hon Kyu (1983). *In litt.*, 8 March.

154. Klots, A.B. (1951). *A Field Guide to the Butterflies of North America, East of the Great Plains*. Houghton Mifflin Company, Boston. 349 pp.

155. Kobayashi, H. and Koiwaya, S. (1978). A new species of *Ornithoptera* (Lepidoptera: Papilionidae) from West Irian. *Transactions of the Himeji Natural History Association (special issue)* 17 pp.

156. Lamas, G. (1983). *In litt.*, 4 April.

157. Larsen, T.B. (1973). Two species of *Allancastria* (Insecta; Rhopalocera) in Lebanon. *Entomologist* 106: 45–52.

158. Larsen, T.B. (1974). *Butterflies of Lebanon*. National Council for Scientific Research, Republic of Lebanon, Beirut. 256 pp.

159. Larsen, T.B. (1983). *In litt.*, 9 March.

160. Larsen, T.B. (1984). *In litt.*, 5 May.

161. Larsen, T.B.(1984). *Butterflies of Saudi Arabia and its Neighbours*. Stacey International, London. 160 pp.

162. Le Cerf, F. (1908). Description d'une variété nouvelle de *Thais cerisyi* God. (Lep.). *Bulletin de la Société Entomologique de France* 1908: 21–22.

163. Lee, C.-L. (1978). The early stages of Chinese Rhopalocera—*Luehdorfia chinensis* Leech (Parnassiidae: Zerynthiinae). *Acta Entomologica Sinensis* 21: 161–163 (in Chinese).

164. Lee, C.-L. (1980). A revision of the Chinese species of *Papilio machaon* L. and their geographical distribution. (sic). *Acta Entomologica Sinica* 23(4): 427–431, 3 pls.

165. Lee, R.F. (1971). A preliminary annotated list of Malawi forest insects. *Malawi Forest Research Institute Research Record* No. 40, 132 pp.

166. Leech, J.H. (1889). Description of a new *Luehdorfia* from Japan. *The Entomologist* 22 (309): 25–27.

167. Legrand, H. (1959). Note sur la sous-espèce *nana* Ch. Oberthür de *Papilio phorbanta* Linné des îles Seychelles [Lep., Papilionidae]. *Bulletin de la Société Entomologique de France* 64: 121–123.

168. Lekagul, B., Askins, K., Nabhitabhata, J. and Samruadkit, A. (1977). *Field Guide to the Butterflies of Thailand*. Association for the Conservation of Wildlife, Bangkok.

169. Lewis, H.L. (1973). *Butterflies of the World*. Harrap, London. 312 pp.

170. Luna de Carvalho (1962). Alguns Papilionideos da Luanda. Subsidios para o estudo da Biologia na Luanda. *Publicaçoes Culturais da Companhia de Diamantes de Angola* 24: No. 60, 165–169.

171. Macfarlane, R. (1983). *In litt.*, 15 March.

172. Martin, Dr. (1912). Ein neuer *Papilio* aus Celebes. *Deutsche Entomologische Zeitschrift "Iris"* 26: 163–5. Plate given separately (1913) *Deutsche Entomologische Zeitschrift "Iris"* 27: Tafel 6.

173. Matsumura, S. (1936). A new genus of Papilionidae. *Insecta Matsumurana* 10: 86.

174. Maza, E., J. de la (1983). *In litt.*, 4 October.

175. Maza, E., J. de la, and Díaz Francés, A. (1979). Notas y descripciones sobre la familia Papilionidae en México. *Revista de la Sociedad Mexicana de Lepidopterologiá* 4(2): 51–56.

176. McCubbin, C. (1971). *Australian butterflies*. Thomas Nelson (Australia) Ltd., Melbourne. xxx + 206 pp.

177. Medicielo, M. (1979). A key to separating members of the genus *Pachliopta* (Lepidoptera: Papilionidae) in the Philippines—excluding Palawan. *The Aurelian* 1(3): 10–11.

178. Mell, R. (1923). Noch unbeschriebene Lepidopteren aus Südchina. II. *Deutsche Entomologische Zeitschrift "Iris"* 1923: 153.

179. Mell, R. (1938). Beiträge zur Fauna Sinica. *Deutsche Entomologische Zeitschrift* 17: 197–345.

180. Mielke, O.H.H. (1983). *In litt.*, 23 March.

181. Miller, J.Y. and Miller, L.D. (1981). Taxonomic notes on some *Graphium* species from the Solomon Islands. *Bulletin of the Allyn Museum* 65: 1–7.

182. Ministry of Law and Justice, India (1972). *The Wildlife (Protection) Act, 1972* and *Revised List of Schedules to the Wildlife Protection Act, 1972, coming into force from 2-10-1980*.

183. Mitchell, G.A. (undated). *The National Butterflies of Papua New Guinea*. Wildlife Branch Department of Natural Resources, Papua New Guinea. 16 pp.

184. Montrouzier, P. (1856). Suite de la faune de l'Ile de Woodlark ou Moiou. *Annales de la Société Phys. Nat., Lyon* : 115–359.

185. Moonen, J.J.M. (1984). Notes on eastern Papilionidae. *Papilio International* 1(3): 47–50.

186. Moreau, E. (1923). Un *Papilio* nouveau de la Guyane Française. *Bulletin de la Société Entomologique de France* 1923: 144 and correction p. 215.

187. Moreau, E. (1924). Description de *Papilio maroni* male de Guyane Française. *Bulletin de la Société Entomologique de France* 1924: 93–94.

188. Munroe, E. (1961). The classification of the Papilionidae (Lepidoptera). *Canadian Entomologist* Supplement 17: 1–51.

189. Munroe, E. and Ehrlich, P.R. (1960). Harmonization of concepts of higher classification of Papilionidae. *Journal of the Lepidopterists' Society* 14: 169–175.

190. Murayama, S.-I. (1978). On some species of Rhopalocera from Southeast Asia with description of a new species and a new subspecies. *Tyo To Ga (Journal of the Lepidopterists' Society of Japan)* 29(3): 153–158.

191. Nabhitabhata, J. (1977). The butterflies of the Mentawi Islands. Unpublished report, 11 pp.

192. National Programme for Environmental Sciences, Working Group for Threatened Animals (1977). Unpublished report in litt., 12 pp.

193. Nekrutenko, Y.P. (1984). *In litt.*, 20 June.

194. Niculescu, E.V. (1961). *Fauna Republicii Populare Române. Insecta XI, 5, Lepidoptera Papilionidae* Edit. Acad. Rom. (Bucharest).

195. Oberlé, P. (ed.) (1981). *Madagascar, un Sanctuaire de la Nature*. Lechevahir S.A.R.L., Paris. 116 pp.

196. Okano, K. (1983). On the data of the type-specimen *Graphium stresmanni*. (sic) *Tokurana (Acta Rhopalocerologica)* 5: 88. (In Japanese).

197. Okano, K. (1983). Some ecological notes of *Teinopalpus*. *Tokurana (Acta Rhopalocerologica)* 5: 94–100.

198. Okano, K. (1983). The revision of classification of the genera of Papilionidae in the world (preliminary report). Part I: with description of a new genus. *Tokurana (Acta Rhopalocerologica)* 5: 1–64, 11 pls.

199. Okano, K. (1984). Color illustration of *Bhutanitis mansfieldi* (Riley, 1940) (Papilionidae): with some notes on the same species. *Tokurana (Acta Rhopalocerologica)* 6/7: 61–65.

200. Owen, D. F. (1983). *In litt.*, 25 February.

201. Panchen, A.L. (1980). Notes on the Behaviour of Rajah Brooke's Birdwing Butterfly, *Trogonoptera brookiana brookiana* (Wallace), in Sarawak. *The Entomologist's Record* 92: 98–102.

202. Panchen, A.L. and Panchen, M.D. (1973). Notes on the butterflies of Corsica, 1972. *The Entomologist's Record* 85: 149–153, 198–202.

203. Parsons, M.J. (1983). *In litt.*, 2 March.

204. Paulian, R. (1951). *Papillons Communs de Madagascar*. Publications de L'Institut de Recherche Scientifique, Tananarive–Tsimbazaza. 90 pp.

205. Paulian, R. (1983). *In litt.*, 10 May.

206. Paulian, R. and Viette, P. (1968). *Faune de Madagascar. XXVII Insectes Lépidoptères Papilionidae*. O.R.S.T.O.M. and C.N.R.S., Paris. 97 pp.

207. Pen, D. (1936). The Papilionidae of south–western Szechwan. *Journal of the West China Border Research Society* 8: 153–165.

208. Pennington, K.M. (1978). *Pennington's Butterflies of South Africa*. (Ed. C.G.C. Dickson). A.D. Donker, Johannesburg. 670 pp + 198 pls.

209. Pérez R., H. (1969). Quetotáxia y morfología de la oruga de *Baronia brevicornis* Salv. (Lepidoptera, Papilionidae, Baroniinae). *Anales del Instituto de Biologiá, Universidad Nacional Autónoma de México* 40, Ser. Zoologia (2): 227–244.

210. Pérez R., H. (1971). Algunas consideraciones sobre la poblacion de *Baronia brevicornis* Salv. (Lepidoptera, Papilionidae, Broniinae), en la región de Mezcala, Guerrero. *Anales del Instituto de Biologiá, Universidad Nacional Autónoma de México*, 42, Ser. Zoologia (1): 63–72.

211. Pérez R., H. (1977). Distribución geografica y estructura poblacional de *Baronia brevicornis* Salv. (Lepidoptera, Papilionidae, Baroniinae) en la República Mexicana. *Anales del Instituto de Biologiá, Universidad Nacional Autónoma de México*, 48, Ser. Zoología (5): 151–164.

212. Pérez R., H. (1983). *In litt.*, 14 May.

213. Perkins, S.F., Perkins, E.M. and Shinginger, F.S. (1968). Illustrated life history and notes on *Papilio oregonius*. *Journal of the Lepidopterists' Society* 22: 53–56.

214. Pinhey, E.C.G. (1965). *The Butterflies of Southern Africa*. Nelson, Johannesburg. 240 pp + 42 pls.

215. Pinhey, E.C.G. and Loe, I.D. *A Guide to the Butterflies of Central and Southern Africa*. Sir Joseph Causton, London and Eastleigh.

216. Pyle, R.M. (1981). *The Audubon Society Field Guide to North American Butterflies* Alfred A. Knopf. Inc, New York. 916 pp.

217. Pyle, R.M. and Hughes, S.A. (1978). Conservation and utilisation of the insect resources of Papua New Guinea. Report of a consultancy to the Wildlife Branch, Dept. of Nature Resources, Independent State of Papua New Guinea. 157 pp., unpublished.

218. Racheli, T. (1979). New subspecies of *Papilio* and *Graphium* from the Solomon Islands, with observations on *Graphium codrus* (Lepidoptera, Papilionidae). *Zoologische Mededelingen* 54(15): 237–240.

219. Racheli, T. (1980). A list of the Papilionidae (Lepidoptera) of the Solomon Islands, with notes on their geographical distribution. *Australian Entomological Magazine*. 7: 45–59.

220. Racheli, T. (1984). Further notes on Papilionidae from the Solomon Islands. *Papilio International* 1: 55–63.

221. Racheli, T. (1984). *In litt.*, 18 June.

222. Racheli, T. (1984). Pers. comm., 25 April.

223. Racheli, T. and Cotton, A.M. (1983). A new subspecies of *Papilio polyctor* Boisduval from South–east Thailand. *Papilio International* 1: 1–8.

224. Racheli, T. and Haugum, J. (1983). On the status of *Papilio heringi* Niepelt 1924. *Papilio International* 1: 37–45.

225. Racheli, T. and Sbordoni, V. (1975). A new species of *Papilio* from Mexico. *Fragmenta Entomologica* 11: 175–183.

226. Remington, C.L. (1968). A new sibling *Papilio* from the Rocky Mountains, with genetic and biological notes. *Postilla* 119: 1–40.

227. Riabov, M.A. (1958). The Lepidoptera (of the Caucasus). In: *Zhivotnyi Mir SSR*. T.5, Moscow–Leningrad, 365–366 (In Russian).

228. Ribbe, C. (1907). Zwei neuen Papilioformen von der Salomon–Insel Bougainville. *Deutsche Entomologische Zeitschrift "Iris"* 20: 59–63.

229. Riley, N.D. (1939). A new species of *Armandia* (Lep. Papilionidae). *The Entomologist*. 72: 206–208, 267.

230. Riley, N.D. (1975). *A Field Guide to the Butterflies of the West Indies*. Collins, London. 244 pp.

231. Robinson, J.C. (1975–6). Swallowtail butterflies of Sabah. *Sabah Society Journal* 6: 5–22.

232. Ross, G. (1964). Life history studies on Mexican butterflies I. *Journal of Research on the Lepidoptera*. 3: 9–17.

233. Rothschild, W. (1895). A revision of the papilios of the eastern hemisphere, exclusive of Africa. *Novitates Zoologicae* 2: 167–463.

234. Rothschild, W. and Jordan, K. (1906). A revision of the American Papilios. *Novitates Zoologicae* 13: 411–752. (Facsimile edition ed. P.H. Arnaud, 1967).

235. Rothschild, W. and Jordan, K. (1901). On some Lepidoptera. *Novitates Zoologicae* 1901: 401–407.
236. Rütimeyer, E. (1969). A new *Papilio* from Colombia and a new sphingid from New Guinea. *Journal of the Lepidopterists' Society* 23: 255–257.
237. Saigusa, S., Nakanishi, A., Shima, H. and Yata, O. (1977). Phylogeny and biogeography of the subgenus *Graphium* Scopoli (Text in Japanese). *Acta Rhopalocerologica* 1: 2–32.
238. Saigusa, T. and Lee, C.-L. (1982). A rare papilionid butterfly *Bhutanitis mansfieldi* (Riley), its rediscovery, new subspecies and phylogenetic position. *Tyo to Ga (Journal of the Lepidopterists' Society of Japan)* 33: 1–24.
239. Salk, P. (1983). Neue beobachtungen zur Biologie von *Parnassius nordmanni* Mén., subsp. *christophi* B.E. (Lep. Papilionidae) *Deutsche Entomologische Zeitschrift* 30: 239–241.
240. Salvin, O. (1893). XIX. Description of a new genus and species of Papilionidae from Mexico. *Transactions of the Entomological Society of London* 1893: 331–2.
241. Samson, C. (1979). Butterflies (Lepidoptera: Rhopalocera) of the Santa Cruz group of islands, Solomon Islands. *The Aurelian* 1(2): 1–19.
242. Samson, C. (1983). Butterflies (Lepidopera: Rhopalocera) of Vanuatu. *Naika* (Journal of the Vanuatu Natural History Society) June 1983: 2–6.
243. Schwartz, A. (1984). *In litt.*, 16 May.
244. Shapiro, A.M. (1975). *Papilio "gothica"* and the phenotypic plasticity of *P. zelicaon* (Papilionidae). *Journal of the Lepidopterists' Society* 79–84.
245. Sheljuzhko, L. (1961). *Parnassius nordmanni* Mén in Kleinasien. *Entomologische Zeitschrift* 71: 33–36.
246. Shirôzu, T. (1960). *Butterflies of Formosa in Colour.* Hoikusha. Osaka, Japan (in Japanese). 483 pp.
247. Shirôzu, T. and Hara, A. (1960). *Early Stages of Japanese Butterflies in Colour.* Hoikusha, Osaka, Japan (in Japanese). Vol. 1, 142 pp.
248. Shirôzu, T. and Hara, A. (1962). *Early Stages of Japanese Butterflies in Colour.* Hoikusha, Osaka, Japan (in Japanese). Vol. 2, 139 pp.
249. Skaife, S.H. (1953). *African Insect Life.* Longmans Green and Co., Cape Town. 387 pp.
250. Smart, P. (1975). *The Illustrated Encyclopedia of the Butterfly World.* Hamlyn, London. 275 pp.
251. Smith, C. (1975). *Commoner Butterflies of Nepal.* Tribhuvan University, Kathmandu, Nepal. 38 pp.
252. Stahl, G. (1979). *Ripponia hypolitus* Cramer. A description of the form 'antiope' from Halmahera, and of a new form from Obi. *Lepidoptera Group of '68 Newsletter* II (5): 135–139.
253. Stanek, V.J. (1969). *The Pictorial Encyclopedia of Insects.* Hamlyn, London. 544 pp.
254. Straatman, R. (1963). A hybrid between *Papilio aegeus aegeus* and *Papilio fuscus capaneus*, with a note on larval foodplants. *Journal of the Lepidopterists' Society* 16: 161–174.
255. Straatman, R. (1969). Notes on the biology and host plant associations of *Ornithoptera priamus urvilleanus* and *O. victoriae* (Papilionidae). *Journal of the Lepidopterists' Society* 23:
256. Straatman, R. (1975). Notes on the biologies of *Papilio laglaizei* and *P. toboroi* (Papilionidae). *Journal of the Lepidopterists' Society.* 29: 180–187.
257. Struhsaker, T.T. (1983). *In litt.*, 30 April.
258. Sullivan, Hon. V.B., Minister for Primary Industries (1974). Fauna Conservation Regulations 1974 and Notifications Made in Pursuance of the Provisions of the Fauna Conservation Act 1974. *Queensland Government Gazette* 246(103): 2269–2308.
259. Talbot, G. (1939). *The Fauna of British India, Including Ceylon and Burma. Butterflies Vol I.* Taylor and Francis Ltd., London, reprint New Delhi 1975. 600 pp.
260. Tanasiychuk, V.N. (1981). Data for the "Red Book" of Insects of the U.S.S.R. *Entomologicheskoye Obozreniye* 60 (3): 168–186. (English translation).
261. Thomas, C. and Cheverton, M. (1982). Cambridge butterfly expedition to Central America, Panama 1979. Final report. 105 pp.
262. Tsukada, E. and Nishiyama, Y. (1982). *Butterflies of the South East Asian Islands. Vol. 1*

Papilionidae. (transl. K. Morishita). Plapac Co. Ltd., Tokyo. 457 pp.

263. Turlin, B. (1983). *In litt.*, 15 September.

264. Turlin, B. (1983). *In litt.*, 1 July.

265. Turlin, B. (1983). Systematique des *Parnassius*. Manuscript *in litt.*, 10 October.

266. Turner, T.W. (1983). *In litt.*, 27 July.

267. Turner, T.W. (1983). The status of the Papilionidae, Lepidoptera of Jamaica with evidence to support the need for conservation of *Papilio homerus* Fabricius and *Eurytides marcellinus* Doubleday. Unpublished report. 14 pp.

268. Tyler, H.A. (1983). *In litt.*, 13 March.

269. Tyler, H.A. (1975). *The Swallowtail Butterflies of North America*. Naturegraph Publishers. viii + 192 pp.

270. U.S. Fish and Wildlife Service (1983). Proposed delisting of Bahama Swallowtail Butterfly and reclassification of Schaus Swallowtail Butterfly from Threatened to Endangered. *Federal Register* 48(168): 39096–39099.

271. Van Someren, V.G.L. (1960). Systematic notes on the associated blue-banded black Papilios of the *bromius-brontes-sosia* complex of Kenya and Uganda, with descriptions of two new species. *Boletim do Sociedade de Estudos da Provincia de Mocambique* 123: 65–78.

272. Van Someren, V.G.L. (1974). List of foodplants of some East African Rhopalocera, with some notes on the early stages of some Lycaenidae. *Journal of the Lepidopterists' Society* 28: 315–331.

273. Van Son G. (1949). *The Butterflies of Southern Africa. Part 1. Papilionidae and Pieridae* Transvaal Museum, Pretoria. 237 pp + 41 pls.

274. Vane–Wright, R.I. (1984). *In litt.*, 3 July.

275. Vázquez G., L. (1953). Observaciones sobre Papilios de México, con descripciones de algunas formas nuevas; una especie nueva para México y localidades nuevas de algunos otros, III. *Anales del Instituto de Biologiá Universidad Nacional Autónoma de México*. 24: 170–175.

276. Vázquez G., L. (1983). *In litt.*, 15 March.

277. Vázquez G., L. and Pérez R., H. (1961). Observaciones sobre la biologia de *Baronia brevicornis* Salv. (Lepidoptera: Papilionidae—Baroniidae). *Anales del Instituto de Biologiá Universidad Nacional Autónoma de México*. 32: 295–311.

278. Vazquez G., L. and Perez, H. (1967). Nuevas observaciones sobre le biologia de *Baronia brevicornis* Salv. Lepidoptera: Papilionidae—Baroniinae. *Anales del Instituto de Biologia Universidad Nacional Autónoma de México*. 37: 195–204.

279. Viette, P. (1980). Mission lépidoptérologique à la Grande Comore (Ocean Indien occidental). *Bulletin de la Société Entomologique de France* 85: 226–235.

280. Viette, P. (1983). *In litt.*, 22 March.

281. Walker, M.V. (1983). *In litt.* to S.M. Wells, 11 April.

282. Watson, A. and Whalley, P.E.S. (1975). *The Dictionary of Butterflies and Moths in Colour*. McGraw–Hill Book Company. 296 pp.

283. Wells, S.M., Pyle, R.M. and Collins, N.M. (1983). *The IUCN Invertebrate Red Data Book*. IUCN, Cambridge and Gland. L + 632 pp.

284. Whitten, A. (undated). *In litt.*

285. Williams, J.G. (1969). *A Field Guide to the Butterflies of Africa*. Collins, London. 238 pp.

286. Wood, G.L. (1982). *The Guinness Book of Animal Facts and Feats*. Third Edition. Guinness Superlatives Ltd., Enfield. 250 pp.

287. Woodhouse, L.G.O. (1950). *The Butterfly Fauna of Ceylon*. Second Edition. Colombo Apothecaries' Co. Ltd., Colombo. 231 + xxxii pp.

288. Wyatt, C. and Omoto, K. (1963). Auf der Jagd nach *Parnassius autocrator* Avin. *Zeitschrift Wiener Entomologische Gesellschaft* 48: 163–170.

289. Yamamoto, A. (1977). A new species of *Graphium* from Mindanao (Lepidoptera: Papilionidae). *Tyô to Ga. (Transactions of the Lepidopterists' Society of Japan)*. 28: 87–88.

290. Yuste, F., Pérez, R., H. and Walls, F. (1972). Compounds of papilionid caterpillars (*Baronia brevicornis* Salv). *Experientia* 28: 1149.

Note added in proof: In a recent publication some *Graphium* species-groups have been revised (Hancock, 1985). Conclusions affecting this text may be summarized as follows:

1. *G. schaffgotschi*, formerly regarded as a second subspecies of *G. taboranus* (species 145), is raised to a full species distributed in Angola, Namibia, Zaire and Zambia.
2. *G. poggianus* (species 150) is now recorded from Tanzania.
3. *G. olbrechtsi* and *G. odin* are now regarded as conspecific with *G. auriger* (species 154–156).
4. *G. weberi* and *G. ucalegonides* are now regarded as conspecific with *G. fulleri* (species 158–160).

291. Hancock, D.L. (1985). Systematic notes on some African species of *Graphium* Scopoli. *Papilio International* 2: 97–103 (plus correction in subsequent issue).

Geographical index to species

This index lists the species of swallowtails found in countries and regions of the world. It is as complete and accurate as possible at the time of going to press, but new data are always emerging and conservation status categories are constantly reviewed and altered. Any new information, amendments or comments on the list will be welcomed.

Under each country or region a list of numbers will be found. These numbers refer to the 573 species recognized in this book and discussed in the main taxonomic list. In addition the number, name and status of any threatened species is printed in full. Full reviews of these threatened species will be found in section 6 of this book. To find the appropriate page number, first refer back to the species list.

Many countries have been listed individually, but to save space and avoid unnecessary repetition some are grouped into regions. These regions usually contain either relatively few species (e.g. North Africa), or the constituent countries have similar swallowtail faunas (e.g. Europe, Middle East), or the distributional information for individual countries within the region is poor (e.g. West Africa, Indochina, Central America). Readers interested in the fauna of one particular country which is not individually listed should note the regional species numbers and refer to the distributions given in the full taxonomic list. Because of the exceptionally large size of the U.S.S.R. and the variation within its swallowtail fauna it has been divided into three separate regions.

EUROPE

Europe has a poor swallowtail fauna which is concentrated in the south and east of the continent. To save repetition it has not been divided into its constituent countries, i.e. Albania, Andorra, Austria, Belgium, Bulgaria, Czechoslovakia, Denmark, Finland, France (including Corsica), Federal Republic of Germany, German Democratic Republic, Gibraltar, Greece, Hungary, Iceland, Ireland, Italy (including Sardinia and Sicily), Liechtenstein, Luxembourg, Malta, Monaco, Netherlands, Norway, Poland, Portugal, Romania, Spain, Sweden, Switzerland, United Kingdom and Yugoslavia. The U.S.S.R. is considered separately.

2, 28, 37, 40, 42, 46, 47, 55, 426–428 (11 species)
40. *Parnassius apollo* RARE (Albania, Andorra, Austria, Bulgaria, Czechoslovakia, Finland, France, F.R.G., G.D.R., Greece, Hungary, Italy (including Sicily), Liechtenstein, Norway, Poland, Romania, Spain, Sweden, Switzerland and Yugoslavia)
427. *Papilio (Papilio) hospiton* ENDANGERED (Corsica and Sardinia)

NORTH AFRICA AND THE MIDDLE EAST

This region has a relatively poor swallowtail fauna and is, therefore, treated as three large subregions.

North Africa
　　　　(Algeria, Canary Is, Egypt, Libya, Morocco and Tunisia)
　　　　47, 55, 428 (3 species)

Arabian Peninsula
　　　　(Bahrain, Kuwait, Oman, Qatar, Saudi Arabia, United Arab Emirates, Yemen and South Yemen)
　　　　55(?), 428, 538, 539 (3 or 4 species)

Middle East
　　　　(Cyprus, Israel, Jordan, Lebanon, Syria and Turkey)
　　　　2, 27, 28, 40, 42, 43, 426, 428 (8 species)
　　　　40.　*Parnassius apollo* RARE (Syria and Turkey)

U.S.S.R.

The U.S.S.R. has been divided into three regions. (total for all regions 35 species).

European U.S.S.R.
　　　　(including Armenia, Azerbaydzhan, Byelorussia, Estonia, Georgia, Latvia, Lithuania, Moldavia and Ukraine)
　　　　2, 27, 28, 40, 42, 44, 46, 55, 428 (9 species)
　　　　40.　*Parnassius apollo* RARE

Soviet Central Asia
　　　　(Kazakhstan, Kirghizia, Tadzhikistan, Turkmenistan and Uzbekistan)
　　　　2, 3, 7, 9, 10, 12, 14, 15, 20, 26, 28, 31, 32, 34, 35(?), 36–3 8, 55(?), 426, 428 (19–21 species)
　　　　15.　*Parnassius autocrator* RARE (Tadzhikistan)

Siberia
　　　　(Asiatic U.S.S.R. excluding Soviet Central Asia)
　　　　16, 24–26, 29, 33, 37, 39–41, 54, 273, 428, 557 (14 species)
　　　　40.　*Parnassius apollo* RARE

MIDDLE ASIA TO INDOCHINA AND JAPAN

Afghanistan
　　　　3, 7, 8, 12, 13, 15, 28, 32, 34–36, 38, 55, 274, 314, 426, 428, 539, 560 (19 species)
　　　　15.　*Parnassius autocrator* RARE

Bangladesh
　　　　121(?), 186, 187, 189, 203(?), 330, 415, 416, 420(?), 428(?), 446, 462, 470, 563, 565(?) (10–15 species)

Bhutan

7, 10, 17(?), 19, 21, 35, 50, 51, 57, 59, 60, 135, 198, 272, 274(?), 284, 285(?), 293, 415(?), 420(?), 421, 428, 458, 462, 470, 478(?), 479, 560(?), 563(?), 565 (22–30 species)
51. *Bhutanitis ludlowi* INSUFFICIENTLY KNOWN
57. *Teinopalpus imperialis* RARE

Burma

50, 57, 59, 60, 114–117, 120, 121, 128, 129, 135, 136, 138, 180, 182, 186–189, 198, 203, 274–278, 284, 285, 291, 293–295, 298, 300, 302, 314, 320, 326, 330, 415, 416, 420–422, 428, 440, 444, 446, 449, 458, 462, 464, 467, 470, 473, 478, 480, 483, 539, 554, 558, 560–563, 565 (68 species)
57. *Teinopalpus imperialis* RARE

China

(including Hong Kong and Xizang Zizhiqu (Tibet))
4, 5, 7, 8, 10–12, 14, 16–22, 24, 29–31, 33–41, 48–50, 52, 54–60, 114–120, 128, 129, 135, 136, 138, 182, 183, 186–189, 198, 203, 271–274, 275(?), 276(?), 278, 280, 281, 283–285, 293, 295(?), 300, 314, 326, 330, 413, 415, 416(?), 420, 421, 422(?), 428, 440, 446–449, 458, 462, 470, 473, 478, 479, 483, 539, 557–559, 561, 563, 565, 567 (99–104 species)
40. *Parnassius apollo* RARE (Xinjiang Uygur (Sinkiang))
48. *Bhutanitis mansfieldi* RARE (Yunnan)
49. *Bhutanitis thaidina* RARE (Yunnan, Sichuan and Shaanxi)
52. *Luehdorfia chinensis* INSUFFICIENTLY KNOWN (Anhui, Hubei, Jiangsu and Jiangxi)
57. *Teinopalpus imperialis* RARE (Hubei, Sichuan)
58. *Teinopalpus aureus* INSUFFICIENTLY KNOWN (Guangdong)

India

(mainland only; for Andaman and Nicobar Islands see Indian Ocean)
6, 7, 8, 10–12, 17, 19–21, 34–36, 38, 50, 55, 57, 59, 60, 114–116, 120, 121, 128, 129, 135, 136, 138, 182, 186–189, 198, 203, 272, 274–276, 278, 284, 285, 293, 294, 300, 302, 304, 314, 325, 326, 330, 415, 416, 420–422, 428, 445, 446, 449, 450, 458, 460, 462, 470, 473, 478, 479, 481, 551, 553, 560, 562, 563, 565, 567 (77 species)
57. *Teinopalpus imperialis* RARE

Indochina

(Kampuchea, Laos, Thailand and Vietnam)
50, 59, 60, 114, 115, 120, 121, 128, 129, 131, 134–138, 180–182, 186–189, 198, 203, 273, 274, 275(?), 276, 277, 279, 285, 291, 293–295, 296(?), 298, 300, 314, 319, 320, 326, 330, 415, 416, 420–422, 444, 446, 449, 462, 464, 467, 470, 471(?), 473, 478–480, 483, 539, 554, 558, 560–563, 565 (66–69 species)

Iran

2, 3, 28, 42, 43(?), 45, 55(?), 426, 428 (7–9 species)

Iraq

2, 28, 42, 43, 55(?), 426, 428 (6 or 7 species)

Japan

(including Ryukyu Is)
24, 29, 30, 33, 53, 54, 187, 203, 273, 314, 428, 429, 440, 446–448, 462, 470, 473, 539, 557, 558 (22 species)

53.   *Luehdorfia japonica* VULNERABLE

North Korea
24, 29, 30, 33, 39, 41, 54, 273, 428, 446–448, 557, 558 (14 species)

South Korea
29, 30, 33, 39, 41, 54, 203(?), 273, 428, 446–448, 470, 557, 558 (14 or 15 species)

Mongolia
16, 20, 22, 24, 26, 29, 33, 37, 39, 40, 428 (11 species)
40.   *Parnassius apollo* RARE

Nepal
10, 17, 20, 21, 35, 57, 116, 117, 121, 128, 135, 136, 182, 187, 189, 198, 203, 272, 274, 284, 285, 294(?), 314, 326, 415, 420, 428, 446, 450, 458, 462, 470, 473, 478, 539, 560, 565, 567 (37 or 38 species)
57.   *Teinopalpus imperialis* RARE

Pakistan
3, 7, 8, 17, 20, 21, 34–36, 38, 55, 426, 428, 560 (14 species)

Sri Lanka
121, 128, 187, 189, 203, 302, 303, 314, 332, 415, 460, 470, 473, 539, 551 (15 species)
303.   *Atrophaneura (Pachliopta) jophon* VULNERABLE

Taiwan
116, 118, 187, 189, 198, 203, 273, 281, 282, 285, 292, 314, 323, 326, 414, 420, 421, 428, 440, 446, 456, 459, 462, 470, 473, 478, 479, 539, 557, 561, 563, 566 (32 species)
414.   *Papilio (Chilasa) maraho* VULNERABLE

SOUTHEAST ASIA TO AUSTRALIA

N.B. No swallowtails have been recorded from New Zealand.

Australia
113, 120, 186, 189, 190, 193, 203, 224, 315, 349–351, 412, 475, 484, 485, 498, 539, 572 (19 species)

Brunei
60, 114–116, 120, 128, 130(?), 134, 136–138, 180, 181, 186–189, 197, 203, 296(?), 298, 314, 316, 320, 321, 330, 416, 422, 444, 453, 467, 470, 473, 478, 484, 554, 564 (35–37 species)
453.   *Papilio (Princeps) acheron* RARE

Indonesia
60, 114, 115, 120, 122–124, 126, 128, 129, 132, 134, 136–138, 141, 142, 180, 181, 185–190, 194–201, 203, 205, 224, 287–291, 296–300, 305–307, 314–316, 318–321, 324, 326, 329–331, 333–344, 347–349, 415–417, 422, 423, 442–444, 451, 452, 455–457, 462, 466, 467, 469–475, 477, 478, 482, 484–486, 489, 493, 496–498, 539, 552, 554, 563, 564, 568–572 (121 species)
195.   *Graphium (Graphium) stresemanni* RARE (Moluccas)

288. *Atrophaneura (Atrophaneura) luchti* RARE (Java)
299. *Atrophaneura (Losaria) palu* INSUFFICIENTLY KNOWN (Sulawesi)
324. *Troides (Troides) prattorum* INDETERMINATE (Buru)
329. *Troides (Troides) dohertyi* VULNERABLE (Talaud Is and Sangihe)
340. *Ornithoptera tithonus* INSUFFICIENTLY KNOWN (Irian Jaya)
341. *Ornithoptera rothschildi* INDETERMINATE (Irian Jaya)
342. *Ornithoptera chimaera* INDETERMINATE (Irian Jaya)
343. *Ornithoptera paradisea* INDETERMINATE (Irian Jaya)
344. *Ornithoptera meridionalis* VULNERABLE (Irian Jaya)
347. *Ornithoptera aesacus* INDETERMINATE (Moluccas)
348. *Ornithoptera croesus* VULNERABLE (Moluccas)
472. *Papilio (Princeps) jordani* RARE (Sulawesi)
568. *Papilio (Princeps) neumoegeni* VULNERABLE (Lesser Sunda Is)

Eastern Malaysia

(Sabah and Sarawak)
60, 114, 115, 120, 126–130, 134, 136–138, 180, 181, 184, 186–189, 197, 203, 296, 298, 314, 316, 320–322, 330, 416, 422, 444, 453, 461(?), 462, 467, 470, 473, 478, 484, 554, 564 (42–43 species)
184. *Graphium (Graphium) procles* INDETERMINATE (Sabah)
322. *Troides (Troides) andromache* INDETERMINATE
453. *Papilio (Princeps) acheron* RARE

Peninsular Malaysia

(including Singapore)
60, 114, 115, 120, 128, 129, 134, 136–138, 180–182, 186–189, 197, 203, 291, 294, 296, 298, 300, 314, 316, 319, 320, 326, 330, 415, 416, 420, 421(?), 422, 444, 462, 467, 470, 471, 473, 478, 480, 484, 539, 554, 563 (46–47 species)
(Malaysia as a whole has 54–56 species)

Papua New Guinea

(including Bismarck Archipelago, Bougainville and associated islands)
120, 142, 186, 189–191, 193, 194, 196, 203–205, 207, 224, 315, 331, 339, 342–346, 349, 423–425, 442, 475, 476, 484, 486, 488, 494, 495, 498, 539, 572 (37 species)
191. *Graphium (Graphium) meeki* RARE (Bougainville)
204. *Graphium (Graphium) mendana* RARE (Bougainville)
342. *Ornithoptera chimaera* INDETERMINATE
343. *Ornithoptera paradisea* INDETERMINATE
344. *Ornithoptera meridionalis* VULNERABLE
345. *Ornithoptera alexandrae* ENDANGERED
424. *Papilio (Chilasa) toboroi* RARE (Bougainville)
425. *Papilio (Chilasa) moerneri* VULNERABLE (Bismarck Archipelago)
495. *Papilio (Princeps) weymeri* RARE (Bismarck Archipelago)

Philippines

114, 115, 120, 126, 127, 133, 134, 136, 139, 140, 180, 181, 186, 187, 189, 196, 197, 202, 203, 286, 298, 308–314, 317, 323, 327, 328, 415, 416, 418, 419, 440, 441, 444, 456, 461, 465, 468, 470, 474, 483, 539, 555, 556, 564 (49 species)

133. *Graphium (Pathysa) idaeoides* RARE
139. *Graphium (Pathysa) megaera* INDETERMINATE (Palawan)
202. *Graphium (Graphium) sandawanum* VULNERABLE (Mindanao)
308. *Atrophaneura (Pachliopta) schadenbergi* VULNERABLE (Luzon, Babuyan)
312. *Atrophaneura (Pachliopta) atropos* INDETERMINATE (Palawan)
418. *Papilio (Chilasa) osmana* VULNERABLE (Leyte and Mindanao)
419. *Papilio (Chilasa) carolinensis* VULNERABLE (Mindanao)
441. *Papilio (Princeps) benguetanus* VULNERABLE (Luzon)
556. *Papilio (Princeps) chikae* ENDANGERED (Luzon)

Solomon Islands
    186, 191, 196, 203, 204, 206, 315, 346, 349, 424, 476, 484, 488, 494, 572 (15 species)
    191. *Graphium (Graphium) meeki* RARE
    204. *Graphium (Graphium) mendana* RARE
    424. *Papilio (Chilasa) toboroi* RARE

TROPICAL AND SOUTHERN AFRICA

This region has been subdivided into individual countries except for western and southern Africa. Madagascar and other islands are listed under Oceanic Islands.

Southern Africa
    (Botswana, Lesotho, Namibia, South Africa, Swaziland and Zimbabwe)
    144–146, 164, 172, 173, 177, 178, 502, 515, 532, 538, 543, 545, 546 (15 species)

Western Africa
    (Benin, Gambia, Ghana, Guinea, Guinea–Bissau, Ivory Coast, Liberia, Mali, Mauritania, Niger, Nigeria, Senegal, Sierra Leone, Togo and Burkina Faso (formerly Upper Volta))
    144, 147, 151, 152, 160, 161, 164, 167, 168, 172, 174, 178, 499, 500, 506, 513, 515, 521, 523, 524, 526, 527(?), 529–531, 533, 534, 538, 544, 545, 548, 549 (31–32 species)
    531. *Papilio (Princeps) antimachus* RARE (Benin (?), Ghana, Guinea, Ivory Coast, Liberia, Sierra Leone and Togo (?))

Angola
    144, 145, 147, 149, 150, 157, 160, 161, 164, 168, 172, 178, 499–501, 512, 513, 515, 521, 525–527, 531, 534, 538, 544, 545 (27 species)
    531. *Papilio (Princeps) antimachus* RARE

Burundi
    144, 149(?), 161(?), 164, 172, 175, 178, 501(?), 503(?), 506, 510, 512, 515, 521, 538, 540, 544, 545, 548(?), 550 (15–20 species)
    540. *Papilio (Princeps) leucotaenia* VULNERABLE

Cameroon
    144, 147, 149, 152, 155, 157–162, 164, 167, 168, 170, 172, 174, 178, 499–501, 506, 510, 513, 515, 521, 523–526, 528, 529, 531, 533, 534, 538, 544, 545, 548 (39 species)

158.   *Graphium (Arisbe) weberi* INSUFFICIENTLY KNOWN
531.   *Papilio (Princeps) antimachus* RARE

Central African Republic
144, 147(?), 149, 151, 152, 155, 157, 160, 161(?), 167(?), 168(?), 172, 178, 499(?), 500, 513, 515, 521, 523, 525–527, 529, 531, 534, 538, 544, 545, 548 (24–29 species)
531.   *Papilio (Princeps) antimachus* RARE

Chad
144, 168, 172, 178, 515, 538, 544(?), 545 (7–8 species)

Congo
144, 147, 149, 152, 155, 157, 159–162, 164, 167, 168, 172, 174, 178, 499–501, 503(?), 506, 513, 515, 521, 523–527, 529, 531, 534, 538, 543–545, 548, 550 (37–38 species)
531.   *Papilio (Princeps) antimachus* RARE

Djibouti
144(?), 164, 172, 178, 515, 538, 545 (6–7 species)

Equatorial Guinea
(including Bioko (formerly Fernando Póo))
144, 149(?), 152(?), 157(?), 160(?), 164, 172, 178, 501, 510, 515, 521(?), 523(?), 524(?), 526, 531, 534(?), 538, 544, 545, 548 (13–21 species)
531.   *Papilio (Princeps) antimachus* RARE

Ethiopia
144, 148, 149, 164, 172, 177, 178, 499, 502, 506, 514, 515, 521(?), 538, 543, 545 (15 or 16 species)

Gabon
144, 147, 149, 152(?), 156, 157, 159–62, 164, 167, 168, 170, 172, 174, 178, 499(?), 501(?), 513, 515, 521, 523, 524, 525(?), 526(?), 531, 534(?), 538, 544, 545 (25–31 species)
531.   *Papilio (Princeps) antimachus* RARE

Kenya
144, 148, 149, 164, 171–173, 176–178, 499, 501, 503, 506, 511, 512, 515, 519–522, 532, 534, 538, 543–545, 550 (30 species)
519.   *Papilio (Princeps) desmondi teita* ENDANGERED

Malawi
144, 145, 146, 148, 164, 171–173, 177, 178, 502–504, 512, 515, 518, 519, 532, 543–545, 547 (22 species)

Mozambique
144, 146, 148, 164, 169, 171–173, 177, 178, 502, 532, 538, 543–545 (16 species)

Rwanda
144, 149(?), 161(?), 164, 172, 175, 178, 501, 503, 506, 510, 512, 515, 521, 531, 538, 540, 544, 545, 548(?), 550 (18–21 species)
531.   *Papilio (Princeps) antimachus* RARE
540.   *Papilio (Princeps) leucotaenia* VULNERABLE

Somalia
144, 164, 171, 172, 177, 178, 514, 515, 538, 543–545 (12 species)

Sudan

144, 148, 149, 164, 172, 178, 500, 501, 506, 512, 515, 521, 525, 526, 534, 538, 544, 545, 548, 550 (20 species)

Tanzania

(including Zanzibar and Pemba Island)
144, 145, 147–149, 161, 164, 165, 171–173, 176–178, 501–506, 511, 512, 515, 518, 519, 521, 532, 538, 543–545, 547, 548, 550 (34 species)
505. *Papilio (Princeps) sjoestedti* RARE

Uganda

144, 147–149, 161, 164, 168, 172, 175, 178, 499, 501, 503, 506, 510, 512, 513, 515, 520–522, 525, 526(?), 529, 531, 534, 538, 540, 544, 545, 548, 550 (31–32 species)
531. *Papilio (Princeps) antimachus* RARE
540. *Papilio (Princeps) leucotaenia* VULNERABLE

Zaire

144, 145, 147, 149–155, 157, 160–162, 164, 167, 168, 170, 172–175, 178, 499–501, 503, 506, 510, 512, 513, 515, 521, 523–527, 529, 531, 532, 534, 538, 540, 543–545, 548 (48 species)
153. *Graphium (Arisbe) aurivilliusi* INSUFFICIENTLY KNOWN
531. *Papilio (Princeps) antimachus* RARE
540. *Papilio (Princeps) leucotaenia* VULNERABLE

Zambia

144, 145, 147, 149, 150, 164, 172, 173, 178, 501–503, 512, 513, 515, 518, 519, 532, 538, 543–545, 547, 548 (23 species)

NORTH AMERICA

Canada

23, 24, 37, 61, 208, 352, 334–356, 359, 391, 428, 430, 432–436
(18 species)

Mexico

1, 62, 66, 69, 71, 81, 87–89, 96, 100, 102, 104, 109, 208, 211, 217, 218, 220, 221, 226–228, 247, 248, 250, 251, 254, 268, 352–358, 371–373, 378, 382–384, 391, 394, 402–404, 407, 409, 430, 432 (52 species)
1. *Baronia brevicornis* RARE
378. *Papilio (Heraclides) esperanza* VULNERABLE

U.S.A.

23, 24, 37, 39(?), 61, 208, 211, 352, 354–359, 382, 384, 385, 388, 391, 407, 428, 430–432, 434, 436–439 (28–29 species)
385. *Papilio (Heraclides) aristodemus ponceanus* ENDANGERED (Florida)

ISLANDS OF THE CARIBBEAN

Bahamas

210, 211, 379, 385, 391 (5 species)

Bermuda
>391 (1 species)

Cayman Islands
>379, 380, 385 (3 species)

Cuba
>65, 210, 211, 225, 379, 383, 385, 386, 388, 391, 399, 400, 432 (13 species)
>386. *Papilio (Heraclides) caiguanabus* INDETERMINATE

Hispaniola (Dominical Republic and Haiti)
>63, 209, 211, 380, 383, 385, 387, 399 (8 species)
>209. *Battus zetides* VULNERABLE
>387. *Papilio (Heraclides) aristor* INDETERMINATE

Jamaica
>64, 211, 374, 379, 381, 388, 399 (7 species)
> 64. *Eurytides (Protesilaus) marcellinus* VULNERABLE
>374. *Papilio (Pterourus) homerus* ENDANGERED

Lesser Antilles
>211, 383, 388 (3 species)

Puerto Rico
>211, 385(?), 399 (2–3 species)

Trinidad and Tobago
>76, 83, 98, 211, 218, 221, 248, 259, 267, 383, 388, 390(?), 394, 407 (13–14 species)

CENTRAL AND SOUTH AMERICA

Due to the relatively small size of its constituent countries, Central America has been treated regionally.

Central America
>(Belize, Costa Rica, El Salvador, Guatemala, Honduras, Nicaragua and Panama)
>62, 66, 68, 69, 71, 76, 81, 83, 87–90, 96–98, 100, 102, 104, 108, 109, 211, 218–221, 227, 228, 247, 248, 250, 251, 254, 260, 268, 354, 356, 357, 360, 367–371, 373, 382–384, 388, 391, 392, 394, 402, 404(?), 407, 409–411, 432 (57–58 species)

Argentina
>70, 71, 73, 74, 76(?), 83, 86, 91, 99, 111, 211, 213, 214, 216, 219, 222, 223, 230, 233, 235, 243, 248, 257–259, 267, 364, 365, 368, 382, 383, 388, 394–397, 407 (36–37 species)
>397. *Papilio (Heraclides) himeros* VULNERABLE

Bolivia
>71, 76, 80, 83, 84, 92, 93, 95, 98, 101, 103, 104, 106, 111, 211, 216, 218, 219, 221, 241, 244, 248, 252, 256, 257, 259, 262, 267, 362, 363, 365, 368(?), 369, 375, 382, 383, 388, 389, 392–394, 396, 398, 410 (43–44 species)
>241. *Parides steinbachi* INSUFFICIENTLY KNOWN
>393. *Papilio (Heraclides) garleppi* INSUFFICIENTLY KNOWN

Brazil

70–80, 82–86, 91, 92, 98, 99, 101, 106, 111, 112, 211, 214, 218, 219, 221–223, 230, 231–237, 239, 240, 243, 244, 246, 248, 249, 252, 255–259, 261–267, 364, 368, 369, 382, 383, 388, 389, 393–398, 407 (74 species)

 91. *Eurytides (Protesilaus) lysithous harrisianus* ENDANGERED
 112. *Eurytides (Eurytides) iphitas* VULNERABLE
 232. *Parides ascanius* VULNERABLE
 236. *Parides hahneli* RARE
 240. *Parides pizarro* INSUFFICIENTLY KNOWN
 249. *Parides burchellanus* VULNERABLE
 393. *Papilio (Heraclides) garleppi* INSUFFICIENTLY KNOWN
 397. *Papilio (Heraclides) himeros* VULNERABLE

Chile

211, 213, 388(?) (2–3 species)

Colombia

67, 68, 71, 76, 80, 83, 84, 90, 92, 93, 96–98, 100, 103–108, 111, 211, 218–221, 244, 247, 248, 251, 252, 254, 255, 259, 260, 264, 267, 268, 360, 362, 363, 367–370, 376, 377, 382, 383, 388, 390–392, 394, 406, 407, 410, 411, 432 (59 species)

Ecuador

71, 75, 76, 80, 83, 84, 92–98, 100, 101, 103, 104, 106, 107, 110, 111, 211, 215, 218, 219, 221, 229, 238, 244, 247, 248, 251–257, 259, 264, 267–269, 360–363, 366, 369, 375–377, 382, 383, 388, 389, 392, 394, 401, 405, 407, 410, 411, 432 (64 species)

French Guiana

71, 76, 80(?), 83, 84, 92, 98, 111, 211, 218, 219(?), 237, 242, 244, 248, 252, 259, 261, 264, 265, 267, 268, 360, 382, 383, 388, 393, 394, 398, 407, 408 (29–31 species)

 242. *Parides coelus* INSUFFICIENTLY KNOWN
 393. *Papilio (Heraclides) garleppi* INSUFFICIENTLY KNOWN
 408. *Papilio (Heraclides) maroni* INSUFFICIENTLY KNOWN

Guyana

71, 76, 80, 83, 84, 92, 98, 111, 211, 218, 219, 237, 244, 248, 252, 255, 259, 261, 264, 265, 267, 268, 360(?), 369, 382, 383, 388, 393, 394, 398, 407 (30–31 species)

 393. *Papilio (Heraclides) garleppi* INSUFFICIENTLY KNOWN

Paraguay

70(?), 71, 72, 73(?), 74, 76, 82–86, 91, 99, 111, 211, 214, 219(?), 222, 223, 230, 248(?), 258, 259, 267, 365(?), 368(?), 382, 383, 388, 394, 395, 407 (26–32 species)

Peru

71, 76, 80, 83, 84, 92–95, 98, 101, 103, 104, 106, 110, 111, 211, 212, 215, 216, 218, 219, 221, 229(?), 238–240, 244, 246, 248, 251–253, 255–257, 259, 262, 264, 267, 362, 363, 366, 369, 375–377, 382, 383, 388, 389, 392, 394, 398, 405, 407, 410, 432 (58–59 species)

 240. *Parides pizarro* INSUFFICIENTLY KNOWN

393.   *Papilio (Heraclides) garleppi* INSUFFICIENTLY KNOWN

Surinam
>71, 76, 80, 83, 84, 92, 98, 111, 211, 218, 219, 221, 237, 244, 248, 252, 259, 261, 264, 265, 267, 268, 360(?), 369, 382, 383, 388, 393, 394, 398, 407 (30–31 species)
>393.   *Papilio (Heraclides) garleppi* INSUFFICIENTLY KNOWN

Uruguay
>211, 222, 223, 364, 368(?), 382, 388, 395 (7 or 8 species)

Venezuela
>67, 71, 76, 80(?), 83, 84, 90, 92, 96, 98, 106, 107(?), 111, 211, 218, 219, 221, 245, 248, 251, 252, 254, 255(?), 259, 260, 267, 268, 362, 370, 376, 382, 383, 388, 390, 392, 394, 398(?), 407, 432 (35–39 species)
>245.   *Parides klagesi* INSUFFICIENTLY KNOWN

INDIAN OCEAN ISLANDS

Including Madagascar, Andaman and Nicobar Islands. For Sri Lanka, see section on Middle Asia to Indochina and Japan.

Andaman and Nicobar Islands
>125, 186, 189, 300–302, 314, 330, 415, 462, 463, 473, 484, 539 (14 species)
>125.   *Graphium (Pathysa) epaminondas* INSUFFICIENTLY KNOWN (Andamans)

Comoros
>166, 507(?), 516, 545 (3–4 species)
>166.   *Graphium (Arisbe) levassori* VULNERABLE
>516.   *Papilio (Princeps) aristophontes* INDETERMINATE

Madagascar
>143, 163, 179, 270, 507, 517, 535–538, 541, 542, 545 (13 species)
>535.   *Papilio (Princeps) morondavana* VULNERABLE
>536.   *Papilio (Princeps) grosesmithi* RARE
>542.   *Papilio (Princeps) mangoura* RARE

Mauritius
>508, 538 (2 species)
>508.   *Papilio (Princeps) manlius* INDETERMINATE

Réunion
>509, 538 (2 species)
>509.   *Papilio (Princeps) phorbanta* VULNERABLE

PACIFIC OCEAN ISLANDS

All islands east of the Solomons, north to Hawaii.

Fiji
>491 (1 species)

Hawaiian Islands
　　440, 539 (2 species)

New Caledonia (including the Loyalty Islands)
　　192, 412(?), 490, 573 (3–4 species)

Vanuatu (*New Hebrides*)
　　203, 484, 485, 487 (4 species)

Western Samoa
　　492 (1 species)

Other Pacific Islands (including Lord Howe, Norfolk, Guam etc.)
　　440, 473, 490, 498, 539 (5 species)

4

Analysis of critical faunas

The aim of this section is to analyse the world distribution of swallowtails so as:

1) to identify countries with particularly rich, valuable or threatened swallow-tail faunas;

2) to present a detailed analysis of the distribution and points of interest in the swallowtail fauna of a single country, Indonesia;

3) to present selective notes on interesting aspects of a few other countries recognized as having important swallowtail faunas.

Similar methods may be applied to any country and to other groups of animals or plants. Baseline data for an analysis may be obtainable from the IUCN Conservation Monitoring Centre.

Introduction

In their work of defining geographical units for conservation Dasmann (7, 8, 9) and Udvardy (28) recognized the need to consider the distribution of species as well as the distribution of ecosystems and biomes. Ideally, all species and all ecosystems should be catalogued, but of course such a task is impossible. Many groups of species remain too poorly known to be catalogued, and under man's influence ecosystems are changing too quickly for their boundaries to be identified. Udvardy finally proposed a system of biogeographical realms, provinces and biomes that included some consideration of the distribution of species, but relied heavily on an analysis of angiosperms, mammals and birds. Molluscs were the only invertebrates that received some attention (28). In terms of biomass, species richness, energy and nutrient flux arthropods dominate the animal life of terrestrial ecosystems from tropical rain forests to antarctic regions, yet they received no study whatsoever. Udvardy was aware of the anomaly, and implied that groups such as arthropods might reveal zoogeographical entities markedly different from those of land vertebrates (28). Certainly they would be expected to reveal patterns in finer detail. To some extent this has been borne out. For example, in a study of the Indo–Australian area it has been shown that although the faunal elements of butterflies and birds are well-matched, those of bats are somewhat simpler because of their more recent evolution and spread (14). In this book we cannot hope to redress the imbalance, but we shall demonstrate that simple analyses of insect groups can give information of value to conservation planners and administrators.

The phylum Arthropoda is a vast assemblage of invertebrates comprising about one million described species from a living total generally estimated at 2–5 million (18). Higher estimates of up to 30 million have been made (11), but have little credence so far. At least three quarters of these species are insects and it is perhaps because of this high diversity, rather than in spite of it, that they have so far received little attention in biogeographical analyses for conservation purposes. The diversity of insects certainly requires that survey work for conservation assessments needs to be highly selective, but there are no grounds for the view that insects (or other invertebrates) are too complex for their use to be contemplated. Certain groups are sufficiently well known ecologically and taxonomically, sufficiently visible and obtainable, to be of great value both for conservation planning from biogeographical analyses and for environmental monitoring on a more local scale (13). There is a vast literature on the zoogeography of insects, from the early observations on patterns by the father of zoogeography, A.R. Wallace (29), to modern numerical and cladistic analyses. Regrettably little of this work concerns itself with conservation, although some is certainly of considerable significance to the subject. It is likely that different groups would be appropriate 'indicators' for the various biogeographical realms as well as for different ecological habitats. For example, the phytophagous Lepidoptera might prove valuable for terrestrial habitats but whereas the milkweed butterflies (Danainae) might be chosen for a study in South East Asia, they would be of little use in South America, where the related glasswings (Ithomiinae) are more diverse. For a number of reasons swallowtail butterflies have great potential for inclusion in faunistic analyses and environmental monitoring: 1) they are large, often brightly patterned and recognizable on the wing, 2) the males can sometimes be baited or decoyed, 3) they inhabit a wide range of latitudes, altitudes and biotopes, 4) there are numerous local, as well as widespread, species, subspecies and forms, 5) they often have specific requirements for foodplants, but the variety of plant families used is wide, particularly in temperate regions.

In Europe and North America in recent years there have been numerous *ad hoc* efforts to protect important insect localities. The butterflies of San Bruno Mountain in San Francisco and the Large Blue (*Maculinea arion*) sites in Britain are good examples (30). In these cases the sites were identified because butterflies were known to be rare and threatened. If conservation is to advance as a practical but scientific discipline methods must be adopted that will identify conservation objectives in advance, not just in response to crises. As the World Conservation Strategy recommends, conservation objectives must be combined with social and economic objectives in formulating development policies (15).

There are very few general surveys or biogeographical studies of insects that have played a major role in identifying key conservation requirements. One of the most detailed is the study by Pyle (22, 23) of the butterflies of Washington State, U.S.A. By superimposing the range limits of 116 species Pyle identified 17 biogeographic provinces. These did not compare closely with the major North American provinces of Dice (10), which were among the first faunistic divisions to be named for the state. A resemblance was found with Dalquest's (6) mammalian provinces, but his western Washington province included all or part of nine of Pyle's butterfly provinces. Different groups of organisms will never reveal identical faunal provinces, but there are clear indications that butterflies offer an excellent level of detail. Pyle's work showed that the proportion of protected areas varied very widely in the butterfly provinces, from 0.2 per cent to 57 per cent. From this information it was possible to recommend focal points for conservation action.

Pioneering work by K.S. Brown Jr. on the distribution of neotropical forest butterflies has played a large part in locating those areas of South American rain forest that survived the arid Pleistocene era, served as refugia for rain forest animals and plants, and are still rich in species (2, 3, 4, 5). The identification of the Peruvian rain forest refugia was based on the distribution of glasswing butterflies (Nymphalidae: Ithomiinae) (4). Lamas (16) extended the study to cover the whole of Peru and a large number of species of Nymphalidae, Papilionidae and Pieridae. As in Pyle's study, the distribution ranges of the butterflies were mapped and biogeographical provinces identified. Twelve provinces were delimited in lowland, lower montane and upper montane forests, comparing favourably with Brown's presumed tropical forest refuge areas for Peru (4). A further 36 provinces were identified in the lowland, lower montane, upper montane and paramo-puna non-forest zones (16). All 48 provinces were clearly distinguishable on the basis of their butterfly faunas, many of which included endemic species or subspecies. A logical extension of the analysis would be to examine the protected areas within the 48 provinces and plan a conservation strategy for the future.

A larger scale but complementary approach has been developed by Ackery and Vane–Wright in their worldwide study of the milkweed butterflies (1). This group of insects is centred on South East Asia and as a consequence of a thorough distributional analysis it was possible to identify "critical faunas" for conservation, based on the whereabouts of narrowly distributed species. The technique is particularly valuable in identifying larger areas of responsibility, such as nations, islands or states.

In the ensuing pages we shall try to combine these two methods in an effort to demonstrate how they may be used to identify priorities in a conservation strategy based on swallowtails. The method is intended to be applicable to any other group of animals or plants, indeed it may prove to be most effective when a range of life forms is analysed together. We shall identify countries with critical swallowtail faunas and try to examine the distribution and vulnerability of swallowtails within a few of those countries. It will not be possible to analyse the distributions as finely as Pyle did for Washington (22, 23) or Lamas did for Peru (16), nor would it be particularly valuable to do so without the identification of a precise need. What we wish to demonstrate and emphasize is that swallowtail and other butterflies can help in identifying conservation priorities, and that the methodology to do so is available.

Methods

This analysis is essentially an assessment of the distribution of endemic species. It is a form of biogeographical analysis but lacks the temporal element, seeking only to describe the distribution of present faunas, not their origins. Ackery and Vane–Wright used physical boundaries to delimit their units of endemism (1) but we choose to adopt political boundaries at the first level of analysis. Hence, whereas Ackery and Vane–Wright have analysed, for example, the faunas of Borneo, Sulawesi and Luzon, our analysis begins with Malaysia, Indonesia and the Philippines. In the second level of the analysis we examine the distribution of swallowtails in regions, provinces or other administrative blocks that comprise those countries with important faunas. The reason for choosing national and provincial boundaries is that these are the units within which conservation is administered. Two exceptions have been made in cases where countries are under the control of a remote government. The French Overseas Departments of French Guiana, Réunion and New Caledonia have been considered individually, as have the Andaman and Nicobar Islands, which are administered by India.

In following the method of analysis it will be useful to refer to Table 4.1. The initial object is to find the smallest number of countries that includes all the 573 species of swallowtails recognized in this work. To do this, the distribution of endemics is analysed and the countries are listed in descending order of richness (Table 4.1, column A). Where more than one country has the same number of endemics the countries are listed in descending order of their total number of species (column F). Countries with endemic species clearly must accept ultimate responsibility for the conservation of those species, but if their endemic faunas could be positively conserved then their non-endemic species might also be protected by the same action. Hence in Indonesia (country number 1) 53 endemic swallowtails are found together with 68 non-endemic species (column B). Thus a total of 121 species have already been included (column E). In the Philippines (country number 2) 21 endemics are found, plus 28 non-endemics. However, 24 of those non-endemics are also found in Indonesia (column C), leaving only four non-endemics and 21 endemics, a total of 25 species (column D) newly accounted for and to be added to the cumulative total (column E). In all, 43 countries contain one or more endemic swallowtails, and by repeating the analysis for each country in turn 561 species may be accumulated. The remaining 12 species are distributed between countries with no endemic swallowtails and are to be found in a minimum of eight countries, making 51 countries in all.

It should be noted that the listing is only one of many possible forms of analysis and should not detract from other considerations. For example, Burma (country number 44) has no endemics but has the fifth largest swallowtail fauna in the world and could clearly play a very important role in conservation. By contrast, Italy (country number 51) has no endemics and a small total fauna of eight species, but two of these are threatened, *Parnassius apollo* (Rare) and *Papilio hospiton* (Endangered), the latter species requiring priority conservation action. *Papilio hospiton* is an example of a butterfly with a very narrow distribution that crosses national borders, in this case Corsica (France) and Sardinia (Italy); another good example is *Papilio leucotaenia* (Vulnerable) which has a narrow range spanning the borders of four central African countries. Clearly this analysis of critical faunas should wherever possible be used in addition to careful and independent national assessments.

We cannot emphasize too strongly that this analysis provides no excuse for disregarding the conservation needs of swallowtail butterflies in unlisted countries.

Results

As Table 4.1 shows, the five countries with the highest swallowtail species endemism are Indonesia (53), Philippines (21), China (15), Brazil (11) and Madagascar (10). If their national conservation plans could ensure the protection of swallowtails these five countries alone could between them account for 309 swallowtail species, almost 54 per cent of the world total. A further five countries, India, Mexico, Taiwan, Malaysia and Papua New Guinea, bring the total to 68 per cent. Increments decrease as further blocks of five countries are added, with 15, 20, 25, 30, 35, 40, and 45 countries respectively including 77, 90, 93, 95, 96, 97 and 99 per cent of the world's swallowtails.

Having identified these 51 countries of particular importance we would ideally have presented an analysis of each, examining the present conservation status of swallowtails within those countries and making recommendations for action. However, this would be a major undertaking beyond our present resources. Instead we offer below a single moderately detailed study, followed by brief notes on selected countries. Should this prove to be of interest to other countries they might choose to include a more detailed analysis in the formulation of their own

Table 4.1.   Analysis of critical swallowtail faunas

	A Endemic species	B Non-endemic species Not found in previous countries	C Non-endemic species Also found in previous countries	D Newly account-able species $(A+B)$	E Cumulative species-list $(D_n + D_{n+1})$	F Total species per country $(A+B+C)$
1. Indonesia	53	68	0	121	121	121
2. Philippines	21	4	24	25	146	49
3. China	15	61	28	76	222	104
4. Brazil	11	63	0	74	296	74
5. Madagascar	10	3	0	13	309	13
6. India	6	8	63	14	323	77
7. Mexico	5	37	10	42	365	52
8. Taiwan	5	0	27	5	370	32
9. Malaysia	4	1	51	5	375	56
10. P.N.G.	4	8	25	12	387	37
11. U.S.A.	4	6	19	10	397	29
12. Cuba	4	3	6	7	404	13
13. Ecuador	3	19	42	22	426	64
14. Colombia	3	10	46	13	439	59
15. Australia	3	2	14	5	444	19
16. Andaman & Nicobar	3	0	11	3	447	14
17. Jamaica	3	0	4	3	450	7
18. Zaire	2	44	2	46	496	48
19. Cameroon	2	2	35	4	500	39
20. U.S.S.R.	2	11	22	13	513	35
21. Tanzania	2	11	21	13	526	34
22. French Guiana	2	0	29	2	528	31
23. Japan	2	0	20	2	530	22
24. Canada	2	0	16	2	532	18
25. Sri Lanka	2	0	13	2	534	15
26. Haiti	2	2	4	4	538	8
27. New Caledonia	2	0	2	2	540	4
28. Comoro Is	2	0	2	2	542	4
29. Peru	1	1	57	2	544	59
30. Bolivia	1	0	43	1	545	44
31. Venezuela	1	0	38	1	546	39
32. Gabon	1	0	30	1	547	31
33. Bhutan	1	0	29	1	548	30
34. Ghana	1	1	22	2	550	24
35. Solomon Is	1	0	14	1	551	15
36. Afghanistan	1	0	18	1	552	19
37. South Africa	1	1	10	2	554	12
38. Iran	1	1	7	2	556	9
39. Vanuatu	1	0	3	1	557	4
40. Mauritius	1	0	1	1	558	2
41. Réunion	1	0	1	1	559	2
42. Fiji	1	0	0	1	560	1
43. Western Samoa	1	0	0	1	561	1
44. Burma	0	2	66	2	563	68
45. Laos	0	2	46	2	565	48
46. Honduras	0	1	33	1	566	34
47. Argentina	0	1	36	1	567	37

Table 4.1.   (cont.)

	A	B	C	D	E	F
	Endemic species	Non-endemic species		Newly account-able species $(A+B)$	Cumulative species-list $(D_n + D_{n+1})$	Total species per country $(A+B+C)$
		Not found in previous countries	Also found in previous countries			
48. Uganda	0	2	30	2	569	32
49. Ethiopia	0	1	15	1	570	16
50. Mozambique	0	1	15	1	571	16
51. Italy	0	2	6	2	573	8

conservation strategies, calling upon IUCN's information resources when needed. Much of the information needed to formulate such an analysis is already contained within this book. Lists of endemics and other species of interest can easily be generated by referring to section 3.

Indonesia   The swallowtails of Indonesia and their distribution between the islands are given in Table 4.2. Further information on these species will be found in the main species list in section 3. Indonesia as a whole has 44 per cent endemism in the swallowtails, a very high level which compares closely with the levels for mammals and birds (17).

In Table 4.3 the relationships between the swallowtail faunas of seven Indonesian provinces are summarized. The geography of Indonesia has made the demarcation of these provinces relatively simple. The range of each province is defined at the bottom of Table 4.2. There are a few departures from Indonesian administrative provinces, notably in the Moluccas (Maluku) and Lesser Sundas, but we have followed the divisions made in the National Conservation Plan for Indonesia (27).

Sumatra shares 42 of its 49 species with Malaysia, 36 of them with Peninsular Malaysia and 34 with Eastern Malaysia. Only two of these species, *Troides aeacus* and *Papilio iswarioides*, are not found on other Indonesian islands. Of the remaining seven, *Graphium cloanthus* has a strange disjunct range mainly in India, Indo–China and China and the other six are Indonesian endemics. Four are found only on Sumatra, namely *Graphium sumatranum*, *Atrophaneura hageni*, *Papilio forbesi* and *Papilio diophantus*. Two species, *Atrophaneura priapus* and *Troides vandepolli*, are shared with Java. Both *Papilio forbesi* and *P. diophantus* are from northern Sumatra, which includes the Gunung Leuser National Park, over 8000 sq. km in extent. *Papilio forbesi* females mimic *A. hageni* which is restricted to the high plateau and probably also occurs in Gunung Leuser as well as other highland areas. *P. forbesi* may be present in the Sumatra Selatan Proposed National Park in the southwest corner of the island. This reserve is over 3500 sq. km in extent and covers the southern end of the Barisan range, which reaches lower altitudes than in the north. Unfortunately there has been extensive damage in certain parts of the reserve, mainly from agriculturalists (23, vol. 2). In general Sumatra's butterfly fauna is likely to be well protected. However, much of its fauna is shared with Peninsular Malaysia and with the increasing extent of logging and the imminent disappearance of virtually all the lowland forest in that state, Sumatra's forests will, in a world context, become increasingly important as conservation zones.

Kalimantan has no known endemic swallowtails, sharing all its species with the Eastern Malaysian states of Sarawak and Sabah. Thirty-five of its 36 species are

Table 4.2. Distribution of swallowtail butterflies within Indonesia

	IN	SM	KA	JA	LS	SW	MO	IJ
Order LEPIDOPTERA								
Family Papilionidae								
Meandrusa payeni	+	+	+	+				
Lamproptera meges	+	+	+	+		+		
Lamproptera curius	+	+	+	+				
Graphium aristeus	+	+	+	+	+		+	+
*Graphium rhesus**	+					+	+	
*Graphium dorcus**	+					+		
*Graphium androcles**	+					+	+	
Graphium euphrates	+					+	+	
Graphium antiphates	+	+	+	+	+	+		
Graphium agetes	+	+	+					
*Graphium encelades**	+					+		
Graphium delesserti	+	+	+	+				
Graphium macareus	+	+	+	+				
Graphium ramaceus	+	+	+					
Graphium megarus	+	+	+	+				
Graphium deucalion	+					+	+	+
Graphium thule	+							+
Graphium arycles	+	+	+	+				
Graphium bathycles	+	+	+	+				
*Graphium meyeri**	+					+	+	
Graphium eurypylus	+	+	+	+	+	+	+	+
Graphium doson	+	+	+	+	+			
Graphium evemon	+	+	+	+				
Graphium agamemnon	+	+	+	+	+	+	+	+
Graphium macfarlanei	+						+	+
Graphium weiskei	+							+
*Graphium stresemanni**	R						R	
Graphium codrus	+					+	+	+
Graphium empedovana	+	+	+	+				
Graphium cloanthus	+	+						
*Graphium sumatranum**	+	+						
*Graphium monticolum**	+					+		
*Graphium milon**	+					+	+	
Graphium sarpedon	+	+	+	+	+	+	+	+
Graphium wallacei	+				+	+		
Cressida cressida	+				+			
*Atrophaneura kuehni**	+					+		
*Atrophaneura luchti**	R		R					
*Atrophaneura hageni**	+	+						
*Atrophaneura priapus**	+	+		+				
Atrophaneura sycorax	+	+		+				
Atrophaneura nox	+	+	+	+				
*Atrophaneura dixoni**	+					+		
Atrophaneura neptunus	+	+	+					
*Atrophaneura palu**	K					K		
Atrophaneura coon	+	+		+				
*Atrophaneura oreon**	+				+			
*Atrophaneura liris**	+				+			
*Atrophaneura polyphontes**	+					+	+	
Atrophaneura aristolochiae	+	+	+	+	+	+		

143

Table 4.2. (cont.)

	IN	SM	KA	JA	LS	SW	MO	IJ
Atrophaneura polydorus	+			+	+		+	+
Trogonoptera brookiana	+	+	+					
*Troides hypolitus**	+					+	+	
Troides cuneifera	+	+		+				
Troides amphrysus	+	+	+	+				
Troides miranda	+	+	+					
*Troides prattorum**	I						I	
Troides aeacus	+	+						
*Troides dohertyi**	V					V		
Troides helena	+	+	+	+	+	+		
Troides oblongomaculatus	+					+	+	+
*Troides criton**	+					+		
*Troides riedeli**	+				+			
*Troides plato**	+				+			
*Troides vandepolli**	+	+		+				
*Troides haliphron**	+				+		+	
*Troides staudingeri**	+				+			
*Ornithoptera goliath**	+						+	+
*Ornithoptera tithonus**	K							K
*Ornithoptera rothschildi**	I							I
Ornithoptera chimaera	I							I
Ornithoptera paradisea	I							I
Ornithoptera meridionalis	V							V
*Ornithoptera aesacus**	I						I	
*Ornithoptera croesus**	V						V	
Ornithoptera priamus	+				+		+	+
Papilio clytia	+				+			
Papilio paradoxa	+	+	+					
*Papilio veiovis**	+						+	
Papilio slateri	+	+	+					
Papilio laglaizei	+							+
Papilio euchenor	+							+
*Papilio gigon**	+					+	+	
Papilio demolion	+	+	+	+	+			
*Papilio lampsacus**	+			+				
*Papilio forbesi**	+	+						
*Papilio oenomaus**	+				+			
*Papilio ascalaphus**	+					+	+	
Papilio rumanzovia	+						+	
*Papilio deiphobus**	+						+	
Papilio memnon	+	+	+	+	+			
*Papilio biseriatus**	+				+			
Papilio iswara	+	+	+					
*Papilio sataspes**	+					+	+	
Papilio helenus	+	+	+	+	+			
Papilio iswaroides	+	+						
*Papilio jordani**	R						R	
Papilio polytes	+	+	+	+	+		+	
Papilio alphenor	+					+	+	
Papilio ambrax	+							+
*Papilio diophantus**	+	+						
Papilio nephelus	+	+	+	+				

Table 4.2. (cont.)

	IN	SM	KA	JA	LS	SW	MO	IJ
Papilio hipponous*	+			+				
Papilio fuscus	+		+			+	+	+
Papilio canopus	+				+			
Papilio albinus	+						+	+
Papilio heringi*	+						+	
Papilio inopinatus*	+				+			
Papilio gambrisius*	+						+	
Papilio tydeus*	+						+	
Papilio aegeus	+							+
Papilio demoleus	+	+			+	+	+	
Papilio blumei*	+					+		
Papilio palinurus	+	+	+					
Papilio paris	+	+		+				
Papilio karna	+	+	+	+				
Papilio neumoegeni*	V				V			
Papilio peranthus*	+			+	+	+		
Papilio pericles*	+				+			
Papilio lorquinianus*	+						+	+
Papilio ulysses	+						+	+

Key:

IN: Indonesia
SM: Sumatra, including Nias, Simeulue, Siberut, Lingga, Bangka, Belitung, Riau, Batu, Enggano, Mentawai
KA: Kalimantan, including Bunguran (Natuna Is)
JA: Java and Bali, including Madura, Kangean and Bawean
LS: Lesser Sundas, including Lombok, Sumbawa, Sumba, Tenggara, Timor, Roti, Wetar, Leti, Moa—Babar, Tanimbar, Flores
SW: Sulawesi, including Talaud, Sangihe, Salayar, Banggai, Tanahjampea, Butung, Tukangbesi I.
MO: Moluccas, including Halmahera, Bacan, Sula, Buru, Ambon, Seram, Obi, Morotai
IJ: Irian Jaya, including Kai, Aru, Misool, Waigeo, Biak, Salawati
Note: Records indicate presence over part or all of the areas listed. Detailed records are available from our database on request. *Graphium procles* (Indeterminate), *Graphium stratiotes*, *Troides andromache* (Indeterminate) and *Papilio acheron* (Rare) are all endemic to Borneo but have not as yet been recorded on the Kalimantan side of the border and are, therefore, not listed above.
*:   Endemics to Indonesia
+:   Records exist for this region
V: Vulnerable in these regions
R: Rare in these regions
I: Indeterminate in these regions
K: Insufficiently Known in these regions

also shared with Sumatra (Table 4.3). Borneo as a whole has four endemics, *Graphium stratiotes*, *Graphium procles* (Indeterminate) *Troides andromache* (Indeterminate) and *Papilio acheron* (Rare). With the possible exception of *G.*

procles, which has a restricted distribution in the Crocker range of Sabah, all of these might well be expected to occur in northern Kalimantan, an area well served by the giant Sungai Kayan–Mentarang Reserve (4000 sq. km) (27, vol. 5). The borders of this reserve are undergoing reassessment and possible alteration as a result of the large human populations inside. A survey of the butterflies could add extra valuable criteria towards the final decisions.

The Javanese fauna also has close links with Sumatra and the mainland, sharing 30 of its 35 species with Malaysia. *Atrophaneura priapus* is found in southern Sumatra as well as Java and its relationship with the Sumatran species *A. hageni* is not at all clear. The Rare Javan endemic *A. luchti* from the mountains of eastern Java differs only slightly from *A. priapus* and there are doubts as to its taxonomic status. Regardless of this, the Ijen, Merapi and Moeleng reserves may be important to the survival of the butterfly. *Troides vandepolli* from north–western Java and western Sumatra is uncommon and protected by Indonesian law. *Papilio lampsacus* is endemic to western Java, occurring only on Gunung (Mt) Gede and Gunung Mas. The Ujung Kulon and Gunung Gede–Pangrango National Parks may effectively protect the habitat of both these species. Their status should be carefully monitored and immediate action taken against habitat disruption, particularly on Gunung Gede.

As might be expected from the biogeographical considerations, the regions lying on the Asian continental plate (the 'Sundaic' regions of Malaysia, Sumatra, Borneo, Palawan, Java and Bali) show strong faunistic similarities in their swallowtails, but quite marked differences begin to develop in the 'Wallacean' regions eastwards from the Lesser Sundas and Sulawesi. Of the 30 species in the Lesser Sundas only 14 are shared with Malaysia. Some species with a mainly Australasian distribution begin to appear, such as *Cressida cressida* and *Papilio canopus*. Of the nine endemics, *Troides riedeli* (Tanimbar Is), *T. plato* (Timor) and *T. staudingeri* (Babar group) are protected by law. Widespread Lesser Sundas endemics include the common *Atrophaneura liris* and *Papilio oenomaus* and the scarcer *P. pericles*. *Atrophaneura oreon* is found only on Sumba, Alor, Flores and Solor, *Papilio biseriatus* only on

Table 4.3. Patterns of species richness and distribution in Indonesian swallowtails

Province[1]	No. of Species	Endemics No.	%	Non-endemics	No. of species shared with						
					1	2	3	4	5	6	7
1. Sumatra	49	4	8	45	—	32	13	9	5	4	
2. Kalimantan	36[2]	0	0	36[2]	35	—	26	12	9	5	5
3. Java (+ Bali)	35	2	6	33	32	26	—	13	9	4	4
4. Lesser Sundas	30	10	33	20	13	12	13	—	10	7	6
5. Sulawesi	38	11	29	27	9	9	9	10	—	19	6
6. Moluccas	37	9	24	29	5	5	4	6	19	—	15
7. Irian Jaya	26	2	8	24	4	5	4	5	6	15	—
8. Indonesia	121[2]	53[3]	44	68[2]							

[1] Limits of provinces are defined in Table 4.2

[2] Not including *Graphium stratiotes*, *Graphium procles*, *Troides andromache* or *Papilio acheron*, all of which are endemic to Borneo but have not been recorded on the Kalimantan side of the border.

[3] 37 species are endemic to only one of areas 1–7; a further 16 species are endemic to Indonesia but occur in more than one of areas 1–7.

Timor, *P. inopinatus* on Romang, Babar, Damar and Tanimbar and *P. neumoegeni* only on Sumba. The last of these species is category Vulnerable and the other three are certainly in need of closer study and assessment. In general the Lesser Sundas are under-represented in terms of protected areas. The Indonesian Conservation Plan (27, vol. 4) includes many recommendations for new reserves and parks, including acquisition of new reserves on Sumbawa, Flores, Timor and Sumba. The sites should be surveyed for local swallowtails.

Sulawesi and its associated islands (see Table 4.2) contain 11 endemic swallowtails, second only to New Guinea and its associated islands, which have 14. The rich endemic fauna of Sulawesi is possibly the result of a very complex geological history which has included the fusion of island arcs from both the Sunda and Australasian continental shelves. In the swallowtails there is clear evidence of a stronger relationship with the Australasian region, 19 species being shared with the Moluccas (including Sula) but only 9 with Kalimantan (Table 4.3). Unfortunately many Sulawesi biotopes have been and continue to be destroyed by deforestation (25). Large areas stretching between Ujung Pandang (Makassar) and the Northern and Central Districts are covered by the unpalatable and useless grass known as lalang (*Imperata cylindrica*) (25), implying severe overexploitation and deterioration of soils.

Two endemic species of *Graphium*, *G. dorcus* and *G. monticolum* are mainly restricted to the high mountains of Sulawesi, *dorcus* in the northern and central areas, rarely southern, *monticolum* mainly central and southern above 1000 m (12). *G. dorcus* is rare and its habitat is not properly known, but *G. monticolum* is quite common (12). Two endemics are widespread in Sulawesi, *Papilio veiovis* and *Papilio blumei*. They are likely to be recorded in a number of parks and reserves. Southern Sulawesi is believed to be the main area for *Graphium encelades*, but the female is unknown and males are very rare. More data are needed to ascertain its conservation status. Two *Atrophaneura* species, *A. dixoni* and *A. palu* (Insufficiently Known), have a mainly central Sulawesi distribution. Both are uncommon, the latter particularly so. The Lore–Lindu Proposed National Park in central Sulawesi may be an important locality for these species. Northern Sulawesi has a particularly vulnerable swallowtail fauna, including the endemics *Atrophaneura kuehni* (very rare, also eastern), *A. dixoni* (also central) and *Papilio jordani* (Rare). The Dumoga–Bone National Park in northern Sulawesi, set up in a joint venture between the Indonesian Government, WWF/IUCN and the World Bank, is likely to be of particular value to conservation of the region's wildlife. In eastern Sulawesi the Morowali Reserve may prove to be of equal value. Sulawesi is the main centre for the widespread species *Papilio ascalaphus*, an Indonesian endemic also found on the Salayar and Sula islands. Just north of Sulawesi lie the Talaud and Sangihe Islands, which have two important endemic swallowtails, *Troides dohertyi* (Vulnerable) and *Papilio hipponous*. *Papilio alphenor* is also found here, as well as in the Philippines and the Moluccas, but not on the Sulawesi mainland. The need for a reserve on Sangihe island is recognized in the Indonesian Conservation Plan and Gunung Sahendaruman has been suggested (27, vol. 6).

The Moluccas and Sulawesi share a number of important Indonesian endemics between them. Some of them are only on Sulawesi and the Sula Islands, not reaching further east, namely *Graphium rhesus*, *G. androcles*, *G. meyeri*, *Papilio gigon*, *P. ascalaphus* and *P. sataspes*. These species would need to be considered in a conservation plan for either the Moluccas or Sulawesi. Human populations are low in the Sula Islands and much of the area is thickly forested. A reserve is urgently needed and

147

has been proposed for Pulau Taliabu. Other Indonesian endemics extend from Sulawesi across to the main Moluccan islands, these including *G. milon*, *Atrophaneura polyphontes* and *Troides hypolitus*.

In addition, there are nine swallowtail species restricted to the Moluccas, out of a total of at least 25 endemic butterflies. Six of these are found on Halmahera and its associated islands, namely *Troides criton* (fairly common, protected by law), *Ornithoptera aesacus* (Obi only, Indeterminate, also protected), *O. croesus* (Vulnerable), *Papilio deiphobus*, *P. heringi* (very poorly known and rare) and *P. tydeus* (common). These islands have the richest fauna and flora in the Moluccas, including 35 (nearly half) of the province's endemic birds. Despite several proposals for establishing reserves in the group, none yet exist or have been approved. Deforestation through commercial timber exploitation is having an increasing impact on Halmahera, and the lack of protected areas is a matter for concern (24). In the southern part of the Moluccas three further endemic species are found. *Graphium stresemanni* (Rare) and *Papilio gambrisius* both occur on Seram and are probably present in the 1800 sq. km Manusela Reserve. The Indonesia Conservation Plan suggests another reserve in the mountains of the Sahuai Peninsula (27, vol. 7). Finally, *Troides prattorum* (Indeterminate) is endemic to Buru, a steeply forested mountainous island once used as a prison. There are no reserves on the island, but the proposed Gunung Kelapat Muda Reserve (27) would almost certainly protect this endemic birdwing. In general the superbly rich butterfly fauna of the Moluccas is very poorly protected. The existing reserve system is quite inadequate but the proposals in the Indonesian Conservation Plan, if accepted, would remedy this regrettable situation.

Irian Jaya has 26 of the 42 species found in New Guinea and its associated islands. Endemics to Irian Jaya and its islands include *Ornithoptera tithonus* (Irian Jaya, Waigeo, Salawati and Misoöl, Vulnerable) and *O. rothschildi*, (Arfak mountains only, western Irian Jaya, Indeterminate) both listed as threatened and considered in detail in section 6. A further three birdwing species, *O. chimaera* (Indeterminate), *O. paradisea* (Indeterminate) and *O. meridionalis* (Vulnerable) are New Guinea endemics also discussed in section 6. There are two endemic *Graphium* species in New Guinea, *G. thule* and *G. weiskei*, the latter being common and widespread. The former is poorly known and rarely collected, but may have been overlooked as it is a mimic of common species of Danainae. Finally, the two endemic *Papilio* species, *P. laglaizei* and *P. euchenor* are widespread and common, although the former, an interesting mimic of the day-flying moth *Alcidis agarthyrsus* (Uraniidae), is quite localized.

Irian Jaya already has a fine network of reserves, including the giant Lorentz Reserve (16 750 sq. km) on the south coast and reserves in western Waigeo, but there are no fully designated national parks. As the Indonesian National Conservation Plan says, Irian Jaya is the richest and most important province in terms of faunal and floral wealth (27, vol. 7). The magnificent birdwing butterflies are a very significant part of the fauna and it is vital to ensure that these are thoroughly protected. There are great opportunities for conservation in the province, but land use conflicts must be quickly resolved. The potential for wildlife industries is very great, to include birdwing ranching as well as crocodile farms and deer ranches.

To summarize, Indonesia has the finest swallowtail fauna in the world, certainly worthy of consideration in the country's conservation planning. On the Sunda shelf the Sumatran fauna is rich, the Kalimantan fauna poorly studied. The Javan endemics are as safe as the island's reserved areas, but those in the Lesser Sundas are poorly protected. Sulawesi has a very rich swallowtail fauna, probably well represented in parks and reserves, but poorly known. The northern Moluccas include important endemic swallowtails that are very poorly protected and require priority action. Irian Jaya has a spectacular fauna that appears to be quite well protected and has great potential for interpretation to local and foreign visitors and entomologists. Indonesia has at least 17 rare and threatened species of swallowtails. The main threat is loss of habitat as a result of agriculture, commercial logging and other forms of human impact.

Philippines With a total of 49 species and 21 endemics, nine of which are threatened (one Endangered, five Vulnerable, one Rare, two Indeterminate), the Philippines has one of the most important swallowtail faunas in the world (Table 4.4). An important factor contributing to this is that the Philippines span a wide latitudinal and biogeographical belt, from Palawan which is the only Philippine island on the Sunda shelf, to Luzon with its Asian mainland connections and Mindanao with submerged island arc connections to Sulawesi. This variety is evidenced by the frequent occurrence of pairs of closely related but distinct species that replace each other in Palawan (and Borneo) and the rest of the Philippines. Examples of such pairs are *Atrophaneura aristolochiae* and *A. kotzebuea*, *Troides plateni* (Palawan endemic) and *T. rhadamantus*, *Papilio helenus* and *P. hytaspes*. Parts of the fauna are still very poorly known, and the fact that new species have been discovered on various islands during the past 20 years implies that still more might be expected. Recently described species include *Graphium sandawanum* (1977, Vulnerable), *Papilio osmana* (1967, Vulnerable) and *P. carolinensis* (1967, Vulnerable), *P. chikae* (1965, Endangered). Three threatened species are considered to be relicts of the Ice Age climate, and they now survive only on high mountains. These are *Graphium sandawanum* (Vulnerable) on Mt Apo, and *Papilio benguetanus* (Vulnerable) and *P. chikae* (Endangered) in the Cordillera Central of Luzon. Such localities are particularly vulnerable to disturbance and slow to recover. Many of the central Philippine islands are almost completely turned over to agriculture: Cebu, for example, has no forest left. In many places inappropriate agricultural methods have degraded the soils to such an extent that only the unpalatable grass *Imperata cylindrica* will grow. Deforestation on Mindanao is of particular international concern and has been aggravated by extensive fires following the drought of 1982–3. Samar still has extensive forests with great potential for conservation.

Table 4.4. Distribution of swallowtail butterflies within the Philippines (after Tsukada and Nishiyama) (22)

	PH	LU	MR	SA	LE	PN	NE	CE	BO	MN	PL
Order LEPIDOPTERA											
Family Papilionidae											
Lamproptera meges	+	+	+	+	+	+	+	+	+	+	+
Lamproptera curius	+										+
Graphium aristeus	+	+					+			+	+
Graphium euphrates	+	+	+								+
Graphium decolor	+	+	+	+	+		+		+	+	+
Graphium idaeoides*	R	R		R	R					R	
Graphium delesserti	+										+
Graphium macareus	+										+
Graphium megaera*	I										I
Graphium stratocles*	+	+	+						+	+	+
Graphium arycles	+										+
Graphium bathycles	+										+
Graphium eurypylus	+	+	+	+	+	+	+	+	+	+	+

Table 4.4. (cont.)

	PH	LU	MR	SA	LE	PN	NE	CE	BO	MN	PL
Graphium doson	+	+	+	+	+	+	+	+	+	+	+
Graphium agamemnon	+	+	+	+	+	+	+	+	+	+	+
Graphium codrus	+	+	+	+	+	+	+	+	+	+	
Graphium empedovana	+										+
Graphium sandawanum*	V									V	
Graphium sarpedon	+	+	+	+	+	+	+	+	+	+	+
Atrophaneura semperi*	+	+	+	+	+	+	+	+	+	+	+
Atrophaneura neptunus	+										+
Atrophaneura schadenbergi*	V	V									
Atrophaneura mariae*	+	+		+	+					+	+
Atrophaneura phegeus*	+			+	+			+		+	
Atrophaneura phlegon*	+		+			+				+	
Atrophaneura atropos*	I										I
Atrophaneura kotzebuea*	+	+	+	+	+	+	+	+	+	+	
Atrophaneura aristolochiae	+										+
Trogonoptera trojana*	+										+
Troides magellanus	+	+		+	+			+	+	+	
Troides plateni*	+										+
Troides rhadamantus*	+	+	+	+	+	+	+	+	+	+	
Papilio clytia	+	+	+	+	+	+	+	+	+	+	+
Papilio paradoxa	+										+
Papilio osmana*	V			V						V	
Papilio carolinensis*	V									V	
Papilio benguetanus*	V	V									
Papilio demolion	+										+
Papilio rumanzovia	+	+	+	+	+	+	+	+	+	+	
Papilio lowi	+										+
Papilio antonio*	+				+					+	
Papilio hystaspes*	+	+	+	+	+	+	+	+	+	+	
Papilio helenus	+										+
Papilio alphenor	+	+	+	+	+	+	+	+	+	+	+
Papilio pitmani	+	+							+	+	+
Papilio demoleus	+	+	+	+	+	+	+	+	+	+	+
Papilio daedalus*	+	+	+	+	+	+	+	+	+	+	+
Papilio chikae*	E	E									
Papilio karna	+										+

Key:

PH: Philippines
LU: Luzon, including Batan Islands, Babuyan islands and Polillo Islands
MR: Mindoro
SA: Samar
LE: Leyte
PN: Panay
NE: Negros
CE: Cebu
BO: Bohol
MN: Mindanao
PL: Palawan, including Balabac and Calamian Islands

Note: Records indicate presence over part or all of the areas listed. Detailed records are available from our database on request.

Table 4.4. (cont.)

*:	Endemics to Philippines
+:	Records exist for this region
E:	Endangered in these regions
V:	Vulnerable in these regions
R:	Rare in these regions
I:	Indeterminate in these regions

The national park system in the Philippines was analysed in a 1982 report that drew attention to the deficiencies in the existing state of affairs (21). Considerable confusion existed on the purposes, objectives and management of national parks and there was evidence of substantial violation of protected area boundaries. Since that report the situation has apparently not improved. Sections of Mt Apo National Park (which has an endemic swallowtail) have been excised and other protected areas are reported to be neglected. It is difficult to obtain precise data on the protected areas of the Philippines, but we understand that the national system is under review. For the sake of the exquisite fauna and flora of the Philippines and of the natural resources of the Filipino people this review should be completed and its recommendations consolidated and enforced at the earliest opportunity. Conflicts between the exploitation and conservation of the natural resources of the Philippines are increasing day by day.

China China is a very large country with a wide range of habitats and has a huge swallowtail fauna of 104 species including 15 endemics (see section 3). Five species are threatened or Insufficiently Known, i.e. *Parnassius apollo* (Rare), *Bhutanitis mansfieldi* (Rare), *Bhutanitis thaidina* (Rare), *Luehdorfia chinensis* (Insufficiently Known) and *Teinopalpus aureus* (Insufficiently Known) (see section 6). China is particularly rich in *Parnassius* and has a fauna of 26 species including the three endemics *P. szechenyii*, *P. cephalus*, *P. przewalskii* and several rare and poorly known species, including *P. acdestis*, *P. loxias*, *P. acco*, *P. hannyngtoni*, *P. actius* and *P. tianschanicus*. All these species have a montane distribution in western and/or northern China, but there is too little information available to make specific conservation recommendations. In the genus *Graphium* three species are very poorly known, *G. mandarinus*, *G. alebion* and *G. leechi*, the last two being endemic. Five species of *Atrophaneura*, *A. daemonius*, *A. plutonius*, *A. mencius*, *A. impediens* and *A. hedistus*, are poorly known and/or rare. Some of these are also restricted to higher altitudes. Of the 25–27 species in the genus *Papilio*, *P. elwesi* (endemic), *P. syfanius* and *P. krishna* are rare and require further study, but most of the others have a wide distribution. Information on the ecology and precise distribution of Chinese swallowtails is very scarce. As a result it is difficult for an outsider to relate China's environmental problems to the conservation of her swallowtails. Suffice it to say that the Chinese swallowtail fauna is of international value and concern, stretching as it does from tropical climates to the snows of the Himalayas. The faunas of the western and northern mountainous regions and plateaux are of particular significance.

In recent years Japanese entomologists have forged links with Chinese in Sichuan and Yunnan, resulting in important collections of *Bhutanitis mansfieldi* and the slightly more common *B. thaidina*. In years to come the accessibility of these important regions will hopefully improve and opportunities for joint programmes of environmental assessment will be possible. It is important that commercial collecting is not encouraged in the absence of impact assessment and monitoring. There is potential for development of seasonal butterfly ranches in western China.

Brazil Brazil has a very rich swallowtail fauna of 74 species, including 11 endemics. Seven species and one subspecies have been given threatened status, three of the species also being found in other countries. Threats directly affecting these taxa are described in section 6. A preliminary analysis of the Brazilian fauna suggests that southern Brazil (from Rio Grande do Sul to São Paulo) has a surprisingly rich fauna of about 39 species; it is not only the Amazon basin that needs consideration. At one time the Atlantic seaboard of eastern Brazil was covered in a strip of semi-deciduous forest, but only 1–2 per cent remains intact. Many species of animals and plants are threatened with extinction, including primates like the Golden-headed Lion Tamarin (*Leontopithecus chrysomelas*) and the Woolly Spider Monkey (*Brachyteles arachnoides*). The beautiful butterfly *Parides ascanius* (Vulnerable) is in the same position. The forests of the Mato Grosso are also rich in species (about 37 in all), while still more records might be expected from Amazonas (about 43 species so far).

Madagascar Madagascar has only 13 swallowtails, but ten of these, (77 per cent) are endemic. The fauna has developed mainly from African stock with one notable exception, *Atrophaneura antenor*, a troidine species with relatives in Asia and South America, but not in Africa. A chain of mountains runs from north to south down the spine of Madagascar, the eastern side is wet and forested, the western side relatively dry with a variety of woodland and savanna ecosystems. Four endemics are mainly distributed in the dry west and central regions, namely *Papilio erithonioides*, *P. grosesmithi* (Rare), *P. morondavana* (Vulnerable) and *Atrophaneura antenor*. The last three species are traded commercially and require monitoring. Rain forest endemics of the eastern side include *Graphium endochus* (uncommon), *Papilio oribazus*, *P. delalandei* and *P. mangoura* (Rare). The other two endemics, *Graphium evombar* and *G. cyrnus* are widespread and common (19, 20). A number of reserves and other protected areas may ensure the perpetuation of these and other Malagasy endemics (see species accounts in section 6), but there is international concern at the alarming rate of degradation of Madagascar's vegetation and soils.

Conclusion

These notes on the swallowtail faunas of the world are necessarily brief. It is recognized that far more detail would be needed precisely to identify centres of species richness and importance within a country, and to plan a system of protected areas around those centres. These few examples must serve to indicate the potential conservation value of investigating swallowtails and other insect groups. Swallowtails not only require species-oriented conservation action, but can also act as indicators of general threats, and as pointers for delimitation of biogeographical boundaries for planning the conservation of whole biotopes, their faunas and floras. Addition of other groups of butterflies, insects or invertebrates (e.g. molluscs) will strengthen and diversify the analysis, particularly when the data are synthesized with a consideration of vertebrates and angiosperms. Critical fauna analyses will serve to direct national and international attention to faunistically important countries, states and provinces, but local knowledge is the basis for more detailed conservation planning. It is axiomatic that baseline data for conservation can only be obtained by the encouragement and completion of field research.

References

1. Ackery, P.R. and Vane–Wright, R.I. (1984). *Milkweed Butterflies: Their Cladistics and Biology*. British Museum (Natural History), London, and Cornell University Press. 448 pp.

2. Brown, K.S., Jr. (1975). Geographical patterns of evolution in Neotropical Lepidoptera. Systematics and derivation of known and new Heliconiini (Nymphalidae: Nymphalinae). *Journal of Entomology* 44: 201–242.

3. Brown, K.S., Jr. (1982). Historical and ecological factors in the biogeography of aposematic neotropical butterflies. *American Zoologist* 22: 453–471.

4. Brown, K.S., Jr. (1977). Geographical patterns of evolution in neotropical Lepidoptera: Differentiation of the species of *Melinaea* and *Mechanitis* (Nymphalidae, Ithomiinae). *Systematic Entomology* 2: 161–197.

5. Brown, K.S., Jr. and Ab'Saber N.A. (1979). Ice-age forest refuges and evolution in the neotropics: Correlation of paleoclimatological, geomorphological and pedological data with modern biological endemism. *Paleoclimas* 5: 1–30.

6. Dalquest, W.W. (1948). *Mammals of Washington*. University of Kansas Natural History Museum, Lawrence. 444 pp.

7. Dasmann, R.F. (1972). Towards a system for classifying natural regions of the world and their representation by national parks and reserves. *Biological Conservation* 4: 247–255.

8. Dasmann, R.F. (1973). A system for defining and classifying natural regions for purposes of conservation. *IUCN Occasional Paper No. 7*: 47 pp.

9. Dasmann, R.F. (1974). Biotic provinces of the world. *IUCN Occasional Paper No. 9*: 57 pp.

10. Dice, L.R. (1943). *The Biotic Provinces of North America*. University of Michigan Press, Ann Arbor. 78 pp.

11. Erwin, T.L. (1982). Tropical forests: their richness in Coleoptera and other arthropod species. *The Coleopterists' Bulletin* 36: 74–75.

12. Haugum, J., Ebner, J. and Racheli, T. (1980). The Papilionidae of Celebes (Sulawesi). *Lepidoptera Group of 1968 Supplement* 9: 21 pp, 1 map, 2 pl.

13. Holloway, J.D. (1980). Insect surveys—an approach to environmental monitoring. *Atti XII Congresso Naz. Ital. Ent.* 12.

14. Holloway, J.D. and Jardine, N. (1968). Two approaches to zoogeography: a study based on the distributions of butterflies, birds and bats in the Indo–Australian area. *Proceedings of the Linnean Society of London* 179: 153–188.

15. IUCN/UNEP/WWF. (1980). *World Conservation Strategy*. IUCN, Gland.

16. Lamas, G. (1982). A preliminary zoogeographical division of Peru, based on butterfly distributions (Lepidoptera, Papilionoidea). In Prance, G.T. (ed.) *Biological Diversification in the Tropics*. Proceedings of the Fifth International Symposium of the Association for Tropical Biology. Columbia University Press, New York. 714 pp.

17. Mackinnon, J. and Wind, J. (1980). *Birds of Indonesia*. FAO, Bogor. xii + 55 pp.

18. Parker, S.P. (1982). *Synopsis and Classification of Living Organisms*. McGraw Hill, New York. 2 vols, 1232 pp.

19. Paulian, R. (1951). *Papillons Communs de Madagascar*. L'Institut de Recherche Scientifique, Tananarive–Tsimbazaza. 90 pp.

20. Paulian, R. and Viette, P. (1968). Insectes Lépidoptères Papilionidae. *Faune de Madagascar* 27, 97 pp, 19 pl. ORSTOM, CNRS, Paris.

21. Pollisco, F.S. (1982). An analysis of the national park system in the Philippines. *Likas–Yaman* 3(12): 56 pp.

22. Pyle, R.M. (1976). The eco-geographic basis for Lepidoptera conservation. Unpublished Ph.D. thesis, Yale University, New Haven, Connecticut 369 pp.

23. Pyle, R.M. (1980(82)). Butterfly eco-geography and biological conservation in Washington. *Atala* 8(1): 1–26.

24. Smiet, F. (1982). Threats to the Spice Islands. *Oryx* 16: 323–328.

25. Straatman, R. (1968). On the biology of some species of Papilionidae from the island of Celebes (East-Indonesia). *Entomologische Berichten* 28: 229–233.

26. Tsukada, E. and Nishiyama, Y. (1982). *Butterflies of the South East Asian Islands. Vol. 1 Papilionidae*. (transl. K. Morishita). Plapac Co. Ltd., Tokyo. 457 pp.

27. UNDP/FAO National Parks Development Project (1981/1982). *National Conservation Plan for Indonesia*. Vols. 1–8. FAO, Bogor.

28. Udvardy, M.D.F. (1975). A classification of the biogeographical provinces of the world. *IUCN Occasional Paper No. 18*: 48 pp.

29. Wallace, A.R. (1876). *The Geographical Distribution of Animals*. Macmillan, London. 2 vols.

30. Wells, S.M., Pyle, R.M. and Collins, N.M. (1983). *The IUCN Invertebrate Red Data Book*. IUCN, Gland and Cambridge, 632 pp.

5

Trade in swallowtail butterflies

The worldwide trade in butterflies is big business, running into tens of millions of dollars annually. The precise amount is controversial; a recent estimate of US $10–20 million per year (21) must be conservative since there have been reports that the export trade from Taiwan alone is $20–30 million (16, 17). Butterflies can be bought from hundreds of dealers throughout the world but particularly in Hong Kong, Japan, Korea, Taiwan, Malaysia, western Europe, the U.S.A., Brazil and Peru. In this section we examine the main areas of commercial interest and ask whether the trade is likely to damage wild populations in a severe or permanent way. All prices quoted are in U.S. dollars.

Introduction

Butterfly collecting began in the 16th and 17th centuries when large numbers of new and exotic species were brought back by explorers to be described, classified and studied by the scientists of the day. It evolved into a popular hobby during the 19th century, pursued by educated aristocrats, middle-class citizens, doctors and clergymen who had the necessary money and time to spare. Some of these became leading authorities on the subject. Large collections were built up, especially by entomologists such as Lord Rothschild who were wealthy enough to send expeditions to remote areas.

The current trade supplies dead specimens to scientists, museums and private collectors, eliminating their need to travel long distances. The number of commercial dealers has risen dramatically over the last three decades (20) and dealers' lists advertise a huge range of butterflies and other insects. During the past five years over 80 per cent of the world's swallowtail species have been made available through international trading organizations, but less than 10 per cent are known to be farmed or ranched (see Table 5.1). Prices range very widely; cheaper species include the common Oriental and Australasian *Graphium sarpedon* (Blue Triangle or Common Bluebottle butterfly) and the Brazilian *Eurytides stenodesmus* at $0.30 or less. At the other end of the scale are rare birdwings like *Ornithoptera alexandrae* (Queen Alexandra's Birdwing) an imperfect male specimen of which was recently advertised for $2850, and *Ornithoptera meridionalis* and *Ornithoptera paradisea* (Paradise Birdwing), specimens of which have been advertised for $7000! All three are restricted to New Guinea where they are protected by law. *O. paradisea* is far more plentiful than *O. meridionalis* and *O. alexandrae*, but certain rare subspecies are highly valued. It is usual for prices to be correlated with rarity, but this is not always clear. Surprisingly, some specimens of *Zerynthia polyxena* (Southern Festoon) from the Alps of Haute-Provence, France, were offered between 1978 and 1984 for the equivalent of about $2500 – $4000 (1, 11). The high price may indicate that the specimens were of a very rare form since this species is widely distributed and rarely advertised for more than $2. Up until the 1970s the highest price paid was for a male *Ornithoptera allottei* (Abbé Allottes Birdwing) from Bougainville which was sold for the equivalent of $1500 at an auction in Paris in 1966 (33). *O. allottei* is only known from about twelve specimens and is thought to be a natural hybrid between *O. victoriae regis* (Queen Victoria's Birdwing) and *O. priamus urvillianus* (Priam's Birdwing) (8).

High prices may also be paid for species which are fairly widely distributed but rare and difficult to obtain. In the foothills of the Himalaya mountains *Teinopalpus imperialis* (Kaiser-I-Hind) is difficult to capture since it is a strong flyer which keeps to the tree tops. In 1955 a collector paid the equivalent of about $250 for a single specimen (33) and the species now costs up to $75 a pair. Species from inaccessible countries rarely appear on the market and when they do they tend to have high price tags. For example, a pair of *Bhutanitis thaidina* from China was advertised by a West German dealer in 1983 for the equivalent of about $225.

The table at the end of this section lists all swallowtail species and shows which of these have legislation protecting them, which are reared in captivity and which are currently in trade. The range of prices demanded for a particular species is shown by a series of price categories. In addition to species the list also includes several rare subspecies that command much higher prices than the nominate form. The information used in the table has been obtained from worldwide lists and correspondence between 1980 and 1985. Higher prices are often asked for female specimens because they are less easily attracted to baits and decoys. For example female specimens of *Graphium latreillanus* are offered at prices much greater than those for males, accounting for the wide range of prices asked for this species. The same consideration applies particularly to females of *Papilio antimachus* and *P. zalmoxis* which are seldom collected. Space in the table does not allow a full consideration of the prices of males and females separately. Where threatened species appear in trade they normally command very high prices, as do most rare species or subspecies.

Some collectors will go to great lengths to obtain rarities, either for their own collections or for sale at a profit. The high prices paid for butterflies in recent years have attracted financiers interested only in investment, with little thought for the insects in the wild (5). The largest and one of the rarest butterflies in the world, *Ornithoptera alexandrae* of Papua New Guinea, is mainly threatened by deforestation but collecting was at one time also a problem. Since 1968 it has been illegal to collect this species but until recently one smuggler was reputedly earning as much as $800 per specimen (16). With *Ornithoptera allottei* now believed to be a hybrid, *Ornithoptera alexandrae* has probably replaced it as the world's most valuable species (although specimens of *Ornithoptera meridionalis* and *Ornithoptera paradisea* have been advertised at higher prices). In 1979 it was estimated that fine specimens of *Ornithoptera alexandrae* were worth up to $2000 (7, 31), but this has since been exceeded. Commercial collecting by expatriates has now been stopped completely in Papua New Guinea and several people have been deported for smuggling (16). *Ornithoptera alexandrae* and some other rare, valuable and coveted species are seriously endangered, mainly by habitat loss rather than trade alone (see section 6). They include *Papilio hospiton* (Corsican Swallowtail) from Corsica and Sardinia, *Papilio homerus* (Homerus Swallowtail) from Jamaica, a female of which was advertised in the U.S.A. in 1984 for $2800, and the uncommon Malagasy species *Papilio morondavana*, females of which are frequently advertised by French dealers for about $120 – $150.

An important development in the last decade, particularly in the United Kingdom, has been the rise of the "butterfly house". These enterprises have concentrated in providing a display of butterflies, particularly exotic species, in "artificial jungles", that is, tropical conditions with appropriate vegetation and climate. Butterfly houses need a constant supply of living specimens for display. Although some butterflies are reared in the houses, most are obtained from their countries of origin as pupae or adults, using the fast, reliable air freight services now available from many parts of the tropics. Good prices are paid for butterflies used in living displays especially when the supply is regular and reliable. Swallowtails are probably in demand because of their size, bright colours and exotic appeal. The main exporting countries are in the tropics where there is a diverse range of beautiful species. Farmed (i.e. bred from parental stock held in captivity) or ranched (i.e. conceived by wild parents but reared from young stages under controlled conditions) specimens of certain species are available, but most material appears to be caught in the wild.

To summarize, there are essentially three different sorts of trade:-

(i) Low value, high volume. Large numbers of common species, often of low quality are used in a range of ornaments and decorations, less as specimens. The trade of Taiwan is typical of this category, with annual sales estimated at 15–500 million butterflies.

(ii) High value, low volume. High quality specimens, sometimes with scientific data including date and place of capture, are sold to museums, students and collectors. Many dealers in Europe, North America and Japan, produce catalogues of specimens in this category. Trade from the insect ranching programme in Papua New Guinea is at the top end of this market (15, 24). Linking (i) and (ii) is the trade in high value, high quality ornamental items. This includes wall mounts and glass domes containing mounted butterflies in life-like settings, or jewellery such as that based on the iridescent *Morpho menelaus* (Blue Morpho) of Brazil.

(iii) The live trade. Fairly low but continuous volume, medium value. Living butterflies, usually pupae or adults, are despatched to butterfly houses for display to the paying public. This trade is probably the fastest growing of the three types.

Countries of origin

The Orient The Oriental region contains an extremely diverse butterfly fauna (see sections 3 and 4). The major threat in the region is loss of habitat resulting from deforestation, agriculture and urbanization. In most areas collecting is small-scale, often by Japanese collectors, but there are large-scale international trading centres in Taiwan and Malaysia, and significant commercial enterprises in Hong Kong and Korea (26).

Taiwan has a butterfly trade which operates on a vast scale involving about 20 000 people including 10 000 collectors (11, 16, 30). Estimates of the number of butterflies traded vary from 15 to 500 million (26). Virtually all butterflies are caught in the wild and sold to the 30 or so factories which process them (19, 30). Many species are imported in bulk from overseas, particularly *Morpho* species from Brazil, *Charaxes* from Africa and birdwings from South East Asia. In most cases the bodies are discarded for pig feed and the wings are glued to paper bodies with bristle antennae and laminated between sheets of clear plastic. They are then used to make bookmarkers, coasters, table mats, wall decorations and even plastic toilet seats! Others are made into pictures or used to decorate hand-bags or purses. Taiwan's export trade was valued at $30 million in 1969 (16) and about $21 million in 1976 (17).

Individual specimens are also sold by a number of dealers in Taiwan, where prices vary according to rarity, size and beauty, from about $0.02 for a specimen of the endemic *Papilio machaon sylvina* to $50 for the endemic *Papilio maraho*. Males of *P. maraho*, now recognized as a Vulnerable species, were advertised in 1983 by a German dealer for the equivalent of about $75. Although an island with an area of only 35 960 sq. km, Taiwan has about 400 butterfly species of which 40 are endemic (19). About 100 species, including 20 crop pests, need no protection, while the remainder may need some form of management (19). Between 50 and 60 of these are very rare (19) and the Entomological Suppliers Association of Great Britain has banned trade in the rare Taiwan endemic *Troides aeacus kaguya* (14). There are no captive-breeding programmes though a few people rear rare papilionids and *Attacus atlas* (Atlas moth) to obtain quality specimens for collectors (19). There is no conclusive evidence that the butterfly industry poses a threat to most butterfly species, but there is now serious concern for *Papilio maraho*. The major threats to Taiwan's butterflies undoubtedly come from habitat destruction for agriculture and forestry. The lowland forests are virtually gone and human population pressure is extending deforestation into more mountainous areas. In densely populated areas urbanization, air pollution and pesticides are taking their toll. There is apparently no environmental legislation in Taiwan, although there is growing conservation concern. Four national parks have been designated, but there are so far no laws to ensure rational exploitation of butterflies and the long-term survival of the industry (11, 16, 26, 28, 30). The import of specimens to Taiwan from overseas has never been measured and it is not known what effect the market has in the countries of origin. More research is needed to assess what proportion of the Taiwan trade is in foreign species, and the impact caused by the demand.

Large numbers of *Trogonoptera brookiana* (Rajah Brooke's Birdwing) are exported from Malaysia (23). There is evidence that the trade exceeds 125 000 specimens annually, all of which are apparently collected in the wild. Under Malaysian law the species is "protected" but not "totally protected", and collecting is permissible with a permit which is easily obtained. The species is also listed on Appendix II of CITES (Convention on International Trade in Endangered Species of Wild Fauna and Flora). This Appendix implies that commercial trading is allowed providing a permit from the country of export is obtained. This can provide a method of monitoring trade levels but there is evidence that most exports of *T. brookiana* are unlicensed. Many dealers list *Trogonoptera brookiana* for sale in bulk quantities of up to 1000 specimens and large numbers have recently been used for artwork (20). In 1983 batches of 100 were on offer by a West German dealer for about $20 and in the U.S.A. a dealer was selling 100 for $30. There have been recent suggestions that the species is being ranched or farmed, but the evidence is circumstantial. Certain dealers in Peninsular Malaysia sell a large proportion of specimens that are in perfect condition, suggesting that they are reared from pupae. The breeding localities, foodplants and habits of the young stages are being kept a closely-guarded secret (9). The main threat to *Trogonoptera brookiana* and other Malaysian butterflies is habitat destruction rather than trade. Nevertheless, the trade in this species should be carefully monitored since the population may be brought to such a low level by loss of habitat that a continuation of heavy collecting could be a serious threat (23).

There is little information on trade in the Philippines and Indonesia although collecting by tourists, private collectors and local people on behalf of dealers is said to be affecting some species, including the Endangered *Papilio chikae* in Luzon (29, also see p. 364). Indonesia has legislation protecting certain species, mainly birdwings, but it is not known how effective this has been. Some of the species are often advertised in dealers' lists.

Butterfly trading based in India and Indochina is now quite extensive and occurs at all levels, from personal collectors to substantial businesses. In South Korea the decline in populations of species such as *Parnassius bremeri*, *Papilio demetrius*, *P. machaon* and *P. macilentus* has been partly attributed to over-collection (18). There is little information from Thailand, but at least one Bangkok business uses local collectors to provide material for the international market. In the mountains around Kathmandu in Nepal collecting at times reaches absurd proportions. There are reports from Phulchoki Mountain that encampments of foreign entomologists wait for hill-topping males of *Teinopalpus imperialis* and kill every specimen that flies in. The Nepal government requires a permit for collecting, but the legislation seems to be openly abused. There is also a substantial market in dead specimens offered by street hawkers in Kathmandu (12). In 1980, the Government of India passed an amendment to the 1972 Wildlife Protection Act listing a large number of butterflies as fully protected. This certainly helps to draw attention to the need for restraint in collecting certain species, but the law is very difficult to enforce. Some foreign tourists visit northern India specifically to collect butterflies known to be fully protected. Capturing material for personal study would no doubt be acceptable, but there are fears that large numbers of specimens find their way to the market-place. There are two large commercial "farms" in the country which have apparently been affected by this legislation but it is thought that neither had a captive breeding programme. The Indian Government is interested in setting up a farming project for common species and two trial farms are planned for north–western and eastern India.

China has one of the world's most important swallowtail faunas, but because of its relative inaccessibility in recent years many species still remain poorly known. However, during the 1980s Japanese entomologists have travelled in the rich collecting areas of Yunnan and Sichuan, bringing back little-known species such as *Bhutanitis mansfieldi* (see p. 192). There is also evidence that local Chinese have been trained to collect specimens and are regularly sending material to Japan. Large numbers of *Bhutanitis thaidina* have been traded in this way (see p. 194). The export of material from China for academic study is greatly to be encouraged, but the build-up of a large commercial operation should be viewed with caution. Many high altitude species are very local and have short flight seasons; they may be vulnerable to over-exploitation. Monitoring of the trade would be advisable. There are opportunities for summer butterfly ranching programmes in western China, worthy of investigation as a source of butterflies for dealers, a seasonal income for local people and a means of protection for the butterflies.

Australasia There is relatively little commercial trade in butterflies in Australia, partly because of the low numbers of amateur enthusiasts in that country. The few commercial enterprises in Australia have perhaps up to 80 per cent of their trade in butterflies originating from outside the country. Endemic swallowtails of particular interest, such as *Graphium macleayanum* and *Protographium leosthenes*, have been intermittently in trade but have not been offered for some years. The endemic birdwings, *Ornithoptera richmondia*, *O. euphorion* and *O. priamus pronomus*, are in demand and also in trade, but not in large numbers. These species, together with the endemic subspecies of the Blue Mountain Butterfly, *Papilio ulysses joesa*, are protected in Queensland; however, the main threat to all these butterflies is destruction of the rain forest habitat rather than commercial collecting. Other

Australian swallowtails are occasionally offered, including *Cressida cressida*, commonly known by the extraordinary name Big Greasy Butterfly because of the way it rapidly loses its wing-scales, producing a greasy effect. The Australian papilionid fauna is a small one and the commercial trade is generally low in volume and high in quality and value.

Papua New Guinea probably runs the best example of a high value, low volume trade, exporting high quality specimens of native butterflies. The spectacular birdwings of the genus *Ornithoptera* are the greatest attraction for collectors and an important natural resource for the country. Mainly because of habitat destruction, but possibly aided by collecting, some species have become very rare, in particular *Ornithoptera alexandrae*. In 1968, the seven rarest *Ornithoptera* species, i.e. *alexandrae*, *allottei*, *chimaera* (Chimaera Birdwing), *goliath*, *meridionalis*, *paradisea* and *victoriae* were protected by a national law to prevent collecting (21).

Before Papua New Guinea's independence in 1975 all collecting was carried out for expatriate dealers by native collectors. They often received as little as $0.20 for butterflies which could fetch up to $250 overseas (7). A few years ago a butterfly dealer visited the island of Nimoa south–east of Papua New Guinea and persuaded villagers to collect hundreds of adults and chrysalids of the spectacular birdwing *Ornithoptera priamus caelestis*, endemic to the Louisiade Archipelago (8). The villagers were paid trifling sums for a haul of several hundred butterflies worth about $5000 (26). Not only were the villagers exploited, but no more birdwings were seen in the vicinity for many years. Eventually they were reintroduced by the Insect Farming and Trading Agency (I.F.T.A.) from a population on the other side of the island and are now in trade (26, 27). Nowadays, only nationals are permitted to profit from the trade, which is co-ordinated by I.F.T.A. at Bulolo. Villagers are encouraged to ranch or, to a lesser extent, collect the commoner unprotected species such as *Ornithoptera priamus* and *Troides oblongomaculatus*. By 1978 over 500 people were involved in this rapidly expanding village industry. All business is handled by I.F.T.A. which pays the villagers all profits less 25 per cent. There has been an increase in the trade in quality specimens from Papua New Guinea. In 1979 villagers were being paid $37 for a box of butterflies but by 1980 this had risen to $50 (21). A hard-working butterfly rancher may earn up to $1200 annually (21). By 1983 the annual trade in insects, mainly butterflies was estimated to be $110 000 and increasing. Approximately 30 per cent of all butterflies reaching the Agency are ranched and the rest are collected as adults in the wild, but 50 per cent of the revenue comes from the usually better quality ranched specimens (21). Swallowtails are more commonly ranched than butterflies of other groups and so the proportion of ranched specimens is higher than 30 per cent of the total.

Ranching is widely encouraged by the I.F.T.A. because it produces fresh undamaged specimens and the habitat enrichment is also a good conservation measure. Under this practice usually only the pupae are kept in captivity. Wild butterflies are free to come and go and the rancher ensures that a proportion of those reared in captivity is released to continue the cycle. Children and potential ranchers are shown which species can easily be ranched, and which plants are attractive as food or for nectar. An average ranch will have a garden of about 0.2 ha surrounded by a thick hedge of *Hibiscus*, *Bougainvillea* or similar nectar-bearing plants. These attract the butterflies but also keep out livestock. Foodplants cultivated inside the fence include *Aristolochia tagala* (Dutchman's Pipe Vine) for the common birdwings, and *Evodia* spp. for *Papilio ulysses* (27). The pupae are removed and usually placed into an emergence cage just before the adults emerge. The ranchers are taught how to judge the correct time and also how to kill and pack into boxes the newly emerged butterflies (12, 15, 24, 27, 31).

A large number of other Papuan butterflies are collected as adults but these tend to be mainly common species such as *Graphium sarpedon* and *G. agamemnon* (Tailed Jay or Green-spotted Triangle butterfly). Adult *Graphium weiskei* (Purple-spotted swallowtails) are collected only as adults because the food plant and larval biology are unknown. This species may be common in some areas but occasionally collection has to be halted for anything up to a year whilst the numbers increase (27). *Ornithoptera victoriae* is common on Bougainville and the I.F.T.A. may in future recommend its removal from the protected list in order to allow ranching. *O. chimaera* and *O. goliath* have also been suggested as good candidates for future ranching projects although their foodplants are more difficult to cultivate. As well as being a valuable village industry in Papua New Guinea, the butterfly ranching project demonstrates that butterfly trade and conservation can be of mutual benefit (27) and the future could be promising for both the trade and the ranching programme.

Africa and Madagascar Large numbers of African *Charaxes* (Nymphalidae), *Papilio* and other genera appear on dealers' lists. Many are only listed as being from Africa, West Africa or East Africa, but the Central African Republic, Madagascar, Malawi and to a lesser extent Congo and Gabon seem to be regular suppliers. Malawi is chiefly a source of *Charaxes* species, whereas both the Central African Republic and Madagascar appear to be suppliers of a wide range of butterflies.

Reports from the Central African Republic indicate that every year for the past fifteen years hundreds of thousands of butterflies have been indiscriminately caught by hundreds of local collectors for sale overseas. Collecting of this sort could have an effect on some species or populations. More information is needed from this and other African countries. Some nations would benefit from a farming or ranching scheme, particularly for *Papilio* and *Charaxes* species.

Madagascar is well known for its unique and endemic fauna, including many butterflies. There is no commercial ranching and specimens are collected by local collectors for expatriate dealers. The collectors are usually untrained and are only interested in large numbers of different species. The resulting indiscriminate collecting may be highly destructive and wasteful since many specimens are damaged and later discarded. Female *Papilio morondavana*, a large black and yellow Vulnerable swallowtail, can command prices up to $150 (see p. 000). Other valuable Malagasy species are the two Rare species *Papilio grosesmithi* (see p. 000) and *P. mangoura* (see p. 000), and the more common *Graphium evombar*. Specimens from Madagascar are subject to a heavy export tax but the fact that some species are common and cheap on many dealers' lists may indicate large-scale smuggling.

The African species commanding the highest prices are often large and spectacular, such as *Papilio antimachus* (African Giant Swallowtail, $20 for a male in 1983 in the USA) and *Papilio zalmoxis* (Great Blue Papilio, $10–15 in 1983 in the USA). Others are very scarce, such as *Papilio sjoestedti* (Kilimanjaro Swallowtail, Rare, see p. 000) from Tanzania, which has been commanding particularly high prices ($212 for a female on one American list in 1984), or *Papilio leucotaenia*, (Cream-banded Swallowtail, Vulnerable, see p. 000) which is restricted to relict forests in Central Africa. Collecting is normally done by local people for overseas dealers. In Africa there is no legislation preventing the collecting of butterflies except in Kenya, where a permit is required to collect Lepidoptera and Coleoptera, and the Province of the Cape of Good Hope, South Africa, where 17 Lepidoptera are protected, none of them swallowtails.

South and Central America Commercial trade in the Caribbean Islands is very variable in extent. There are many endemic taxa of swallowtails in the Caribbean, particularly endemic subspecies, but on the whole these have been little exploited. Some of the commercial collecting in Jamaica has caused unease, particularly the exploitation of the endemic, and very valuable, *Papilio homerus* (p. 297). One commercial enterprise is established in the Dominican Republic, but most trading activity in the Caribbean Islands seems to be by visiting dealers from elsewhere, particularly the U.S.A. Apart from the trade in swallowtails from Jamaica and Hispaniola, highly-priced specimens of *Battus devilliers* (from the Bahamas) have recently been offered and other species from various parts of the Caribbean are advertised from time to time. On the whole, however, the volume of trade is low, is concentrated on the high value end of the market, and is not to be compared with the volume of trade in butterflies from Peru, Brazil or even Mexico.

Butterfly "farms" exist in Costa Rica and some species of *Morpho* (Morphidae) and some swallowtails are ranched in Brazil (21, 25, 32). The beautiful iridescent blue *Morpho* butterflies are much in demand for high quality jewellery and other decorative purposes and 50 million a year are used in Brazil alone (6, 26). Although the law requires that morphos in trade must have been farmed, there is evidence that farms or ranches only meet part of the trade and many specimens continue to be caught in the wild (21). Survey work is needed to assess the impact of this trade. In Brazil one butterfly, *Parides ascanius*, is completely protected by law and another, *Eurytides lysithous harrisianus*, has been proposed for listing. In Honduras there is a low value, high volume trade similar to that in Taiwan, but on a smaller scale (26). In Peru there is a considerable insect industry, with large numbers of swallowtails coming out of the classic Tingo Maria and Rio Satipo regions as well as other areas. A major butterfly collecting company reportedly operates in Colombia (20). Mexico has a large export trade in butterflies, all of which is required to be licensed under Mexican legislation. Butterflies originating in Bolivia, Colombia, Costa Rica, Ecuador and Venezuela are also not uncommon on dealers' lists.

North temperate regions In general the trade in temperate species is for serious collectors rather than the ornamental market. Butterflies of the Palearctic region are often in demand and although some farms exist, most species are collected as young stages or adults in the field. Eastern Palearctic species are particularly difficult to obtain and China has species of *Papilio* and *Parnassius* that would be suitable for farming or ranching. There is extensive European legislation to prevent and monitor butterfly collecting and trade. The EEC is party to CITES and thus monitors trade in birdwings and *Parnassius apollo* in all EEC countries. Some or all of the native Papilionidae are protected in Austria, Czechoslovakia, Finland, France, East Germany (DDR), Luxembourg, Netherlands, Poland, Switzerland, Great Britain and the U.S.S.R. (Lithuania). Turkey has a complete ban on collecting and export of butterflies (14). In the U.S.A. the Floridan subspecies *Papilio aristodemus ponceanus* (Schaus' Swallowtail) is currently proposed as endangered under the 1972 Endangered Species Act, but *Papilio andraemon bonhotei* (Bahama Swallowtail) is proposed for delisting as a result of its improved circumstances (see p. 301).

Butterfly trade and conservation

The impact of private collectors The major threat to butterflies is loss of habitat caused by the actions of humans (2). Urbanization, deforestation, overgrazing, agricultural intensification, expansion of subsistence agriculture and atmospheric pollution all take their toll. Nevertheless, there are circumstances in which private collectors may have a serious impact, notably when populations are already severely reduced by other factors (22).

Such was probably the fate of *Lycaena dispar* (the Large Copper, Lycaenidae) which became extinct in Britain, on the edge of its range, in 1847/8. Habitat destruction by drainage was the main cause of the demise, but collectors may have dealt the final blow. More recently, Canadian naturalists expressed concern over excessive collecting of butterflies in the Great Lakes area. Species that were already threatened by habitat loss and pollution became the target of collectors interested by their increased rarity value. As a result, *Artogeia virginiensis* (the West Virginia White, Pieridae) has been listed as endangered and protected in Ontario (2), but subsequent investigations indicate that the species may be more widespread than was at first realized (4).

Even butterflies with a seemingly wide range may have vulnerable isolated populations with slight pattern variations coveted by collectors. In 1926 a collector tried to wipe out an entire local population of *Parnassius apollo* (Apollo butterfly) in the Italian Alps to increase the value of specimens already possessed (3).

Recently some other examples of irresponsible exploitation of butterflies by collectors have come to light. In the Far East, Japanese collectors are particularly implicated. Package tours to collect butterflies have been accused of causing serious reductions in the butterfly populations of some areas of the Himalayas (10). In Luzon, Japanese collectors have been known to offer cameras for specimens of *Papilio chikae*, a beautiful iridescent swallowtail (9), and a female specimen on a West German dealer's list in 1983 was offered for the equivalent of about $150. It is now quite common for lepidopterists to travel to the tropics on collecting tours. Collecting a few specimens for a private collection can do no harm, but it seems to be common practice to collect large numbers of specimens for sale to dealers. The money obtained may be offset against the cost of the travel. Reports indicate that Japanese collectors frequently travel to Indonesia, the Philippines or other parts of Asia with the sole purpose of collecting large numbers of butterflies. The recent rediscovery of *Atrophaneura palu* in western Sulawesi has resulted in a number of highly-priced specimens appearing in the European trade and yet no data on the habitat or ecology of the species have been published for the benefit of the scientific community.

There is a danger that collectors may be unable to recognize when they are depleting butterfly stocks below the threshold of recovery, particularly when they only visit the breeding areas for short periods of time or employ unsupervised assistants. Many species of butterfly can reproduce exponentially if conditions are right, but they may also decrease exponentially when conditions are bad. This means that a seemingly common species may very suddenly plummet into rarity or even extinction.

The impact of commercial collectors Given suitable conditions and resources most butterflies are able to reproduce sufficiently to prevent permanent reduction of their populations by collectors. However, the evidence that millions of specimens are now involved in the international trade in ornaments is disturbing. In Taiwan the butterfly populations are reputedly remaining constant despite the enormous pressure on them (26), but there is also evidence that more material is imported to Taiwan from Africa and South America, and that the Taiwan trade is in decline. There is clearly a need for an independent assessment. Similarly the main collecting areas in South America and Africa require further research, not only to protect the butterflies, but also to ensure long-term rational utilization of the resources that are the livelihood of many people.

As a result of their limited size, island populations are often particularly vulnerable to excessive collecting as well as habitat destruction. For the giant birdwings of New Guinea, Indonesia and the Solomon Islands the problem is exacerbated because they generally lay no more than 30 eggs per generation. In such cases trade does need to be controlled and in Papua New Guinea the butterfly ranching programme is a good example of how this can be done successfully.

There is some controversy over the effect of removing large numbers of males from butterfly populations. Males tend to be collected more than females because they are often brighter and more easily captured on hilltops or at baits of urine or rotten fruit. One male may be capable of fertilizing several females, but there is little scientific information on the degree of male depletion a population can withstand. In Brazil where 50 million male morpho butterflies enter the trade annually, a survey was undertaken to determine the proportion of males and females in nature (6). Results showed a ratio of ten or even several hundred males to every female in most species and it was concluded that the number of specimens taken by commercial collecting would not threaten populations of the species taken. These sex ratios have been widely disputed since the habits of females make them much more difficult to capture (13) and the sex ratio of captive-bred butterflies has not been shown to depart significantly from 1:1. More research is needed on this important aspect of butterfly reproduction and its implications to management.

Collecting and legislation CITES lists the birdwing butterflies (*Ornithoptera*, *Trogonoptera*, and *Troides*) and *Parnassius apollo* on Appendix II. Since these listings there has been a significant reduction in advertised trade of some species, although CITES Appendix II seeks only to regulate trade, not to prevent it. A total of 87 countries are now party to CITES.

Many examples of legislation have been cited in the above sections. National legislation or voluntary restrictive codes are increasingly necessary in regions where habitat destruction is so extensive that the relatively minor effects of collecting may have a serious impact (13). Such laws or codes should encourage scientific study for management purposes and can be used to encourage public awareness of butterfly conservation needs. In the U.S.A., the Endangered Species Act not only prevents collecting but also specifies a recovery programme and allows for the protection of habitat which is critical to the survival of the species. Legislation against collectors can only be successful if habitat is set aside to conserve wild populations. Emphasis should be put on protected areas as well as protected species. This course of action requires thorough survey work in order to delineate critical habitat, and expert ecological studies to ensure adequate long-term management.

In countries with large areas of natural habitat, private collectors are most unlikely to have a significant impact on butterfly populations. Even when butterflies are protected within national parks there would generally be no harm in permitting amateur collecting on a small scale. It is particularly important that future generations of entomologists are not dissuaded from their studies by unnecessarily severe restrictions.

Commercial collecting has not been independently assessed in any of the main tropical centres of origin. There are insufficient data to judge whether monitoring or regulation of the trade is warranted. Research into the origin and extent of world trade in ornamental butterflies is urgently required.

Conclusions Private collecting of butterflies can be an instructive hobby and is important for research into ecology, population dynamics, genetics and taxonomy. It brings a great deal of pleasure to many people and does not usually threaten butterfly species. However, in a small number of cases irresponsible over-collection can cause a permanent decline. The latter is particularly true if the species has a very small range, has naturally low populations and low reproductive rate, or has already been severely reduced by other impacts.

Commercial collecting can be an important source of income and should not be dismissed as necessarily harmful. If populations are harvested in a sustainable manner, then both conservation and commercial interests can be satisfied. Habitat destruction is the main cause of decline in butterfly populations, but there is a danger that commercial collecting levels that were sustainable in the past may become damaging in the ever-decreasing areas of suitable habitat. Such may soon be the case for *Trogonoptera brookiana*, which lives only in the primary lowland forests of Malaysia and Indonesia. Responsible commercial collectors should take all possible steps to conserve the habitats in which they collect. This is clearly essential for both the trade and the species. Where possible, captive breeding should be encouraged, to provide high quality specimens for the trade, to provide local employment and to ease the pressure on wild populations. At present, less than 10 per cent of swallowtail species are known to be farmed or ranched (see Table 5.1). Commercial organizations should invest in research to raise this figure.

Legislation against collecting and trade is unlikely to preserve a species unless parallel measures to protect its habitat are also enforced. Preliminary assessments are needed in order to decide whether extensive monitoring of international butterfly commerce is advisable.

References

1. Anon. (1978). Extract from *Nice Matin* in litt.
2. Anon. (1982). Save the butterflies. *Nature Canada* 11(4): 46.
3. Bourgogne, J. (1971). Un témoignage de plus sur la destruction de la nature (Papilionidae). *Alexanor* 7: 50.
4. Brownell, V.R. (1981). *The West Virginia white butterfly (Artogeia virginiensis Edwards) in Canada: a status report*. Nongame Program, Wildlife Branch, Ontario Ministry of Natural Resources. 55 pp + appendices.
5. Campbell, G. (1976). Why not try a flutter on butterflies? *Daily Telegraph* 13 March.
6. Carvalho, J.C.M., and Mielke, O.H.M. (1972). The trade of butterfly wings in Brazil and its effects upon survival of the species. *Proceedings 19th International Congress of Entomology (Moscow)* 1: 486–488.
7. Cherfas, J. (1979). How to raise protection money. *New Scientist* 6 December.
8. D'Abrera, B. (1975). *Birdwing Butterflies of the World*. Lansdowne Press, Melbourne. 260 pp.
9. D'Abrera, B. (1982). *Butterflies of the Oriental Region. Part 1. Papilionidae and Pieridae*. Hill House, Victoria, Australia. xxxi + 244 pp.
10. Dasgupta, M.A. (1979). Save our butterflies. Letter to the Editor. *Illustrated Weekly of India* 23 September.
11. Donahue, J.P. (1984). *In litt.*, 24 May.
12. Feeny, P. (1984). Pers. comm., 1 August.
13. Heath, J. (1981). Insect conservation in Great Britain (including "A code for insect collecting". *Beiheft Veröffentlichung Naturschutz Landschaftspflege Baden–Württemberg* 21: 219–223.
14. Heath, J. (1981). *Threatened Rhopalocera (Butterflies) in Europe*. Nature and Environment series (Council of Europe) no. 23, 157 pp.
15. Hutton, A.F. (1978). Conservation and utilization of the insect resources of Papua New Guinea. Unpublished report.

16. Inskipp, T. and Wells, S. (1979). *International Trade in Wildlife*. Earthscan, London. 104 pp.
17. Jackman, B. (1976). Bye-bye birdwing. *Sunday Times* 12 September.
18. Kyu, K.H. (1982). Status survey of Alpine butterfly fauna, South Korea. *WWF Yearbook 1982*. World Wildlife Fund, Gland, Switzerland. 492 pp.
19. Marshall, A.G. (1982). The butterfly industry of Taiwan. *Antenna* 6: 203–4.
20. Nagano, C.D. (1984). The International Trade in Butterflies. Unpublished manuscript. 27 pp.
21. National Research Council (1983). *Butterfly Farming in Papua New Guinea*. Managing Tropical Animal Resources Series. National Academy Press, Washington, D.C.
22. Owen, D.F. (1974). Trade threats to butterflies. *Oryx* June.
23. Owen, D.F. (1976). Rajah Brooke's birdwing. *Oryx* 13: 259–261.
24. Parsons, M. (undated). *Insect Farming and Trading Agency Farming Manual*. Insect Farming and Trading Agency, Bulolo, P.N.G.
25. Pyle, R.M. (1979). How to conserve insects—for fun and necessity. *Terra* 17(4): 18–22.
26. Pyle, R.M. (1981). Butterflies: now you see them. *International Wildlife* 11(1): 4–11.
27. Pyle, R.M. and Hughes, S.A. (1978). Conservation and utilisation of the insect resources of Papua New Guinea. Unpublished report to the Wildlife Branch, Dept. of Nature Resources, Independent State of Papua New Guinea. 157 pp.
28. Severinghaus, S.R. (1977). The butterfly industry and butterfly conservation in Taiwan. *Atala* 5(2): 20–23.
29. Tsukada, E. and Nishiyama, Y. (1982). *Butterflies of the South East Asian Islands. Vol. 1 Papilionidae*. (transl. K. Morishita). Plapac Co. Ltd., Tokyo. 457 pp.
30. Unno, K. (1974). Taiwan's butterfly industry. *Wildlife* 16: 356–359.
31. Vietmeyer, N.D. (1979). Butterfly ranching is taking wing in Papua New Guinea. *Smithsonian* 10(2): 119–135.
32. Wiltshire, E.P. (1959). First impressions of the tropical forests of southeastern Brazil and their Lepidoptera. *Journal of the Lepidopterists' Society* 13(2): 79–88.
33. Wood, G.L. (1982). *The Guinness Book of Animal Facts and Feats*. Third Edition. Guinness Superlatives Ltd., Enfield. 250 pp.

Table 5.1: Swallowtail butterflies in trade, 1980–1985

Data in this table are drawn from a selection of trade literature acquired from dealers worldwide. We would like to be informed of other species known to be traded and of species already listed here but found to be traded at prices markedly different from those given. The numbers next to each species are the same as those used in the species list (section 3) and may be used for cross-reference.

Key:

IUCN   conservation status categories (column CAT)
     E   Endangered
     V   Vulnerable
     R   Rare
     I   Indeterminate
     K   Insufficiently Known

Legislation (column LAW)
Lists all species which are protected by law in part of their range. In addition, *Parnassius apollo* and all species of the three birdwing genera *Trogonoptera*, *Troides* and *Ornithoptera* are listed on appendix 2 of the Convention on International Trade in Endangered Species of Wild Fauna and Flora (CITES).

Ranching or farming (column R/F)
Shows traded specimens which are ranched, reared or farmed in captivity. Most will be ranched.

Price categories (columns A–H)
Exchange rates have fluctuated widely during the preparation of this table. The price equivalents in the various ranks are therefore very approximate and the table can only be used as a guide. Source data are held at the Conservation Monitoring Centre and are available to interested parties.

UK	USA	GERMANY	FRANCE	JAPAN
A. Up to £0.60	Up to $1	Up to DM3	Up to Fr6	Up to Y240
B. £0.60–1.20	$1–2	DM3–6	Fr6–12	Y240–480
C. £1.20–5	$2–7.50	DM6–20	Fr12–50	Y480–1800
D. £5–20	$7.50–30	DM20–80	Fr50–200	Y1800–7200
E. £20–50	$30–75	DM80–200	Fr200–500	Y7200–18000
F. £50–100	$75–150	DM200–400	Fr500–1000	Y18000–36000
G. £100–1000	$150–1500	DM400–4000	Fr1000–10000	Y36000–360000
H. £1000+	$1500+	DM4000+	Fr10000+	Y360000+

Table 5.1

	CAT	LAW	R/F	A	B	C	D	E	F	G	H
Order: LEPIDOPTERA											
Family: PAPILIONIDAE											
Subfamily: BARONIINAE											
1 *Baronia brevicornis*	R			+	+						
Subfamily: PARNASSINAE											
Tribe: PARNASSIINI											
2 *Archon apollinus*						+	+				
3 *Hypermnestra helios*							+				
4 *Parnassius szechenyii*										+	
5 *Parnassius cephalus*											
6 *Parnassius maharaja*											
7 *Parnassius delphius*		*					+	+	+		
8 *Parnassius stoliczkanus*		*					+				
9 *Parnassius patricius*								+			
10 *Parnassius acdestis*											
11 *Parnassius imperator*		*									
12 *Parnassius charltonius*		*					+				
13 *Parnassius inopinatus*							+				
14 *Parnassius loxias*											
15 *Parnassius autocrator*	R										
16 *Parnassius tenedius*							+				
17 *Parnassius acco*		*									+
18 *Parnassius przewalskii*											
19 *Parnassius hannyngtoni*		*									
20 *Parnassius simo*											
21 *Parnassius hardwickii*						+	+				
22 *Parnassius orleans*											
23 *Parnassius clodius*					+						
24 *Parnassius eversmanni*							+	+	+		
25 *Parnassius felderi*											
26 *Parnassius ariadne*											
27 *Parnassius nordmanni*								+			
28 *Parnassius mnemosyne*		*		+	+	+	+				
29 *Parnassius stubbendorfi*						+	+				
30 *Parnassius glacialis*				+	+	+					
31 *Parnassius apollonius*						+					
32 *Parnassius honrathi*					+		+				
33 *Parnassius bremeri*							+				
34 *Parnassius jacquemontii*		*			+		+				
35 *Parnassius epaphus*		*					+				
36 *Parnassius actius*							+				
37 *Parnassius phoebus*		*		+	+	+					
38 *Parnassius tianschanicus*						+	+				
39 *Parnassius nomion*						+	+				
40 *Parnassius apollo*	R	*	○	+	+	+					
Tribe: ZERYNTHIINI											
41 *Sericinus montela*				+	+						
42 *Allancastria cerisy*				+	+						
43 *Allancastria deyrollei*					+						
44 *Allancastria caucasica*											
45 *Allancastria louristana*											

Table 5.1 (cont.)

	CAT	LAW	R/F	A	B	C	D	E	F	G	H
46 *Zerynthia polyxena*		*	o	+	+	+					+
47 *Zerynthia rumina*		*	o		+	+	+				
48 *Bhutanitis mansfieldi*	R										
49 *Bhutanitis thaidina*	R							+	+	+	
50 *Bhutanitis lidderdalii*		*					+				
51 *Bhutanitis ludlowi*	K										
52 *Luehdorfia chinensis*	K										
53 *Luehdorfia japonica*	V			+	+	+					
54 *Luehdorfia puziloi*					+	+					
Subfamily: PAPILIONINAE											
Tribe: LEPTOCIRCINI											
55 *Iphiclides podalirius*		*	o				+				
56 *Iphiclides podalirinus*											
57 *Teinopalpus imperialis*	R	*					+		+		
58 *Teinopalpus aureus*	K										
59 *Meandrusa sciron*		*				+	+				
60 *Meandrusa payeni*				+	+	+	+				
61 *Eurytides marcellus*					+	+	+				
62 *Eurytides epidaus*				+	+						
63 *Eurytides zonaria*											
64 *Eurytides marcellinus*	V						+			+	
65 *Eurytides celadon*						+					
66 *Eurytides philolaus*				+	+	+					
67 *Eurytides anaxilaus*											
68 *Eurytides xanticles*							+				
69 *Eurytides oberthueri*											
70 *Eurytides bellerophon*					+	+					
71 *Eurytides agesilaus*				+							
72 *Eurytides orthosilaus*					+				+		
73 *Eurytides helios*					+	+					
74 *Eurytides stenodesmus*				+	+						
75 *Eurytides earis*											
76 *Eurytides telesilaus*			o	+							
77 *Eurytides aguiari*											
78 *Eurytides embrikstrandi*											
79 *Eurytides travassosi*											
80 *Eurytides molops*				+		+					
81 *Eurytides macrosilaus*											
82 *Eurytides nigricornis*			o								
83 *Eurytides protesilaus*				+		+					
84 *Eurytides glaucolaus*											
85 *Eurytides asius*			o	+		+					
86 *Eurytides microdamas*											
87 *Eurytides thymbraeus*					+						
88 *Eurytides belesis*				+	+						
89 *Eurytides branchus*				+		+					
90 *Eurytides ilus*											
91 *Eurytides lysithous*			o		+	+					
ssp. *harrisianus*	E	*									
92 *Eurytides ariarathes*				+							
93 *Eurytides harmodius*				+		+					

169

Table 5.1 (cont.)

	CAT	LAW	R/F	A	B	C	D	E	F	G	H
94 *Eurytides trapeza*					+	+					
95 *Eurytides xynias*				+	+						
96 *Eurytides phaon*				+	+						
97 *Eurytides euryleon*						+					
98 *Eurytides pausanias*				+		+					
99 *Eurytides protodamas*			o				+				
100 *Eurytides marchandi*				+	+						
101 *Eurytides thyastes*			o	+	+	+					
102 *Eurytides calliste*											
103 *Eurytides leucaspis*				+	+						
104 *Eurytides lacandones*				+		+					
105 *Eurytides dioxippus*						+					
106 *Eurytides serville*				+							
107 *Eurytides columbus*				+		+					
108 *Eurytides orabilis*											
109 *Eurytides salvini*						+					
110 *Eurytides callias*				+	+						
111 *Eurytides dolicaon*			o	+							
112 *Eurytides iphitas*	V						+				
113 *Protographium leosthenes*					+						
114 *Lamproptera meges*				+							
115 *Lamproptera curius*				+	+						
116 *Graphium eurous*				+		+					
117 *Graphium mandarinus*						+	+	+			
118 *Graphium alebion*						+	+				
119 *Graphium tamerlanus*							+				
120 *Graphium aristeus*		*		+	+	+					
121 *Graphium nomius*					+						
122 *Graphium rhesus*				+		+					
123 *Graphium dorcus*									+		+
124 *Graphium androcles*						+	+				
125 *Graphium epaminondas*	K					+				+	
126 *Graphium euphrates*						+					
127 *Graphium decolor*				+							
128 *Graphium antiphates*				+							
129 *Graphium agetes*				+							
130 *Graphium stratiotes*							+				
131 *Graphium phidias*											
132 *Graphium encelades*				+		+					
133 *Graphium idaeoides*	R						+	+	+		
134 *Graphium delesserti*				+	+						
135 *Graphium xenocles*				+		+					
136 *Graphium macareus*				+	+						
137 *Graphium ramaceus*				+							
138 *Graphium megarus*		*		+	+						
139 *Graphium megaera*	I										
140 *Graphium stratocles*						+					
141 *Graphium deucalion*				+	+						
142 *Graphium thule*									+		
143 *Graphium endochus*					+	+					
144 *Graphium angolanus*				+	+		+				
145 *Graphium taboranus*						+					

Table 5.1 (cont.)

	CAT	LAW	R/F	A	B	C	D	E	F	G	H
146 *Graphium morania*					+						
147 *Graphium ridleyanus*				+		+	+				
148 *Graphium philonoe*						+	+				
149 *Graphium almansor*				+	+						
ssp *carchedonius*							+				
ssp *kigoma*							+	+			
150 *Graphium poggianus*											
151 *Graphium adamastor*				+	+		+				
152 *Graphium agamedes*				+	+		+				
153 *Graphium aurivilliusi*	K										
154 *Graphium olbrechtsi*											
155 *Graphium odin*				+	+		+				
156 *Graphium auriger*											
157 *Graphium hachei*					+						
158 *Graphium weberi*	K										
159 *Graphium fulleri*											
160 *Graphium ucalegonides*					+	+					
161 *Graphium ucalegon*				+	+						
162 *Graphium simoni*											
163 *Graphium cyrnus*				+	+	+			+		
164 *Graphium leonidas*				+	+	+					
165 *Graphium pelopidas*											
166 *Graphium levassori*	V										
167 *Graphium tynderaeus*				+	+						
168 *Graphium latreillanus*				+		+			+	+	
169 *Graphium junodi*											
170 *Graphium nigrescens*											
171 *Graphium polistratus*						+					
172 *Graphium policenes*				+	+						
173 *Graphium porthaon*						+					
174 *Graphium illyris*						+	+				
175 *Graphium gudenusi*						+					
176 *Graphium kirbyi*						+					
177 *Graphium colonna*						+					
178 *Graphium antheus*				+	+			+			
179 *Graphium evombar*				+	+			+			
180 *Graphium arycles*		*		+			+				
181 *Graphium bathycles*				+	+						
182 *Graphium chiron*				+							
183 *Graphium leechi*											
184 *Graphium procles*	I										
185 *Graphium meyeri*				+		+					
186 *Graphium eurypylus*		*		+	+	+					
187 *Graphium doson*				+	+	+					
188 *Graphium evemon*		*		+							
189 *Graphium agamemnon*			○	+	+	+					
190 *Graphium macfarlanei*			○		+	+					
191 *Graphium meeki*	R						+				
192 *Graphium gelon*							+				
193 *Graphium macleayanum*						+					
194 *Graphium weiskei*					+	+	+				
195 *Graphium stresemanni*	R							+	+	+	

Table 5.1 (cont.)

	CAT	LAW	R/F	A	B	C	D	E	F	G	H
196 *Graphium codrus*			o			+	+				
197 *Graphium empedovana*						+	+	+			
198 *Graphium cloanthus*				+	+		+				
199 *Graphium sumatranum*						+					
200 *Graphium monticolum*						+					
201 *Graphium milon*				+	+	+					
202 *Graphium sandawanum*	V								+	+	
203 *Graphium sarpedon*				+		+					
204 *Graphium mendana*	R		o								
205 *Graphium wallacei*			o			+					
206 *Graphium hicetaon*			o				+				
207 *Graphium browni*						+					
Tribe: TROIDINI											
208 *Battus philenor*				+							
209 *Battus zetides*	V										
210 *Battus devilliers*									+	+	
211 *Battus polydamas*				+		+					
212 *Battus streckerianus*											
213 *Battus archidamas*						+					
214 *Battus polystictus*			o	+		+					
215 *Battus philetas*						+					
216 *Battus madyes*				+	+	+					
217 *Battus eracon*											+
218 *Battus belus*				+		+					
219 *Battus crassus*				+	+						
220 *Battus laodamas*				+		+					
221 *Battus lycidas*				+		+					
222 *Euryades duponchelii*							+				+
223 *Euryades corethrus*						+	+				+
224 *Cressida cressida*					+	+	+				
225 *Parides gundlachianus*											
226 *Parides alopius*							+				
227 *Parides photinus*				+	+						
228 *Parides montezuma*				+	+	+					
229 *Parides phalaecus*											
230 *Parides agavus*			o		+						
231 *Parides proneus*					+						
232 *Parides ascanius*	V	*						+			
233 *Parides bunichus*						+					
234 *Parides diodorus*											
235 *Parides perrhebus*				+	+	+					
236 *Parides hahneli*	R		o						+	+	
237 *Parides mithras*					+						
238 *Parides chabrias*							+				
239 *Parides quadratus*											
240 *Parides pizarro*	K										
241 *Parides steinbachi*	K										
242 *Parides coelus*	K										
243 *Parides tros*							+	+			
244 *Parides aeneas*					+	+	+				

Table 5.1 (cont.)

		CAT	LAW	R/F	A	B	C	D	E	F	G	H
245	*Parides klagesi*	K										
246	*Parides orellana*											
247	*Parides childrenae*						+					
248	*Parides sesostris*				+	+	+					
249	*Parides burchellanus*	V										
250	*Parides polyzelus*				+		+					
251	*Parides iphidamas*				+	+						
252	*Parides vertumnus*				+	+	+	+				
253	*Parides cutorina*											
254	*Parides lycimenes*					+	+					
255	*Parides phosphorus*					+		+				
256	*Parides drucei*					+						
257	*Parides erlaces*				+	+	+					
258	*Parides nephalion*			○	+	+	+					
259	*Parides anchises*					+	+					
260	*Parides erithalion*				+		+					
261	*Parides panthonus*				+		+					
262	*Parides aglaope*						+					
263	*Parides castilhoi*											
264	*Parides lysander*				+	+	+					
265	*Parides echemon*				+			+				
266	*Parides zacynthus*			○	+	+		+				
267	*Parides neophilus*				+		+					
268	*Parides eurimedes*				+	+						
269	*Parides timias*				+							
270	*Atrophaneura antenor*						+	+				
271	*Atrophaneura daemonius*											
272	*Atrophaneura plutonius*		*									
273	*Atrophaneura alcinous*			○	+		+	+				
274	*Atrophaneura latreillei*		*		+	+	+					
275	*Atrophaneura polla*		*					+				
276	*Atrophaneura crassipes*		*									
277	*Atrophaneura adamsoni*						+	+				
278	*Atrophaneura nevilli*		*									
279	*Atrophaneura laos*											
280	*Atrophaneura mencius*											
281	*Atrophaneura impediens*											
282	*Atrophaneura febanus*				+							
283	*Atrophaneura hedistus*											
284	*Atrophaneura dasarada*						+	+				
285	*Atrophaneura polyeuctes*			○	+		+					
286	*Atrophaneura semperi*						+	+				
287	*Atrophaneura kuehni*						+	+				
288	*Atrophaneura luchti*	R										
289	*Atrophaneura hageni*							+	+			
290	*Atrophaneura priapus*							+		+		
291	*Atrophaneura sycorax*							+	+	+		
292	*Atrophaneura horishanus*				+	+						
293	*Atrophaneura aidoneus*							+				
294	*Atrophaneura varuna*				+	+						
295	*Atrophaneura zaleucus*							+	+			
296	*Atrophaneura nox*				+	+						

Table 5.1 (cont.)

		CAT	LAW	R/F	A	B	C	D	E	F	G	H
297	*Atrophaneura dixoni*						+	+				
298	*Atrophaneura neptunus*					+	+	+				
299	*Atrophaneura palu*	K							+			
300	*Atrophaneura coon*		*			+	+	+	+			
301	*Atrophaneura rhodifer*								+			
302	*Atrophaneura hector*				+	+	+					
303	*Atrophaneura jophon*	V	*									
304	*Atrophaneura pandiyana*								+			
305	*Atrophaneura oreon*						+					
306	*Atrophaneura liris*						+					
307	*Atrophaneura polyphontes*					+	+					
308	*Atrophaneura schadenbergi*	V										
309	*Atrophaneura mariae*				+		+					
310	*Atrophaneura phegeus*						+					
311	*Atrophaneura phlegon*					+	+					
312	*Atrophaneura atropos*	I						+				
313	*Atrophaneura kotzebuea*				+							
314	*Atrophaneura aristolochiae*			o	+	+	+	+				
315	*Atrophaneura polydorus*			o	+	+	+	+				
316	*Trogonoptera brookiana*		*	o	+	+	+	+				
317	*Trogonoptera trojana*						+	+				
318	*Troides hypolitus*		*				+	+	+			
319	*Troides cuneifer*						+	+	+	+		
320	*Troides amphrysus*		*				+	+		+		
321	*Troides miranda*		*					+	+	+	+	
322	*Troides andromache*	I	*					+			+	
323	*Troides magellanus*						+	+				
324	*Troides prattorum*	I							+	+	+	
325	*Troides minos*					+						
326	*Troides aeacus*					+	+	+				
327	*Troides plateni*						+					
328	*Troides rhadamantus*		*				+					
329	*Troides dohertyi*	V	*				+	+				
330	*Troides helena*		*			+	+		+			
331	*Troides oblongomaculatus*			o		+	+	+	+			
332	*Troides darsius*					+						
333	*Troides criton*		*				+	+				
334	*Troides riedeli*		*					+				
335	*Troides plato*		*									
336	*Troides vandepolli*		*					+	+		+	
337	*Troides haliphron*		*				+	+				
338	*Troides staudingeri*		*									
339	*Ornithoptera goliath*		*						+	+	+	
340	*Ornithoptera tithonus*	K	*					+	+	+	+	
341	*Ornithoptera rothschildi*	I	*					+	+		+	
342	*Ornithoptera chimaera*	I	*					+	+			
	ssp. *charybdis*										+	
343	*Ornithoptera paradisea*	I	*						+	+	+	+
344	*Ornithoptera meridionalis*	V	*								+	+
345	*Ornithoptera alexandrae*	E	*							+		+
346	*Ornithoptera victoriae*		*					+		+		
	ssp. *regis*										+	

Table 5.1 (cont.)

	CAT	LAW	R/F	A	B	C	D	E	F	G	H
347 *Ornithoptera aesacus*	I									+	+
348 *Ornithoptera croesus*	V					+	+	+		+	
349 *Ornithoptera priamus*		*	○			+	+	+	+		
ssp. *admiralitatis*		*	?			+	+				
ssp. *arruana*		*				+	+				
ssp. *boisduvali*		*					+				
ssp. *bornemanni*		*				+	+				
ssp. *caelestis*		*	○			+	+	+	+		
ssp. *demophanes*		*				+	+				
ssp. *hecuba*		*									
ssp. *miokensis*		*	?				+				
ssp. *poseidon*		*	○			+					
ssp. *priamus*		*				+		+			
ssp. *pronomus*								+			
ssp. *teucrus*		*				+					
ssp. *urvilliana*		*	○			+	+				
350 *Ornithoptera euphorion*		*					+	+			
351 *Ornithoptera richmondia*							+				
Tribe: PAPILIONINI											
352 *Papilio glaucus*			○		+	+					
353 *Papilio alexiares*											
354 *Papilio multicaudatus*				+		+					
355 *Papilio rutulus*					+	+					
356 *Papilio eurymedon*					+	+					
357 *Papilio pilumnus*						+	+				
358 *Papilio palamedes*					+		+				
359 *Papilio troilus*				+		+					
360 *Papilio ascolius*							+				
361 *Papilio neyi*											
362 *Papilio zagreus*					+	+					
363 *Papilio bachus*					+	+					
364 *Papilio hellanichus*									+		
365 *Papilio scamander*			○		+	+		+			
366 *Papilio xanthopleura*											
367 *Papilio birchalli*											
368 *Papilio cleotas*			○			+					
369 *Papilio menatius*				+	+	+	+				
370 *Papilio phaeton*											
371 *Papilio victorinus*				+	+						
372 *Papilio garamas*						+					
373 *Papilio abderus*						+					
374 *Papilio homerus*	E									+	+
375 *Papilio warscewiczi*						+					
376 *Papilio cacicus*						+					
377 *Papilio euterpinus*											
378 *Papilio esperanza*	V										
379 *Papilio andraemon*						+					
380 *Papilio machaonides*									+		
381 *Papilio thersites*								+			
382 *Papilio astyalus*			○	+	+						
383 *Papilio androgeus*				+	+	+				+	

Table 5.1 (cont.)

	CAT	LAW	R/F	A	B	C	D	E	F	G	H
384 *Papilio ornythion*				+			+				
385 *Papilio aristodemus*								+			
ssp. *ponceanus*	E	*									
386 *Papilio caiguanabus*	I										
387 *Papilio aristor*	I										
388 *Papilio thoas*			○	+	+	+					
389 *Papilio cinyras*				+							
390 *Papilio homothoas*						+					
391 *Papilio cresphontes*				+							
392 *Papilio paeon*				+	+	+					
393 *Papilio garleppi*	K							+			
394 *Papilio torquatus*				+		+					
395 *Papilio hectorides*			○	+	+						
396 *Papilio lamarchei*											
397 *Papilio himeros*	V										
398 *Papilio hyppason*					+	+					
399 *Papilio pelaus*											
400 *Papilio oxynias*											
401 *Papilio epenetus*											
402 *Papilio erostratus*								+			
403 *Papilio erostratinus*											
404 *Papilio pharnaces*				+							
405 *Papilio chiansiades*				+	+						
406 *Papilio dospassosi*											
407 *Papilio anchisiades*			○	+	+						
408 *Papilio maroni*	K										
409 *Papilio rogeri*				+		+					
410 *Papilio isidorus*				+	+						
411 *Papilio rhodostictus*											
412 *Papilio anactus*						+	+				
413 *Papilio elwesi*								+			
414 *Papilio maraho*	V							+			
415 *Papilio clytia*		*		+	+	+					
416 *Papilio paradoxa*		*		+	+	+					
417 *Papilio veiovis*						+	+				
418 *Papilio osmana*	V										
419 *Papilio carolinensis*	V										
420 *Papilio agestor*				+	+		+				
421 *Papilio epycides*		*		+		+	+				
422 *Papilio slateri*		*		+		+					
423 *Papilio laglaizei*						+	+	+			
424 *Papilio toboroi*	R					+					
425 *Papilio moerneri*	V										
426 *Papilio alexanor*						+					
427 *Papilio hospiton*	E	*									
428 *Papilio machaon*		*	○	+	+	+					
429 *Papilio hippocrates*						+					
430 *Papilio zelicaon*					+						
431 *Papilio indra*						+			F		
432 *Papilio polyxenes*				+	+	+					
433 *Papilio brevicauda*											
434 *Papilio oregonia*						+					

Table 5.1 (cont.)

	CAT	LAW	R/F	A	B	C	D	E	F	G	H
435 *Papilio kahli*							+				
436 *Papilio nitra*						+					
437 *Papilio coloro*							+				
438 *Papilio bairdi*							+				
439 *Papilio joanae*											
440 *Papilio xuthus*			○	+	+						
441 *Papilio benguetanus*	V						+				
442 *Papilio euchenor*						+	+				
443 *Papilio gigon*						+					
444 *Papilio demolion*				+	+	+					
445 *Papilio liomedon*		*									
446 *Papilio protenor*			○	+	+	+					
447 *Papilio demetrius*			○	+							
448 *Papilio macilentus*				+		+					
449 *Papilio bootes*		+			+	+					
450 *Papilio janaka*						+					
451 *Papilio lampsacus*							+	+			
452 *Papilio forbesi*					+	+	+				
453 *Papilio acheron*	R							+			
454 *Papilio oenomaus*							+				
455 *Papilio ascalaphus*						+					
456 *Papilio rumanzovia*					+	+					
457 *Papilio deiphobus*					+	+	+				
458 *Papilio alcmenor*				+	+	+					
459 *Papilio thaiwanus*				+	+						
460 *Papilio polymnestor*						+					
461 *Papilio lowi*					+	+					
462 *Papilio memnon*			○	+	+	+	+				
463 *Papilio mayo*		*						+			
464 *Papilio noblei*							+				
465 *Papilio antonio*											
466 *Papilio biseriatus*											
467 *Papilio iswara*				+	+						
468 *Papilio hystaspes*						+					
469 *Papilio sataspes*						+					
470 *Papilio helenus*			○	+	+	+					
471 *Papilio iswaroides*					+	+					
472 *Papilio jordani*	R										
473 *Papilio polytes*			○	+	+	+					
474 *Papilio alphenor*				+							
475 *Papilio ambrax*			○		+	+					
476 *Papilio phestus*						+					
477 *Papilio diophantus*						+	+	+			
478 *Papilio nephelus*				+	+	+					
479 *Papilio castor*				+							
480 *Papilio mahadeva*				+		+		+			
481 *Papilio dravidarum*							+				
482 *Papilio hipponous*						+					
483 *Papilio pitmani*					+		+				
484 *Papilio fuscus*		*	○	+	+	+					
485 *Papilio canopus*											
486 *Papilio albinus*			○		+	+					

Table 5.1 (cont.)

		CAT	LAW	R/F	A	B	C	D	E	F	G	H
487	*Papilio hypsicles*						+					
488	*Papilio woodfordi*			○	+		+	+				
489	*Papilio heringi*											
490	*Papilio amynthor*											
491	*Papilio schmeltzii*						+					
492	*Papilio godeffroyi*											
493	*Papilio inopinatus*							+	+			
494	*Papilio bridgei*			○			+	+	+			
495	*Papilio weymeri*	R		○			+	+				
496	*Papilio gambrisius*						+	+				
497	*Papilio tydeus*							+				
498	*Papilio aegeus*			○	+	+	+					
499	*Papilio cynorta*				+		+	+				
500	*Papilio plagiatus*						+	+	+			
501	*Papilio zoroastres*				+							
502	*Papilio echerioides*						+					
503	*Papilio jacksoni*					+	+					
504	*Papilio fuelleborni*											
505	*Papilio sjoestedti*	R									+	+
506	*Papilio rex*						+					
507	*Papilio epiphorbas*				+	+	+					
508	*Papilio manlius*	I										
509	*Papilio phorbanta*	V	*									
510	*Papilio charopus*					+	+					
	ssp. *juventus*								+			
511	*Papilio hornimani*						+					
512	*Papilio mackinnoni*				+	+		+				
513	*Papilio sosia*				+							
514	*Papilio aethiopsis*						+					
515	*Papilio nireus*				+	+	+					
516	*Papilio aristophontes*	I										
517	*Papilio oribazus*				+	+	+					
518	*Papilio thuraui*								+			
519	*Papilio desmondi*						+		+			
	ssp. *teita*	E										
520	*Papilio interjecta*											
521	*Papilio bromius*				+	+	+					
522	*Papilio chrapkowskii*											
523	*Papilio zalmoxis*						+	+				
524	*Papilio gallienus*				+	+						
525	*Papilio mechowi*				+	+						
526	*Papilio zenobius*						+					
527	*Papilio mechowianus*						+					
528	*Papilio andronicus*											
529	*Papilio zenobia*				+	+	+					
530	*Papilio maesseni*											
531	*Papilio antimachus*	R					+	+		+		
532	*Papilio ophidicephalus*						+					
533	*Papilio menestheus*					+	+					
534	*Papilio lormieri*				+	+						
535	*Papilio morondavana*	V						+	+	+	+	
536	*Papilio grosesmithi*	R					+	+	+	+		

Table 5.1 (cont.)

	CAT	LAW	R/F	A	B	C	D	E	F	G	H
537 *Papilio erithonioides*				+	+	+		+			
538 *Papilio demodocus*				+	+	+					
539 *Papilio demoleus*			o	+	+						
540 *Papilio leucotaenia*	V						+	+			
541 *Papilio delalandei*				+	+	+					
542 *Papilio mangoura*	R					+	+				
543 *Papilio constantinus*						+					
544 *Papilio phorcas*				+	+	+					
545 *Papilio dardanus*				+	+	+	+				
546 *Papilio euphranor*								+			
547 *Papilio pelodorus*							+				
548 *Papilio hesperus*				+	+	+					
549 *Papilio horribilis*								+			
550 *Papilio nobilis*							+	+			
ssp. *crippsianus*								+		+	
551 *Papilio crino*					+	+	+				
552 *Papilio blumei*						+	+				
553 *Papilio buddha*		*						+			
554 *Papilio palinurus*						+	+				
555 *Papilio daedalus*						+	+				
556 *Papilio chikae*	E									+	+
557 *Papilio maackii*			o			+	+				
558 *Papilio bianor*			o	+	+	+					
559 *Papilio syfanius*								+			
560 *Papilio polyctor*						+	+	+			
561 *Papilio dialis*				+	+						
562 *Papilio elephenor*		*									
563 *Papilio paris*			o	+	+	+	+				
564 *Papilio karna*						+	+	+			
565 *Papilio arcturus*						+	+				
566 *Papilio hoppo*				+	+	+					
567 *Papilio krishna*									+		
568 *Papilio neumoegeni*	V						+	+	+		
569 *Papilio peranthus*						+	+				
570 *Papilio pericles*						+	+				
571 *Papilio lorquinianus*						+	+	+			
572 *Papilio ulysses*		*	o	+	+	+	+				
573 *Papilio montrouzieri*							+	+			

6

Reviews of threatened species

The following 78 reviews summarize our present knowledge of those swallowtail species for which some degree of threat has been established or is suspected. The order of appearance is given on the Contents pages. A list of the 78 species sorted into their threatened categories is given in Appendix A at the end of this section (p. 369).

The criteria used in selecting species for full review have been outlined in section 1, How to use this book. There it was emphasised that Red Data Book categories are constantly being updated and changed. There are always species that could arguably be in a higher category, while others could be in a lower one. To be categorized as threatened (i.e. Endangered, Vulnerable, Rare or Indeterminate) a species must be at risk on a world scale, hence the absence of an RDB category for a particular species implies nothing about localized threats. Species not given a full review may nevertheless be threatened over part of their range, and the full species list in section 3 (p. 33) should be consulted for details.

As noted in section 1, there are very many species that have not been given an RDB category but for which monitoring and further information are still required. Those species, of which there are almost 100, are listed in Appendix B (p. 371).

We would be grateful if readers could bring new information and shortcomings in these reports to our attention.

Baronia brevicornis Salvin, 1893   RARE

Subfamily   BARONIINAE

Summary *Baronia brevicornis* (the Baronia) is the most primitive living swallowtail, a 'living fossil' of great scientific interest. The species is confined to Mexico and although it is well distributed there, its populations should continue to be monitored, studied and protected.

Description *Baronia brevicornis* is a dark brown tailless butterfly with variable ochreous markings (Plate 1.1). The wingspan is 55–65 mm, and the female is larger than the male but otherwise similar in appearance (2). Both sexes are rather variable in colour and some individuals, particularly females, are much darker than others (3). The antennae are very short (11). Although the adults have the fore-tibial spur which is characteristic of the Papilionidae, the wing venation is very uncharacteristic in having both anal veins on the hindwing, like Pieridae and Nymphalidae (8, 12). It must therefore be assumed that *Baronia brevicornis* is primitive and diverged before the time when the vein was lost by other members of the Papilionidae (12).

Male: UFW ground colour brown with yellow–ochre oval spots forming three arches on the wing.

UHW with yellow–ochre markings which include two large patches, one covering two-thirds of the cell, the other a triangle below the cell, and a discal row of spots.

LFW/LHW paler than the upper surface. The subapical spots of the forewing, and all of the hindwing spots are silver.

The female is larger than the male, with more extensive yellow–ochre markings.

The mature larva has a black, yellow or green (15) head and a green body with small tubercles which may be either black or yellow (6, 8, 12). The body has a continuous yellow dorsal stripe and white transverse lines on each segment (6, 8, 12). In common with all other Papilionidae, the caterpillar possesses a defensive osmeterium (4, 6, 8, 11, 12).

Distribution This species is restricted to areas where its foodplant occurs in the Sierra Madre del Sur in south-west Mexico. It has been recorded from localities in the States of Jalisco, Oaxaca, Chiapas, Guerrero, Morelos, Puebla, Michoacán, and Colima (6, 9, 10, 12, 14). Its occurrence is very local.

This distribution is somewhat enigmatic for such a primitive species, being far removed from South-East Asia, the region which, on grounds of having the greatest diversity of swallowtails, has often been presumed to be their centre of origin (12). However, in a recent assessment Hancock has succinctly re-evaluated the data and proposed from fossil and geological evidence that North America holds the key to the ancestry of the Papilionidae (5). *Praepapilio colorado*, a fossil papilionid from middle Eocene deposits in Colorado, U.S.A., shares some of the primitive features found in *Baronia* (5) and might easily be included within that genus (1). At the time when the family was differentiating, Mexico was separated from South America by a considerable body of water and the western Sierra Madre was a peninsula, presumably with *Baronia brevicornis* inhabiting the tip (12). The impact of physiographic, climatic and vegetational factors on the distribution of *Baronia brevicornis* has been assessed, and the presence of disjunct populations as far south as Oaxaca and Chiapas is believed to be the result of southward movements of the lepidopteran fauna during the Pleistocene Ice Ages (10). The present extent of the range of *Baronia brevicornis* is limited to alititudes of 505–1335 m (10), where its foodplant grows.

Habitat and Ecology *Baronia brevicornis* was described in 1893 (11) but it was not until 1961 that Vázquez and Pérez elucidated some of its fascinating life history (10, 14). The habitat is thickets of deciduous scrub composed mainly of the very unusual foodplant *Acacia cochliacantha* (Leguminosae) (not *A. cymbispina*, as is often quoted) (10). The *Acacia* is known locally as 'cubata' and the thickets are called 'cubatera' (14). Other plants in the thickets are *Bursera* spp. (Torchwood), *Ipomoea* spp. (Morning Glory), *Mimosa polyantha* and *Neobuxbaumia mezcalaensis* (12). The *Acacia* trees are bare until the first rains come and, as the leaves grow out, the butterflies emerge. Females lay their eggs singly, few on each tree (14). The butterfly is said to be single-brooded, and the larvae dig themselves into loose ground for pupation (10, 14). The adults have a short flight range and the males are territorial, defending their areas from intrusion by assailing intruders from a vantage point high in the branches of an *Acacia* tree (7). At Morelos during July, both males and females come down to drink on the sandy banks of small streams (7).

At suitable localities and in the right season *Baronia* may be quite numerous (7, 10). The population in the Mezcala region of Guerrero State has been studied for ten years (9). Questions elucidated in that study include the influence of rainfall on adult emergence and the duration of egg-laying periods, larval stages and adult populations (9). Population parameters have been described, notably the drop in populations during the 1960s and the impact of natural predators on the species (9). The eggs of *Baronia brevicornis* are parasitized by chalcid wasps of the family Trichogrammatidae. The caterpillars suffer little predation, possibly because of the osmeterium (see below), but some are taken by bugs of the family Pentatomidae. Adults are devoured by spiders in the Argiopidae and by robber flies (Asilidae) which catch them in flight (9).

As noted above, *Baronia brevicornis* is the 'living fossil' of the Papilionidae. Because of its primitive characteristics and unusual distribution and feeding habits, it is of very great scientific interest. Studies of the osmeterial secretions of *Baronia* showed them to be mainly composed of two aliphatic acids, isobutyric acid and 2-methyl butyric acid (4). When disturbed, the larvae extrude the osmeterial gland and try to wipe it against the attacker. The secretion is strongly odorous and visably coats the erected gland (4). In *Papilio machaon* at least, the same acids are strongly repellent to ants (4). Other studies have demonstrated the presence of cholesterol, three saturated hydrocarbons and two monounsaturated acids in the caterpillar bodies, although the siginificance of these discoveries is not clear (16).

Threats *Baronia brevicornis* is under no immediate threat, but with such a limited distribution it must be considered to be rare on a world basis. Being of such great evolutionary interest, the populations attract considerable attention from collectors and scientists worldwide. The human population of Mexico is growing at an alarming rate and this may eventually lead to degradation of the *Acacia* thickets in which the butterflies breed.

Conservation Measures The Mexican Government should recognize the international importance of *Baronia brevicornis* and, whilst ensuring that populations are protected from excessive exploitation, should encourage further research and study by local and expatriate scientists. The extent of *Acacia* thickets within national park boundaries is unknown. An assessment of these, and their utilization by *Baronia* would be a necessity for its long-term protection from habitat destruction. A colony of *Baronia* does occur within the Cañon del Sumidero National Park near Tuxta Gutiérrez, Chiapas (7).

References

1. Brown, K.S., Jr. (1982). Historical and ecological factors in the biogeography of aposematic neotropical butterflies. *American Zoologist* 22: 453–471.
2. D'Abrera, B. (1981). *Butterflies of the Neotropical Region. Part I. Papilionidae and Pieridae*. Lansdowne Editions, Melbourne, xvi + 172 pp.
3. Diaz Francés, A. and Maza E. J. de la, (1978). Guía illustrada de las mariposas Mexicanas. Parte 1. Familia Papilionidae. *Sociedad Mexicana de Lepidopterologiá A.C. Publicaciones Especiales* 3: 15 pp.
4. Eisner, T., Pliske, T.E., Ikeda, M., Owen, D.F., Vazquez, L., Perez, H., Franclemont, J.G. and Meinwald, J. (1970). Defense mechanisms of arthropods. XXVII. Osmeterial secretions of papilionid caterpillars (*Baronia, Papilio, Eurytides*). *Annals of the Entomological Society of America* 63: 914–915.
5. Hancock, D.L. (1983). Classification of the Papilionidae (Lepidoptera): a phylogenetic approach. *Smithersia* 2: 1–48.
6. Igarashi, S. (1979). *Papilionidae and Their Early Stages*. Vol 1: 219 pp., Vol 2: 102 pp. of plates. Kodansho, Tokyo. (In Japanese).
7. Maza E. J. de la (1984). *In litt.*, 3 January.
8. Pérez R., H. (1969). Quetotáxia y morfología de la oruga de *Baronia brevicornis* Salv. (Lepidoptera Papilionidae Baroniinae). *Anales del Instituto de Biología Universidad de México*. 40, Ser. Zool. (2): 227–244.
9. Pérez R., H. (1971). Algunas consideraciones sobre la población de *Baronia brevicornis* Salv. (Lepidoptera, Papilionidae, Baroniinae) en la región de Mezcala, Guerrero. *Anales del Instituto de Biología Universidad de México* 42, Ser. Zool. (1): 63–72.
10. Pérez R., H. (1977). Distribución geográfica y estructura poblacional de *Baronia brevicornis* Salv. (Lepidoptera, Papilionidae, Baroniinae) en la República Mexicana. *Anales del Instituto de Biología Universidad de México* 48, Ser. Zool. (5): 151–164.
11. Salvin, O. (1893). Description of a new genus and species of Papilionidae from Mexico. *Transactions of the Entomological Society of London* 1893 (4): 331–332.
12. Tyler, H.A. (1975). *The Swallowtail Butterflies of North America* Naturegraph, California. viii + 192 pp., 16 pl.
13. Tyler, H.A. (1983). *In litt.*, 13 March.
14. Vázquez G., L. and Pérez R., H. (1961). Observaciones sobre la biología de *Baronia brevicornis* Salv. (Lepidoptera: Papilionidae—Baroniinae). *Anales del Instituto de Biología Universidad de México*. 32: 295–311.
15. Vázquez G., L. and Pérez R., H. (1967). Nuevas observaciones sobre le biología de *Baronia brevicornis* Salv. Lepidoptera: Papilionidae—Baroniinae. *Anales del Instituto de Biología Universidad de México*. 37: 195–204.
16. Yuste, F., Perez, H. and Walls, F. (1972). Compounds of papilionid caterpillars (*Baronia brevicornis* S.). *Experientia* 28: 1149.

Parnassius autocrator Avinoff, 1913   RARE

Subfamily PARNASSIINAE   Tribe PARNASSIINI

Summary *Parnassius autocrator* flies at high altitudes in the Hindu Kush of Afghanistan and the Pamir mountains of Tadzhikskaya S.S.R. It is extremely local in its distribution and possibly threatened by degradation of high mountain pastures in the U.S.S.R. The status of the species in Afghanistan is presently unknown.

Description *Parnassius autocrator* was originally described as a subspecies of *P. charltonius* Gray. Bryk agreed with this (3) and Munroe omitted the species from his list (6), but recent authors consider it to be a full species (4, 5, 8, 9, 11). *Parnassius autocrator* is somewhat unusual in appearance, lacking the red spots so characteristic of many other *Parnassius* species, but with yellow marks on the hind wings (8). The sexes differ in that the female has a wide orange patch on the UHW, while the male has only a wavy yellow–orange line (1) (Plates 1.3 and 1.4).

UFW/UHW chalky white with shadowy black markings and transparent apices. Hind wings with a row of black submarginal lunules and a bright yellow postdiscal bar. The undersides are similar.

Distribution Restricted to a narrow range in the Pamir Mountains of south-eastern Tadzhikistan (Tadzhikskaya) S.S.R. in the U.S.S.R. and the Hindu Kush in north-eastern Afghanistan (1, 4, 11). Very few specimens have been taken in the U.S.S.R. and it is not known to complete its life cycle there (7).

Habitat and Ecology *Parnassius autocrator* is a member of the *charltonius* species group (5), whose young stages feed on Fumariaceae. The foodplant of *P. autocrator* is *Corydalis adiantifolia* (and possibly *C. hindukushensis*), a very local plant that grows on steep, rocky slopes and cliffs at altitudes around 3000 m and above (4, 11). In north-eastern Afghanistan the butterflies are found in the narrow steep-sided valleys where the foodplant occurs (11). They fly at altitudes between 2800 and 4000 m, but are most numerous between 3200 and 3500 m (11), sometimes in the company of *P. charltonius* (4, 11). The adults emerge around the second week of July and are on the wing for four weeks. Females flutter around the foodplant or settle on stones or thistle flowers, taking off at the slightest disturbance (11). The males are more active, soaring about a metre from the ground up and down the slopes, mainly between 09.00 and 15.00 hrs (11).

The climate at these altitudes in the Hindu Kush is dry and hot in summer with deep snow in winter. The landscape in summer is bare and dry with thorny, succulent and aromatic shrubs growing on rocky bare earth and scree (11). This is in great contast to the meadows in which alpine parnassians such as *P. phoebus* fly. The U.S.S.R. populations of *P. autocrator* are said to fly in pastures (9) but the composition and extent of these habitats are unknown.

Threats Listed in the U.S.S.R. Red Data Book (2) and stated by Tanasiychuk (9) to be an extremely rare species in the U.S.S.R., threatened with extinction because of degradation of high mountain pastures. No further detail is given and the statement is taken to be rather speculative. However, the habitat and foodplants of *Parnassius autocrator* may be damaged by heavy grazing of sheep and goats. The species is known to be extremely local in north-eastern Afghanistan, but no threats are known (11). The region is at present inaccesible, but there are reasons to fear that environmental degradation may be an increasing problem.

Conservation Measures No protected areas are known in this region and there is very little information on the extent of human impact on the fauna and flora of the high pastures where the butterfly lives. The main part of the butterfly's range is clearly in Afghanistan and conservation efforts should be directed to that country. More data are required on the precise distribution, biology and habitat requirements of *Parnassius autocrator*. When the region is once again open to scientists it will be necessary to develop a conservation plan.

References

1. Ackery, P.R. (1975). A guide to the genera and species of Parnassiinae (Lepidoptera: Papilionidae). *Bulletin of the British Museum (Natural History). Entomology* 31: 71–105.
2. Bannikov, A.G. and Sokolov, V.I. (eds) (1984) *The Red Data Book of the USSR. Rare and Threatened Species of Animals and Plants.* Lesnaya Prom. Press, Moscow. (In Russian).
3. Bryk, F. (1935). Lepidoptera, Parnassiidae. Part 2. (Subfam. Parnassiinae). *Tierreich* 65: 51 + 790 pp., 698 figs.
4. Eisner, C. and Naumann, C.M. (1980). Parnassiana Nova LVII. Beitrag zur Ökologie und Taxonomie der Afghanischen Parnassiidae (Lepidoptera). *Zoologische Verhandelingen* 178: 35 pp. + 9 pl.
5. Hancock, D.L. (1983). Classification of the Papilionidae (Lepidoptera): a phylogenetic approach. *Smithersia* 2: 1–48.
6. Munroe, E. (1961). The classification of the Papilionidae (Lepidoptera). *Canadian Entomologist* Supplement 17: 1–51.
7. Nekrutenko, Y.P. (1984). *In litt.*, 20 June.
8. Smart, P. (1975). *The Illustrated Encyclopedia of the Butterfly World.* Hamlyn, London. 275 pp.
9. Tanasiychuk, V.N. (1981). Data for the "Red Book" of Insects of the U.S.S.R. *Entomologicheskoye Obozreniye* 60 (3): 168–186.
10. Turlin, B. (1984). *In litt.*, 1 June.
11. Wyatt, C. and Omoto, K. (1963). Auf der Jagd nach *Parnassius autocrator* Avin. *Zeitschrift der Wiener Entomologischen Gesellschaft* 48: 163–170.

Parnassius apollo (Linnaeus, 1758)   RARE

Subfamily PARNASSIINAE   Tribe PARNASSIINI

Summary *Parnassius apollo*, the Apollo butterfly, was the first insect to be included under the Convention on International Trade in Endangered Species of Wild Fauna and Flora (CITES). Prone to local subspeciation in montane and northern Eurasia, the Apollo has many named regional populations. A number of these are Extinct or Endangered although others are still numerous.

Description The Apollo is a medium-sized (50–80 mm wingspan) butterfly with rounded, chalky white wings, with black spots, grey markings and transparent areas lacking scales. The pattern, density and intensity of marking varies according to locality, but the hindwings always have striking scarlet spots. The antennal shaft is grey with darker rings (1, 23).

Distribution *Parnassius apollo* was formerly widely distributed in Europe and Asia, although its range is now somewhat depleted. In northern Europe it has been recorded from Norway, Denmark, Sweden and Finland across to Latvia and central Siberia in the U.S.S.R. In central and southern Europe it has been recorded in Spain, France, Switzerland, Italy, Liechtenstein, Austria, West Germany, Netherlands (vagrants?), East Germany, Poland, Czechoslovakia, Hungary, Yugoslavia, Greece, Bulgaria, Turkey, Romania, and Syria, eastwards into the Ukraine, Bol'shoy Kavkaz (Caucasus) and Siberia in the U.S.S.R., China (Sinkiang) and Mongolia (3, 4, 5, 8, 18, 19, 21, 22, 23, 35). Populations are often isolated, disjunct and local.

Habitat and Ecology This butterfly, a relict of the glacial epoch, occurs in subalpine situations between 750 and 2000 m in the Alps and associated ranges, but near sea level in the northern parts of its range. The larvae feed exclusively on stonecrops (*Sedum* spp.) (24). There is one generation per year, with over-wintering in the egg stage. Its population biology differs from one colony to another and individuals may be numerous or extremely rare. The Apollo seems to be able to withstand a certain degree of grazing of its habitats, which are commonly rocky and relatively xeric (5). Major alterations affecting the host plant are not tolerated (21, 23). Additional details for specific races are available in the literature (12, 26, 28, 30, 37, 39, 40).

Threats *Parnassius apollo* is reported to have declined and become rare or endangered in all or part of the following countries: Bulgaria, Czechoslovakia, Finland (26, 27), France, West Germany (2, 5, 6, 35), East Germany (extinct), Greece, Italy, Liechtenstein (4), Netherlands, Norway, Poland (11), Romania, Spain (38), Sweden (25), Switzerland (9, 17) and the U.S.S.R. (Ukraine, Carpathians and Crimea) (21, 22). The species has long been prized by collectors, who aim to possess as many of the different variants as possible. The trade therefore comprises a substantial value and volume (34). Rare or very limited subspecies may command impressive prices. There is disagreement over the importance of over-collecting in bringing about the observed declines. Collecting has been considered a potential threat in Finland given the current depleted condition of populations (26, 27). Over-collecting is reported to have occurred in Spain (34), Silesia (26, 27) and Italy, where an effort was made to extirpate purposefully a rare, local race of Apollo in order to enhance the value of specimens already in hand (7). Market collecting has also been held partly responsible for the Polish decline of the species (12). None of these assertions contain numerical data. One lepidopterist, finding seven Apollos killed along 1 km of road, suggested that vehicles were a greater mortality factor than collectors (36), and vehicles on a motorway system near Bozen, South Tyrol (Italy) nearly wiped out a local race of Apollo (34). However, most observers agree that it is not collectors or cars, but 'Tannensoldaten' (ranks of conifers in plantations) that threaten Europe's parnassians. Examples of afforestation interfering with Apollo survival have been cited in Bavaria (10, 33), Poland (12), Spain (20) and Switzerland (21). Succession of suitable habitat to scrubland is another recurring threat, for instance in Poland and Switzerland (21). Climatic change (Finland) and acid rain (Norway) are among purported causes of the widespread withdrawal of *P. apollo* from much of Fennoscandia (21), although agriculture has also been blamed in Finland (27) and the causes in Sweden are not properly understood (25). Urbanization has been given as another reason for decline in Finland and tourist development is implicated in Bulgaria (21). On the whole, habitat change seems to be a far more important threat to Apollo than collecting. Whatever the major cause or causes, the effects are dramatic, at least locally: in France, two subspecies are Extinct, one Endangered and seven Vulnerable (16, 21); in East Germany, the insect is Extinct (21) and in Poland the decline is widespread and continuing: between 1950 and 1965 three colonies disappeared from Pieniny National Park and many other Polish populations are considered Endangered (12–16, 30, 31, 32, 40).

Conservation Measures *Parnassius apollo* was the first invertebrate to be included in Appendix II of the Convention on International Trade in Endangered Species of Wild Fauna and Flora (CITES), requiring monitoring of imports and exports to and from signatory states. Laws exist (21) at the national, regional or local level to protect *P. apollo* in the following countries: Austria, Czechoslovakia, Finland, France, Greece, Netherlands, Switzerland, West Germany (since the 19th century (34)) and Poland. However, these laws and regulations usually address protection of individuals (i.e. collecting restrictions) instead of habitats, and may have little effect. For example, despite legal protection accorded in 1952, the Apollo has continued to decline dramatically in Poland (11, 12, 30, 32). *P. apollo* does exist in a number of national parks and other protected areas, such as Tatra and Pieniny National Parks in Poland, but this does not always lead to its conservation (15) since specific management measures have seldom been taken. An attempt has been made to reintroduce *P. apollo* into the Pieniny Mountains in Poland (16, 31); the results are not yet known. Publicity items have been produced stressing the need for conservation of *P. apollo*, such as a poster issued by the Fedération Française des Sociétés de Sciences Naturelles, depicting four of the threatened French races of the butterfly.

Many authors have called for specific measures on behalf of *P. apollo*. In Poland the status of the butterfly inside and outside the national parks should be assessed in order to identify the factors detrimental to its survival, and the government should implement the aims of the 1952 law protecting the butterfly (15, 31). Reserves for the Apollo are needed in the mountains of Crimea (39). The Spanish Lepidoptera Red Data Book (38) offers suggestions for conserving the butterfly in the Pyrenees, and strong measures (including a new national park for this and other wildlife) are recommended in a later Spanish appraisal (20). Artificial breeding of two generations per year (instead of the usual one) has been suggested as a means of reinforcing populations or enhancing reintroduction attempts in Scandinavia (28, 29). Reintroductions should be attempted only with larvae or pupae, to avoid emigration of the adults (29). Past attempts to release butterflies have not succeeded. All countries where the species occurs should sponsor detailed surveys of Apollo colonies, and take steps to establish reserves and implement management measures where possible. Reforestation with conifers should be avoided and scrub succession should be arrested in important Apollo habitats. The provisions of CITES should be enforced, to determine levels of commercial trade in the butterfly and its possible impact on populations. Attention should also be given to *Parnassius phoebus* and *P. mnemosyne*, which may also be threatened in Europe (21, 26, 27, 34).

As a large, conspicuous and very attractive butterfly, the Apollo has a certain value as a tourist attraction in some alpine areas.

References

1. Ackery, P.R. (1975). A guide to the genera and species of Parnassiinae (Lepidoptera: Papilionidae). *Bulletin of the British Museum (Natural History). Entomology* 31: 71–105.
2. Anon. (1976). Rote Liste bedrohter Tiere in Bayern (Wirbeltiere und Insekten). *Schriften der Natur. Lans. Bayern* 3: 1–12.
3. Bernardi, G., Nguyen, T. and Nguyen, T.H. (1981). Inventaire, cartographie et protection des Lépidoptères en France. *Beiheft Veröffentlichungen Naturschutz Landschaftspflege Baden–Württemberg* 21: 59–66.
4. Biedermann, J. (1982). Lebensraum für Insekten. *Liechtensteiner Umweltbericht* June 1982, 4–5.
5. Blab, J. and Kudrna, O. (1982). *Naturschutz Aktuell, Hilfsprogramm für Schmetterlinge*. Kilda–Verlag, Greven. 135 pp.
6. Blab, J., Nowak, E. and Trautmann, W. (1977). Rote Liste der Gefährdeten Tiere und Pflanzen in der Bundesrepublik Deutschland. *Naturschutz Aktuell* 1: 1–67.
7. Bourgogne, J. (1971). Un témoignage de plus sur la destruction de la nature (Papilionidae). *Alexanor* 7: 50.
8. Bryk, F. (1935). Lepidoptera. Parnassiidae pars II (Subfam. Parnassiinae). *Tierreich* 65: Li + 790 pp., 698 figs.
9. Burckhardt, D., Gfeller, W. and Muller, H.U. (1980). *Geschützte Tiere der Schweiz*. Schweiz. Bund. für Naturschutz, Basel. 223 pp.
10. Christensen, G. (1975). Wer rottet aus? *Entomologische Zeitschrift* 85: 246–48.
11. Dabrowski, J.S. (1975). Some problems in the preservation of butterflies in Poland. *Atala* 3: 4–5.
12. Dabrowski, J.S. (1980). The protection of the Lepidopterofauna - the latest trends and problems. *Nota Lepidopterorum* 3: 114–118.
13. Dabrowski, J.S. (1980). O stanie zagrozenia lepidopterofauny w niektorych parkach narodowych Polski. *Wiadomosci Entomol.* 1: 143–149.
14. Dabrowski, J.S. (1980). Mizeni biotopu jasone cervenookeho—*Parnassius apollo* (L.) v Polsku a nutnost jeho aktivni ochrany (Lepidoptera, Papilionidae). *Casopis Slezskeho Muzea Opava (A)* 29: 181–185.
15. Dabrowski, J.S. (1981). Remarks on the state of menacing the lepidopteran fauna in National Parks. Part 11 (general): the National Park in Tatra. *Zeszyty Naukowe Uniwersytetu Jagiellonskiego, Prace Zool* 27: 77–100.
16. Dabrowski, J.S. and Palik, E. (1979). Uwagi o stanie zagrozenia w parkach narodowych, Czesc 1: Zmiany zachodzace we wspolczesnej lepidopterofaunie Pieninskiego Parku Narodowego, ze szczegolnym uwzglednieniem zanikania gatunku *Parnassius apollo* (L.), (Lepidoptera: Papilionidae), Dokumentacja n/t na zlec. Kom. Nauk.: '*Czlowiek i Srodowisko*' *PAN*, 1–38 (maszynopis).
17. Gfeller, W. (1975). Geschützte Insekten in der Schweiz. *Mitteilungen der Schweizerischen Entomologischen Gesellschaft* 48: 217–213.
18. Gómez Bustillo, M.R. and Fernández-Rubio, F. (1974). *Mariposas de la Península Ibérica (tomo 1)*. Servicio de Publicaciones del Ministerio de Agricultura, Madrid. 198 pp.

19. Gómez Bustillo, M.R. and Fernández-Rubio, F. (1974). *Mariposas de la Península Ibérica (tomo 2)*. Servicio de Publicaciones del Ministerio de Agricultura, Madrid. 258 pp.
20. Gómez-Bustillo, M.R. (1981). Protection of Lepidoptera in Spain. *Beiheft Veröffentlichungen Naturschutz Landschaftsflege Baden-Württemberg* 21: 67–72.
21. Heath, J. (1981). Threatened rhopalocera (butterflies) in Europe. *Council of Europe, Nature and Environment Series* No. 23, 157 pp.
22. Heath, J. and Leclerq, J. (Eds) (1981). *European Invertebrate Survey. Provisional Atlas of the Invertebrates of Europe, Maps 1–27*. Institute of Terrestrial Ecology, Monks Wood and Faculté des Sciences Agronomiques, Gembloux.
23. Higgins, L.G. and Riley, N.D. (1980). *A Field Guide to the Butterflies of Britain and Europe*. 4th ed. Collins, London. 384 pp.
24. Igarashi, S. (1979). *Papilionidae and their early stages*. 2 vols., 219 pp., 102 pl. Kodansho, Tokyo. (In Japanese).
25. Janzon, L.-A. and Bignert, A. (1979). Apollofjärilen i Sverige, *Fauna Flora, Uppsala* 74: 57–66.
26. Mikkola, K. (1979). Vanishing and declining species of Finnish Lepidoptera *Notulae Entomologicae* 59: 1–9.
27. Mikkola, K. (1981). Extinct and vanishing Lepidoptera in Finland. *Beiheft Veröffentlichungen Naturschutz. Landschaftspflege Baden-Württemberg* 21: 19–22.
28. Nikusch, I. (1981). Die Zucht von *Parnassius apollo* Linnaeus mit jährlich zwei Generationen als Möglichkeit zur Erhaltung bedrohter Populationen. *Beiheft Veröffentlichungen Naturschutz Landschaftspflege Baden-Württemberg* 21: 175–176.
29. Nikusch, I. (1982). First trials to save threatened populations of *Parnassius apollo* by transplantation to new suitable biotypes. Third European Congress of Lepidopterology, Cambridge. In press.
30. Palik, E. (1966). On the process of dying out of *Parnassius apollo* Linné. *Tohoku Koncho Kenkyu* 2: 45–47.
31. Palik, E. (1980). The protection and reintroduction in Poland of *Parnassius apollo* (Linnaeus) (Papilionidae). *Nota Lepidopterorum* 2: 163–164.
32. Palik, E. (1981). The conditions of increasing menace for the existence of certain Lepidoptera in Poland. *Beiheft Veröffentlichungen Naturschutz Landschaftspflege Baden-Württemberg* 21: 31–33.
33. Pfaff, G. (1935). Wer rottet aus? *Entomologische Zeitschrift* 49: 105–107.
34. Pyle, R.M. (1976). *The eco-geographic basis for Lepidoptera conservation*. Part iii. A review of world Lepidoptera conservation. Yale Univ. PhD. Thesis, Publ. Univ. Microfilm Int. 369 pp.
35. Rowland-Brown, H. (1913). *Parnassius apollo* in Germany. *Entomologist* 46: 289–290.
36. Schmiedel, R. (1934). Tragt der sammler die schuld am Ruckgang unserer Insektenfauna? *Entomologische Rundschau* 51: 174–177.
37. Svenson, I. (1981). Changes in the Lepidoptera fauna of Sweden after Linnaeus. *Beiheft Veröffentlichungen Naturschutz. Landschaftspflege Baden-Württemberg* 21: 23–30.
38. Viedma, M.G. de and Gómez-Bustillo, M.R. (1976). *Libro Rojo de los Lepidópteros Ibéricos*. Instituto Nacional para la Conservación de la Naturaleza, Madrid. 120 pp.
39. Yermolenko, V.M. (1973). On protection of useful, relict and endemic insects of the Ukrainian Carpathian mountains and mountains of Crimea. In: *On Insect Protection*. Armenian Academy of Sciences Symposium, Yerevan. Pp. 29–35.
40. Zukowski, R. (1959). Extinction and decrease of the butterfly *Parnassius apollo* on Polish territories. *Sylwan* 103: 15–30.

This review has been adapted from *The IUCN Invertebrate Red Data Book*, whose authors and contributors are gratefully acknowledged.

Bhutanitis mansfieldi (Riley, 1939)   RARE

Subfamily PARNASSIINAE   Tribe ZERYNTHIINI

Summary Until recently *Bhutanitis mansfieldi* was known from only two female specimens and one male collected in 1918 from Yunnan Province, south–western China. In spring 1981 a team of Japanese mountaineers rediscovered this enigmatic and beautiful rarity on Mt Gonga in Sichuan. More information on its conservation status is needed.

Description *Bhutanitis mansfieldi* was originally described from a single female specimen (4) (Plate 1.5). A second female was known to have been collected (2), but its whereabouts remained a mystery until March 1983, when it was rediscovered in the British Museum (Natural History) (3). In 1982 a Japanese insect dealer obtained a very battered male specimen, illustrated in (3). Also in 1982 Saigusa and Lee described a second subspecies, *pulchristata*, from 11 males and three females taken in Sichuan (5). The sexes are similar, with a forewing length of about 40 mm.

UFW ground colour pale yellow with eight black, transverse bands (which extend into the UHW) including a broad outer margin.

UHW elongated, with a scalloped edge, a deep notch at the anal angle and three tails, of which the outermost is long and club-shaped and the others shorter. There is a broad, black distal band, a large bright red patch extending from the anal margin, two blue eye-spots, and four pale yellow submarginal lunules (1, 3, 4).

LFW/LHW differs only slightly from the upperside. One of the discal spots is red, and the submarginal lunule between the first two tails is orange.

Distribution The type specimen of this species was found by M.J. Mansfield in a collection of butterflies made by a British botanist, G. Forrest, who collected extensively in Yunnan, China. The specimen lacked data, but was assumed to have originated in that province. No further material was collected until spring 1981 when two members of a mountaineering party sent to Mt Gonga in Sichuan by the Hokkaido Alpine Association unexpectedly collected 14 specimens near Xinxing (2200 m a.s.l.) (5). A further party of Chinese collectors sponsored by a pool of Japanese entomologists was then sent to Xinxing in 1983, where they caught about 30 specimens early in the season (6). Snow was still present in patches. In view of these records from Sichuan, there may be cause to question the assumption that Yunnan was the provenance of the type material. There are no records from Laos, Vietnam or Burma, which border onto Yunnan.

Habitat and Ecology In common with *Bhutanitis thaidina*, *B. mansfieldi* may feed on species of *Aristolochia* (Aristolochiaceae), a genus of climbing vines. If this is so, the habitat may be in valley forests suitable for supporting the vine. However, there would seem to be every possibility that *Bhutanitis* may also have adapted to a more temperate alpine genus or family of plants, as have most of the Parnassiini. *Luehdorfia*, which is believed to be derived from *Bhutanitis*, uses *Asarum* as a foodplant, another genus within the Aristolochiaceae but with a more temperate distribution. There is no information on breeding biology or young stages, although it is known that the adults fly early in the season, before *B. thaidina* (6). There is a great need for further study of both these beautiful species.

Much of Yunnan and Sichuan is rugged, highland country. Yunnan is mainly a plateau region at an altitude of about 2000 m in south–western China while Sichuan, lying north of Yunnan, includes some lower-lying regions. The climate is temperate alpine, dry and bright in winter, wet in summer. The terrain is mountainous with fertile valleys. Despite being rich in minerals (particularly Yunnan), the region is so remote from administrative centres that it has remained under-developed and very poorly known to both local and foreign entomologists.

Threats Now that a precise locality has been discovered there is clearly a need to plan and execute conservation measures. Any collecting in the region should be moderated and sustainable. There is no precise information on environmental threats in the region of Xinxing, nor on the provision for protected areas nearby. More information is needed.

Conservation Measures No measures have been taken so far, although the rediscovery of the butterfly by Japanese entomologists is a great step forward. The Japanese have shown that they have the resources and the lines of communication to find the butterfly, they should also be encouraged to allocate further resources to ensuring the long-term survival and protection of this species.

References

1. Ackery, P.R. (1975). A guide to the genera and species of Parnassiinae (Lepidoptera: Papilionidae). *Bulletin of the British Museum (Natural History). Entomology* 31: 71–105.
2. Mansfield, M.J. (1941). *Bhutanitis mansfieldi* Riley not unique. *Entomologist* 74: 44.
3. Okano, K. (1984). Color illustration of *Bhutanitis mansfieldi* (Riley, 1940) (Papilionidae): with some notes on the same species. *Tokurana (Acta Rhopalocera)* 6/7: 61–65.
4. Riley, N.D. (1939). A new species of *Armandia* (Lep. Papilionidae). *The Entomologist* 72: 206–208, 267.
5. Saigusa, T. and Lee, C. (1982). A rare papilionid butterfly *Bhutanitis mansfieldi* (Riley), its rediscovery, new subspecies and phylogenetic position. *Tyo to Ga* 33: 1–24.
6. Turlin, B. (1984). *In litt.*, 1 June.

Bhutanitis thaidina Blanchard, 1871　　　　　　　　　　　　RARE

Subfamily　PARNASSIINAE　　　　　　　　Tribe　ZERYNTHIINI

Summary *Bhutanitis thaidina* has a fairly wide distribution in south–western China and Tibet, but was until recently known from very few specimens. Further documentation of the precise localities of the known populations, their exploitation by local collectors, and the life history of this beautiful species is needed.

Description *Bhutanitis thaidina* is a medium-sized three-tailed butterfly with a forewing length of 42–48 mm. The sexes are similar in appearance, the female slightly larger than the male (1, 3).

UFW yellow ground colour reduced to eight transverse lines separating broad bronze–black bands from the costal margin to the inner margin (1).

UHW elongated with a scalloped outer margin and three tails of which the outer is long and club-shaped and the others shorter. The dark forewing bands continue into the discal region of the hindwing as streaks and spots. There is a broad, black distal band with a large red patch, four orange–yellow submarginal lunules and three light blue postdiscal spots (1).

LFW/LHW paler than the upperside, with broader yellow lines and a smaller, paler red patch on the LHW.

Distribution *Bhutanitis thaidini* occurs in south–western China where it has been recorded from the upper Jinsha Jiang (Yangtzekiang) River in Yunnan and Sichuan (Szechwan) provinces, from Kangding (Tatsienlu) in Sichuan, and localities in Shaanxi (Shensi) province (1, 3, 4). Records also exist for Mou-pin in Tibet (Xizang Zizhiqu) (2).

Habitat and Ecology According to Bryk (3) the foodplant of *Bhutanitis thaidina* is *Aristolochia* sp. (Aristolochiaceae), a genus of scrambling vines. This requires confirmation from field studies. The genus *Luehdorfia*, derived from *Bhutanitis*, has changed its feeding habits to the more temperate aristolochiaceous genus *Asarum*. No further details are available on the breeding biology or habitat of the species.

The range of this species is mainly high altitude plateaux (over 2000 m) with a rugged and highly dissected terrain. The natural vegetation would be variable, but the foodplants are likely to be found in forested country.

Threats As in the case of *Bhutanitis mansfieldi*, *B. thaidina* is too poorly known for environmental threats to be properly documented. Highly priced specimens occasionally appear on dealers' lists, but no information is given on their precise sources. It has been reported that teams of Chinese collectors trained and sponsored by Japanese entomologists and dealers travelled to various parts of Sichuan during the collecting seasons of 1981, 1982 and 1983. Numerous specimens were taken near the town of Kangding, where the species is very local but common during the flight season (5). Collecting on this scale is most unlikely to pose an immediate threat to the butterfly, but there is a need to monitor the trade in order to ensure sustainable exploitation of the populations.

Conservation Measures Unfortunately Tibet and south–western China are very difficult of access, but there is a great need for more research into the

Lepidoptera of these regions. It is now clear that the specimens rarely appearing in the market are freshly caught, probably through the arrangements of Japanese entomologists. The entomologists who have developed lines of access and communication to these remote parts of China should be encouraged to ensure assessments of the impact of their collecting. It is apparent that *Bhutanitis thaidina* is fairly widespread and there is perhaps potential for a local industry in butterfly ranching along the lines of that in Papua New Guinea. Research into the breeding biology and management of *Bhutanitis* would be an essential preliminary step. There would certainly be a valuable market for this and many other Chinese species (see section 5 of this book). It would be essential to monitor and regulate any trade in a sustainable manner, both to ensure the welfare of the butterfly and to safeguard the ranching programme itself. Now that populations are beginning to be located, the Chinese authorities might also ensure that adequate habitat is protected in national parks and reserves. There is a need for information on protected areas on this region.

References

1. Ackery, P.R. (1975). A guide to the genera and species of Parnassiinae (Lepidoptera: Papilionidae). *Bulletin of the British Museum (Natural History). Entomology* 31: 71–105.
2. Bouchard, M.E. (1871). Remarques sur la faune de la principauté thibétaine du Mou-pin. *Comptes Rendus Hebdomadaires des Séances de L'Academie de Science, Paris* 72: 807–813.
3. Bryk, F. (1935). Lepidoptera. Parnassiidae pars II (Subfam. Parnassiinae). *Tierreich* 65: Li + 790 pp., 698 figs.
4. Pen, D. (1936). The Papilionidae of south–western Szechwan. *Journal of the West China Border Research Society* 8: 153–165.
5. Turlin, B. (1984). *In litt.*, 1 June.

Bhutanitis ludlowi Gabriel, 1942   INSUFFICIENTLY KNOWN

Subfamily PARNASSIINAE   Tribe ZERYNTHIINI

Summary *Bhutanitis ludlowi* is known only from the type series, three males and one female collected in the Trashiyangsi Valley in Bhutan in 1933 and 1934. Its presence and status there need to be ascertained.

Description *Bhutanitis ludlowi* is as beautiful as *B. mansfieldi* and *B. thaidina*, as well as being rather larger, with a forewing length of about 59 mm (male) or 63 mm (female). The sexes are similar (1, 3) (Plate 1.6).

UFW ground colour dull black crossed by eight transverse, off-white lines.

UHW elongated with scalloped outer margin and three tails, a long, spatulate, outermost tail, and two smaller ones. The transverse lines continue from the UFW, and a large velvety black and red patch, four dark grey lunules, and three large, blue–grey spots are present.

LFW/LHW with more pronounced transverse lines and yellow–ochre submarginal lunules (1, 3).

Distribution *Bhutanitis ludlowi* has only been recorded from the Trashiyangsi Valley in north–eastern Bhutan (1). The type series of three males and two females is unique. No other specimens have been found since the original collections in 1933 and 1934 (3).

Habitat and Ecology There is no published information on the habitat or ecology of *Bhutanitis ludlowi*. It may use *Aristolochia* as a foodplant, as does *Bhutanitis thaidina* (1), but its habitat may be too high for these vines. The specimens were collected at an altitude of 2300–2500 m (3). Information on the vegetation at this altitude is given in reference (2) and summarized in the review of *Teinopalpus imperialis*.

Threats No information is available specifically concerning the butterfly. Bhutan is certainly the best forested area in the Himalaya, but even here the resource is being eroded by unregulated grazing, forest fires and high consumption of wood for fuel and construction. Such degradation will ultimately affect all wildlife, including butterflies, and is extremely difficult to reverse in the Himalayan conditions of climate and topography.

Conservation Measures The presence of *Bhutanitis ludlowi* in the Trashiyangsi Valley needs to be confirmed. Appropriate measures for its conservation and protection may then be developed. Bhutan has a good conservation record, with about 20 per cent (9213 sq. km) of its area protected as wildlife sanctuaries and reserved forests. The giant Jigme Dorji wildlife sanctuary (7900 sq. km) extends along the northern border of Bhutan and may well include *B. ludlowi* habitat within its boundaries. A butterfly survey of the sanctuary would be extremely valuable.

References

1. Ackery, P.R. (1975). A guide to the genera and species of Parnassiinae (Lepidoptera: Papilionidae). *Bulletin of the British Museum (Natural History). Entomology* 31: 71–105.

2. FAO/UNEP (1981). *Tropical Forest Resources Assessment Project. Forest Resources of Tropical Asia.* FAO, Rome. 475 pp.

3. Gabriel, A.G. (1942). A new species of *Bhutanitis* (Lep. Papilionidae). *Entomologist* 75: 189.

Luehdorfia chinensis Leech, 1893   INSUFFICIENTLY KNOWN

Subfamily PARNASSIINAE   Tribe ZERYNTHIINI

Summary *Luehdorfia chinensis* is a poorly known species restricted to certain eastern provinces in China. Its taxonomic status is uncertain and its conservation status is obscured by the inaccessibility of its habitat and the lack of information on its biology.

Description The taxonomic status of *Luehdorfia chinensis* is not entirely certain. It was not listed by Munroe (6), but has since been variously treated as a separate species (3, 4), and as a subspecies of *L. japonica* (1) or *L. puziloi* (2). Recently however, the young stages have been described and the species is now generally considered to be distinct (5). *L. chinensis* is similar to *L. japonica* in appearance (see next review), but differs in having a red submarginal band and yellow marginal spots on the UHW. Igarashi (1) gives two subspecies, *chinensis* and *lenzeni*.

Distribution *Luehdorfia chinensis* has a restricted distribution in the provinces of Anhui (Anhwei), Hubei (Hupeh), Jiangsu (Kiangsu), and Jiangxi (Kiangsi) in eastern China (1).

Habitat and Ecology The foodplant of *Luehdorfia* is *Asarum* (Aristolochiaceae), a genus of about 100 species of perennial herbs with vestigial or no petals to the flowers. Further details of the young stages and breeding biology are given by Lee (5) and Igarashi (4) (in Chinese and Japanese respectively).

Threats There is no information on threats to this poorly known species. However, the eastern lowlands of China are highly developed and support high human populations. If the habitat is open deciduous woodland, as it is for *L. japonica* (see next review), there is cause for concern because the level of destruction of such formations is likely to have been high. Listing as Insufficiently Known is appropriate until further details emerge.

Conservation Measures More details are required concerning the habitat, biology and conservation status of *Luehdorfia chinensis*.

References

1. Ackery, P.R. (1975). A guide to the genera and species of Parnassiinae (Lepidoptera: Papilionidae). *Bulletin of the British Museum (Natural History). Entomology* 31: 71–105.
2. Bryk, F. (1935). Lepidoptera. Parnassiidae pars II (Subfam. Parnassiinae). *Tierreich* 65: Li + 790 pp., 698 figs.
3. Hancock, D.L. (1983). Classification of the Papilionidae (Lepidoptera): a phylogenetic approach. *Smithersia* 2: 1–48.
4. Igarashi, S. (1979). *Papilionidae and Their Early Stages.* Vol 1: 219 pp., Vol 2: 102 pp. of plates. Kodansho, Tokyo. (In Japanese).
5. Lee, C.-L. (1978). The early stages of Chinese Rhopalocera—*Luehdorfia chinensis* Leech (Parnassiidae: Zerynthiinae). *Acta Entomologica Sinensis* 21: 161–163. (In Chinese).
6. Munroe, E. (1961). The classification of the Papilionidae (Lepidoptera). *Canadian Entomologist* Supplement 17: 1–51.

Luehdorfia japonica Leech, 1889   VULNERABLE

Subfamily PARNASSIINAE   Tribe ZERYNTHIINI

Summary *Luehdorfia japonica* is found in open, deciduous woodland in western Honshu, Japan. Suitable habitat is disappearing through changes in management and deforestation for development. Research on the management of the butterfly is needed, together with identification of appropriate reserves and protected areas.

Description *Luehdorfia japonica* is a relatively small butterfly with a forewing length of 30–35 mm, the female slightly larger than the male (8, 9) (Plate 1.2). The sexes are otherwise alike.

UFW pale yellow with a black base and outer margin, and black bands across the wing (1, 5, 8, 9).

UHW also pale yellow with black bands, a medium length tail, a deep notch at the anal angle, and a scalloped outer margin. The outer third of the wing is black with orange submarginal patches and blue postdiscal spots. There is a small, blue anal eye-spot with a large, crimson patch above it (1, 5, 8, 9).

LFW/LHW virtually the same as the upper, differing only on the hindwing. The crimson patch near the anal angle is extended as a band to the costal margin, and the orange submarginal patches are also merged into a continuous band (1, 5, 8).

The young stages have been described by Igarashi (5) and Shirozu and Hara (11) (in Japanese).

Distribution *Luehdorfia japonica* is restricted to the western part of Honshu (3), the main island of Japan. The northern limit is Yuri in Akita Prefecture, the eastern limit is the Tama Hills in the vicinity of Tokyo and the western limit is around Hagi city in Yamaguchi Prefecture (8). The subspecies, *L. j. formosana* Rothschild, 1918, listed by Shirozu (10), appears to be either a doubtful record or is now extinct. *L. puziloi* also flies in Japan, but the two species are almost entirely separated. Where they do occur together, in Akita, Yamagata, Nagano, Niigata and Yamanashi Prefectures, hybrids are occasionally found (8). These areas are known as the "*Luehdorfia* line".

Habitat and Ecology In Japan at least, the habitat of *Luehdorfia japonica* is open formations of deciduous broad-leaved forest at middle altitudes (2). Such forest is a succession stage towards dense, climax forest with a closed canopy and is therefore naturally unstable (2). The climax forest is unsuitable for the butterfly.

The foodplants are small perennial herbs from the family Aristolochiaceae, *Asarum nipponicum*, *A. tamaense*, *A. blumei*, *A. caulescens* and *A. sieboldi* (1, 5). The adult butterflies emerge early in the spring (March in the west, April generally and May/June in mountainous or northern areas) and seek nectar in the flowers of *Erythronium japonicum* and *Viola*. The eggs hatch within several days of laying (4). The larvae feed on *Asarum* and grow rapidly, changing into pupae at the end of May or beginning of June (4). The pupae may be found attached to the stems of shrubs or herbs near the ground (8). They pass about ten months in diapause (4).

The pupal diapause in *Luehdorfia japonica* has been the subject of research programmes by Japanese teams and is now known to be complicated. Two types of diapause seem to be experienced in one generation. One is summer diapause maintained by the long day and high temperatures in summer and broken by the short day (below 15 hours) and moderate temperatures (14–25°C) in autumn; the other is winter diapause terminated by the winter cold (4, 6, 7). Consequently, the differentiation of the adult resumes in autumn and continues at a low rate during the winter, reaching completion in spring.

Threats The deciduous broadleaved forests of Japan were formerly managed as a source of timber and charcoal and the constant thinning encouraged *Luehdorfia japonica* (2). In recent times however, the forests are not used as a resource. They are either left to reach dense climax woodland or else they are cleared for residential developments and amenities such as golf courses. The resulting depletion of suitable habitat for *L. japonica* is causing continuous reductions in its range.

Conservation Measures There are still many good localities for *Luehdorfia japonica* in western Honshu, but the constant and accelerating loss of its habitat requires conservation action. There are no permanent reserves for the butterfly and suitable localities should be identified and surveyed. Once protected, habitats will need to be carefully managed in order to maintain suitable conditions. A research programme into the management of the butterfly is desirable. The butterfly is not difficult to rear artificially and suitable localities could be kept well stocked. Pupae are already occasionally advertised for sale. More data on the habitat and biology of *Luehdorfia japonica* are needed, but the programme of research into its pupal diapause has given a good basis for future management studies.

References

1. Ackery, P.R. (1975). A guide to the genera and species of Parnassiinae (Lepidoptera: Papilionidae). *Bulletin of the British Museum (Natural History). Entomology* 31: 71–105.
2. Ae, S.A (1983). *In litt.*, March 18.
3. Bryk, F. (1935). Lepidoptera. Parnassiidae pars II (Subfam. Parnassiinae). *Tierreich* 65: Li + 790 pp.
4. Hidaka, T., Ishizuka, Y. and Sakagami, Y. (1971). Control of pupal diapause and adult differentiation in a univoltine papilionid butterfly, *Luehdorfia japonica*. *Journal of Insect Physiology* 17: 197–203.
5. Igarashi, S. (1979). *Papilionidae and Their Early Stages*. Vol 1: 219 pp., Vol 2: 102 pp. of plates. Kodansho, Tokyo. (In Japanese).
6. Ishii, M. and Hidaka, T. (1982). Characteristics of pupal diapause in the univoltine papilionid *Luehdorfia japonica* (Lepidoptera, Papilionidae). *Kontyu* 50: 610–620.
7. Ishii, M. and Hidaka, T. (1983). The second pupal diapause in the univoltine papilionid, *Luehdorfia japonica* (Lepidoptera: Papilionidae) and its terminating factor. *Applied Entomology and Zoology* 18: 456–463.
8. Kawazoe, A. and Wakabayashi, M. (1976). *Colored Illustrations of the Butterflies of Japan*. Hoikusha, Osaka, Japan. 422 pp.
9. Leech, J.H. (1889). Description of a new *Luehdorfia* from Japan. *The Entomologist* 22(309): 25–27.
10. Shirôzu, T. (1960). *Butterflies of Formosa in Colour*. Hoikusha. Osaka, Japan (In Japanese). 483 pp.
11. Shirôzu, T. and Hara, A. (1960). *Early Stages of Japanese Butterflies in Colour*. Hoikusha, Osaka. Vol. 1, 142 pp. (In Japanese).

Teinopalpus imperialis Hope, 1843   RARE

Subfamily PAPILIONINAE   Tribe LEPTOCIRCINI

Summary *Teinopalpus imperialis*, the Kaiser-I-Hind, is a superb green, black and orange, tailed butterfly from the forests of the Himalaya, Burma and China. Although widely distributed, it is very local and rare. Some of the forests still retain their original character, but human population pressure is increasing their degradation through fire, deforestation, grazing and fuelwood collecting. As well as supporting unique wildlife, the forests play a vital role in controlling the flow of water to the lowlands.

Description *Teinopalpus imperialis*, the Kaiser-I-Hind, is a large and beautiful, tailed butterfly with sexes that differ in size and coloration and were originally described as two separate species (Plate 1.7). There are two (or perhaps three) subspecies. The wingspan of the nominate subspecies is 90–120 mm, (male forewing length 52 mm, female 60–65 mm); the female *T. i. imperatrix* de Nicéville, is even larger with a wingspan of 130 mm, a forewing length of 70 mm and five distinct tails (2, 4, 5, 17). *Teinopalpus behludinii* (15) from Sichuan is almost certainly referable to *T. imperialis*, but the type specimen is apparently untraceable (12, 15) and no other specimens are known. New material is needed to establish *behludinii* as a good species. Until this is forthcoming it may be assumed that *behludinii* is a subspecies or population of *imperialis*.

Male: UFW with a black ground colour densely suffused with bands of green scales. UHW scalloped, with a long golden-tipped tail. The green bands of the UFW continue, but the distal region has a bright orange patch and a series of submarginal lunules in orange and bright green.

LFW basal area densely covered with green scales, the distal two-thirds yellow–brown with narrow black bands. LHW similar to UHW but with broader yellow markings (2, 4, 5, 17).

Female: UFW basal area resembling the male but the outer two-thirds of the wing has broad, indistinct alternate grey and green bands. UHW deeply scalloped with two long black tails and a shorter one between them. The yellow discal patch of the male is replaced by a much larger dark grey patch, with yellow markings beneath.

LFW/LHW similar to the male but with the yellow–brown colour replaced by grey and with a broad ochreous submarginal band on the LFW (2, 4, 5, 17).

Distribution *Teinopalpus imperialis* is found at high altitudes from Nepal to southern Burma and in the Chinese provinces of Sichuan (as *T. behludinii*) and Hubei (11). As Mell (11) suggested, records from Yunnan are to be expected. The records from Guangdong are referable to *T. aureus* (see next review). *T.i. imperialis* occurs between 2000 and 3500 m in Nepal, across north–eastern India and Bhutan to northern Burma and at lower altitudes (over 1000 m) in China (11). In India it has been recorded from Sikkim, the Darjeeling region of northern West Bengal, the Khasia Hills, north of Cherrapunji in Meghalaya, and the states of Assam and Manipur in the north–east. It may also prove to be present in Arunachal Pradesh and Nagaland, completing the circle around the headwaters of the Brahmaputra River. *T.i. imperatrix* occurs only in Burma, where its distribution records indicate an extension to lower altitudes than the nominate subspecies. It is known from the Shan States in the north, southwards to the Ataran River in Karen State, and Mergui in

Upper Tenasserim State (2, 4, 5, 17). Although there have been no published records of *Teinopalpus imperialis* in the highlands of western Thailand, its presence there is a possibility. In China *Teinopalpus imperialis* is possibly quite widespread, perhaps occurring in the montane plateau of Yunnan and Sichuan, eastward to the highland parts of Hubei (11). Unfortunately there have been no reports from China for almost 50 years (11, 15).

Habitat and Ecology The Kaiser-I-Hind is restricted to mountainous wooded districts, where it flies high up in the canopy (17). It is said to descend nearer the ground from 8.00 to 11.00 a.m. on sunny days, when it can be attracted by baiting (17). Its flight is very rapid, particularly in the male, and the best localities for observation are open places and mountain tops surrounded by forests (10). The butterflies fly up through the forest to the hilltops in search of a mate. Often the males will defend mating territories on the summits, fighting off other suitors. In cloudy or misty weather the butterflies rest on low bushes, concealed by the ventral camouflage of the wings (8).

Adults have not been seen to visit either flowers or water (8). Okano (14) gives details of the foodplants and young stages of *Teinopalpus* (in Japanese). The caterpillar is green with a large head and the pupa is smooth, green and horned (13, 17). Larvae pupate at the end of September and adults emerge in April–May in Sikkim, May–July in Manipur (8). There are reports that the species is double-brooded, at least in India (18). The foodplants are *Daphne* spp., laurel-like small trees or shrubs in the Thymeleaceae (8, 13, 14). The habitat in which *Teinopalpus imperialis* flies is mainly broad-leaved evergreen forest, sometimes mixed with conifers and often well-stocked with various species of oak (*Quercus*) (3, 16). The foodplants are generally a component of the understorey.

Threats The butterfly is very local and is never reported to be abundant, although it is fairly widespread (17). For its survival it appears to be dependent upon the natural forests that harbour its foodplant *Daphne* in the Himalaya, Burma and China. There is no evidence of its spread into degraded forest, the regenerative stages of shifting cultivation, or agricultural land. Nothing has been published concerning threats specific to the butterfly, but there is a growing literature describing the disturbing extent of deforestation over the entire area of the butterfly's range.

In Nepal deforestation has largely resulted from rising human population and migration into the Terai, or southern lowland strip, where broad-leaved forests grow. Loss of these lowland forests is running at an estimated 80 000 ha per year, almost 5 per cent of the 1980 total forest cover (6). The migrants come from the hill areas, which are more difficult to cultivate. Virtually all of central Nepal is now deforested although certain spots known to contain *Teinopalpus*, such as Phulchoki mountain, still remain wooded.

In India the Himalayan forests are possibly being deforested more slowly than in Nepal, although the available data are acknowledged as being unreliable (6). Few data are available for specific regions, but the isolated montane area of Meghalaya seems to be particularly seriously deforested. These hills have a very high rainfall and Cherrapunji, with an average rainfall of 1150 cm per year, is one of the wettest places on earth (1). The traditional form of agriculture was shifting cultivation, or 'jhumming', but over-exploitation has led to severe soil erosion (1). Such degradation in the lower altitude regions has undoubtedly led to more severe environmental pressures and forest degradation in the high altitude central core of Meghalaya. In

Darjeeling the forests are reported to be virtually all gone and in Sikkim also large areas are denuded and eroded (7).

In Bhutan the forests are said to be subjected to heavy biotic interference, particularly grazing and fires (6). In Burma deforestation is relatively limited, but again the main pressure on forests is rising human population and the increased need for fuelwood, grazing and land for shifting cultivation (6). Few data are available for the Chinese part of the butterfly's range, but specimens have been recorded as low as 1000–1400 m (11), where habitat destruction is likely to have been severe.

Information on the degradation of the Himalayan forests is scarce and often out of date. It is thus difficult to obtain an accurate impression of human impact on the butterfly's habitat, but the indications are that its range must be shrinking rapidly. As well as supporting important flora and fauna, not to mention human communities, these forests have important climatic and environmental influences. Most of the important rivers of northern India originate in the Himalaya. The huge River Brahmaputra has its headwaters in the habitat of *Teinopalpus imperialis*. Biotic interference is often serious in the lower zones of hardwood forest, declining higher up. Fires may be devastating at any altitude, and grazing and fuelwood collection are on the increase. In general most of the high altitude coniferous forests seem to have retained their original character, particularly in the Himalaya, but increasing population pressure may eventually lead to intensified use of the land (6). In this eventuality great skill will be needed if environmental degradation is to be avoided.

Despite the wide range of *T. imperialis*, some of the reports of over-collecting must be a matter for concern. For example, in July 1984 on Phulchoki (Pulchoki) mountain near Kathmandu in Nepal, every specimen flying in was being caught by foreign collectors (7). Although such an isolated event is unlikely to have a serious impact on the butterfly species, other hills may be receiving the same treatment. Moreover, in both Nepal and India collecting is illegal without a licence. Excessive collecting in this way reduces the natural beauty of the area.

Conservation Measures Specific measures for the Kaiser-I-Hind are difficult to propose at this stage. Collecting could pose a problem in certain localized areas and the legislation already enacted in India and Nepal should be enforced more effectively. The best way to preserve the Kaiser-I-Hind is rational utilization and protection of the forests in which it lives. The Government of Nepal is preparing a National Conservation Strategy that will be of benefit to all wildlife (9). In India such strategies would have to be on a state-wide basis, since each state is virtually autonomous in its forestry practice (6). The presence of *Teinopalpus imperialis* in protected areas should be noted and research towards management measures encouraged.

References

1. Agarwal, A., Chopra, R. and Sharma, K., Eds. (1982). *The State of India's Environment 1982. A Citizens' Report*. Centre for Science and Environment, New Delhi. 192 pp.
2. Antram, C.B. (1924). *Butterflies of India*. Thacker, Spink & Co., Calcutta and Simla. 226 pp.
3. Champion, H.G. and Seth, S.K. (1968). *A Revised Study of the Forest Types of India*. Publication Division, Government of India, Delhi.
4. D'Abrera, B. (1982). *Butterflies of the Oriental Region. Part 1. Papilionidae and Pieridae*. Hill House, Victoria, Australia. xxxi + 244 pp.
5. Evans, W.H. (1932). *The Identification of Indian Butterflies*. Bombay Natural History Society. 2nd ed., revised. 454 pp.

6. FAO/UNEP (1981). *Tropical Forest Resources Assessment Project. Forest Resources of Tropical Asia.* FAO, Rome. 475 pp.
7. Feeny, P. (1984). Pers. comm., 1 August.
8. Ghosh, S.K. and Mandal, D.K. (1983). Review of Kaiser-I-Hind butterfly, *Teinopalpus imperialis* Hope. Unpublished manuscript. 3 pp.
9. His Majesty's Government of Nepal/IUCN (1983). *National Conservation Strategy for Nepal. A Prospectus.* IUCN, Gland. 36 pp.
10. Jordan, K. (1909). In Seitz, A. *Macrolepidoptera Fauna Indo–Australasia* 9: 17–109.
11. Mell, R. (1938). Beiträge zur Fauna Sinica. *Deutsche Entomologische Zeitschrift* 17: 197–345.
12. Moonen, J.J.M. (1984). Notes on eastern Papilionidae. *Papilio International* 1(3): 47–50.
13. Munroe, E. (1961). The classification of the Papilionidae (Lepidoptera). *Canadian Entomologist* Supplement 17: 1–51.
14. Okano, K. (1983). Some ecological notes on *Teinopalpus*. Tokurana 5: 94–100. (In Japanese).
15. Pen, D. (1936). The Papilionidae of south–western Szechwan. *Journal of the West China Border Research Society* 8: 153–165.
16. Stainton, J.D.A. (1972). *Forests of Nepal*. John Murray, London. 181 pp.
17. Talbot, G. (1939). *The fauna of British India, including Ceylon and Burma. Butterflies vol I.* Taylor and Francis Ltd., London, reprint New Delhi 1975. 600 pp.
18. Wankhar, D.M. (1984). *In litt.*, 11 April.

Teinopalpus aureus Mell, 1923   INSUFFICIENTLY KNOWN

Subfamily PAPILIONINAE   Tribe LEPTOCIRCINI

Summary *Teinopalpus aureus*, sometimes known as the Golden Kaiser-I-Hind, is a superb green and golden tailed butterfly from China. No specimens have been seen for many years and there is a great need for more information on the distribution, ecology and conservation status of the species.

Description *Teinopalpus aureus* is a large and beautiful butterfly with a forewing length of 45–55 mm. Its taxonomic status is uncertain because the butterfly is so poorly known. It was described as a subspecies of *T. imperialis* (3), but Munroe (4) and Hancock (2) list it as a full species. The female is unknown.

UFW black with dense, bright green scaling and a black band with a bright yellow–green distal border. UHW with a large, golden–yellow discal patch, concentric bands of black–blue, orange and green, and a similar long, yellow-tipped tail. The underside is similar, but somewhat paler (1).

Distribution There is very little information on the distribution of *T. aureus*. The type locality is in south–east China where it has been recorded from montane forests near Lianping in the north of Guangdong (Kwangtung) province (3). In addition, there is a colour photograph of this species in the British Museum (Natural History) labelled 'Haut–Donnai, Annam', suggesting it may also be found in Vietnam (1).

Habitat and Ecology *Teinopalpus aureus* is a montane species, probably flying in habitats above 1000 m, in a region that would once have had broad-leaved and coniferous evergreen forests at such altitudes. The species is so poorly known that there are no published details of its present habitat, breeding biology or foodplant requirements. However, it is known to fly in late March and April (3). The closely-related *Teinopalpus imperialis* (see previous review) feeds on *Daphne*, a laurel-like small tree or shrub in the Thymeleaceae.

Threats The extent of montane land in the vicinity of Lianping, the locality for the specimens in the British Museum (Natural History), appears from maps to be very limited. In all probability the species is very rare but fairly widespread, although it seems strange that such a striking butterfly has not been recorded more often. If, like *Teinopalpus imperialis*, *T. aureus* flies in montane evergreen forest, then its survival in southern China is likely to be severely imperilled by the extent of deforestation there.

Conservation Measures *Teinopalpus aureus* is extremely poorly known, but is likely to be threatened by deforestation of its montane habitats. It may already be extinct and searches are needed to establish its present distribution. Once found, protected areas and ecological study will be needed to ensure its future survival.

References

1. D'Abrera, B. (1982). *Butterflies of the Oriental Region. Part 1. Papilionidae and Pieridae.* Hill House, Victoria, Australia. xxxi + 244 pp.
2. Hancock, D.L. (1983). Classification of the Papilionidae (Lepidoptera): a phylogenetic approach. *Smithersia* 2: 1–48.

3. Mell, R. (1938). Beiträge zur Fauna Sinica. *Deutsche Entomologische Zeitschrift* 17: 197–345.
4. Munroe, E. (1961). The classification of the Papilionidae (Lepidoptera). *Canadian Entomologist* Supplement 17: 1–51.

Eurytides (Protesilaus) marcellinus (Doubleday, 1845)   VULNERABLE

Subfamily PAPILIONINAE   Tribe LEPTOCIRCINI

Summary *Eurytides marcellinus*, the Jamaican Kite Swallowtail, is endemic to Jamaica and participates in spectacular migrations across the island. Its populations fluctuate widely and its main stronghold is threatened by loss of the foodplant *Oxandra lanceolata* as a result of cultivation.

Description This species is easily distinguished from other Jamaican species by its small size (forewing length 30–35 mm), long slender tails and pattern of longitudinal bands of black and delicate pale blue–green (1, 5) (Plate 1.8). Its nomenclature has been somewhat changeable at both the specific (1) and generic levels (3, 4). The most recent revision places it in the revived genus *Protesilaus*, as part of the *marcellus* species group (1, 3, 4) but here we retain the name *Eurytides*. The sexes are believed to be similar (5).

UFW/UHW black with bold pale blue–green stripes and submarginal spots and a bright red anal spot.

LFW/LHW similar but with brown ground colour and a wide red stripe on the hindwings. Tails about 10 mm long (5, 6, 2).

The eggs are spherical, pale brown when laid, becoming black as they approach hatching (7). The mature larva is blue–green dorsally and grey ventrally with an orange head; the pupa dark brown with no anterior prominences (7).

Distribution Endemic to Jamaica, with a former range which includes the parishes of St Elisabeth in the south–west, St Andrew and Kingston, and St Thomas in the south–east (1). The species is still locally common at Roselle (probably synonymous with Rozelle Falls) in the south–east but is rare elsewhere (7).

Habitat and Ecology Prior to 1968 the species was found only sporadically. In 1968 T.W. Turner found the main breeding colony at Roselle and raised specimens from eggs (1). The preferred habitat is a wooded area with the calciphile foodplant *Oxandra lanceolata*, a 10–15 m straight-trunked tree in the family Annonaceae, known locally as the Black lancewood or Okra. At least two broods of the butterfly occur each year, in May–July and September. Adults are abundant for only one week in May or June and are otherwise rare. Pupal diapause may occur. According to D'Abrera (2), *Eurytides marcellinus* is known to maintain very low populations for long periods, occasionally undergoing a population explosion and appearing all over the island. The males are strongly migratory and participate in spectacular westerly migrations in some years; the purpose of these movements is not clear. In the absence of the foodplant from most of the island, such mass movements are doomed to reproductive failure.

Threats The breeding habitat is very restricted and is being severely deforested for cultivation. The larval foodplants are unable to survive the disturbance (7). The breeding ground is not officially protected.

Conservation Measures No long-term conservation measures have been taken so far, but the scientific community in Jamaica has shown considerable concern and started a dialogue with the owners of the 2 sq. km breeding ground. About a quarter of a hectare has recently been set aside as a reserve where it is hoped the larval foodplant can be preserved. An adjacent landowner is prepared to provide land for the foodplant to be planted and funds are urgently required for this project. Since the tree is straight-trunked, it may even prove to be a valuable plantation species.

Further measures required include a survey of the distribution of *Oxandra*, identification of *Oxandra* to local landowners and a plea for help in protecting the trees, establishment of *Oxandra* seedlings in nearby properties in order to extend the range of the plant and the butterfly, and an investigation of predators of the caterpillars and pupae with a view to preventing heavy losses (7). The butterfly can be raised in captivity but the adults are virtually impossible to contain (7).

The Natural Resources Conservation Department of Jamaica should be encouraged to take some responsibility for the protection of *Eurytides marcellinus*, and to take steps to create permanently protected breeding grounds.

References

1. Brown, F.M. and Heinemann, B. (1972). *Jamaica and its Butterflies*. Classey, London. 478 pp.
2. D'Abrera, B. (1981). *Butterflies of the Neotropical Region. Part I. Papilionidae and Pieridae*. Lansdowne Editions, Melbourne. xvi + 172 pp.
3. Hancock, D.L. (1983). Classification of the Papilionidae (Lepidoptera): a phylogenetic approach. *Smithersia* 2: 1–48.
4. Munroe, E. (1961). The classification of the Papilionidae (Lepidoptera). *Canadian Entomologist* Supplement 17: 1–51.
5. Riley, N.D. (1975). *A Field Guide to the Butterflies of the West Indies*. Collins, London. 244 pp.
6. Rothschild, W. and Jordan, K. (1906). A revision of the American Papilios. *Novitates Zoologicae* 13: 411–752. (Facsimile edition ed. P.H. Arnaud, 1967).
7. Turner, T.W. (1983). The status of the Papilionidae, Lepidoptera of Jamaica with evidence to support the need for conservation of *Papilio homerus* Fabricius and *Eurytides marcellinus* Doubleday. Unpublished report. 14 pp.

Eurytides (Protesilaus) lysithous harrisianus (Swainson, 1822) ENDANGERED

Subfamily PAPILIONINAE   Tribe LEPTOCIRCINI

Summary *Eurytides lysithous* is a kite swallowtail which mimics various species of *Parides* swallowtails in different parts of its range. The strikingly patterned subspecies *E. l. harrisianus*, Harris' Mimic Swallowtail, resembles the Vulnerable species *P. ascanius* (also in this volume). Common in the nineteenth century, it was still found regularly around Rio de Janeiro, Brazil, until the 1940s. Nearly all known colonies have been destroyed by development and only a single known locality remains, in which the unpalatable model *P. ascanius* may already be extinct. *E. l. harrisianus* has recently been proposed to be added to Brazil's list of animals threatened with extinction.

Description All members of Munroe's *lysithous* species-group are mimetic of Troidini or Heliconiinae (9). Recently, Hancock (7) (following D'Almeida (6)) has taken a new look at the group and reduced it to 15 species, renaming it after the most primitive member *Eurytides asius*. New data resulting from biological and morphological studies of the species indicate that still further taxonomic analysis at the subgeneric level is needed (2). *Eurytides lysithous* is a highly variable species (5) and *E. l. harrisianus* looks and behaves very much like *Parides ascanius*. It is a medium-sized black swallowtail with narrow tails and, in its commonest form (*platydesma*), both wings are crossed by a broad white band (Plate 2.1). The hindwing is variably marked with large red spots which appear similar to the rose-red patch on *P. ascanius*. Unlike the model, the mimic possesses a red streak at the base of the wings underneath. The antennae, tongue and tails are short for a swallowtail. The immature stages are identical to those of other subspecies of *G. lysithous* (3, 6).

The taxonomic position of the genus is much disputed. Munroe (9) placed the *lysithous* species-group in the genus *Eurytides* subgenus *Protesilaus*. In a recent revision, Hancock has raised *Protesilaus* to generic status and transferred several *Eurytides* species-groups into this new genus (8). Other authorities consider *Eurytides* to be a synonym of *Graphium*, a name normally reserved for Old World species. Hancock has also treated *E. chibcha*, *hipparchus* and *kumbachi* as aberrations of other species and removed them from Munroe's *lysithous* species-group, renaming it the *asius* species-group. In addition, *harrisianus* has been considered to be a form (5) and even a full species (7), as well as a subspecies. These arguments do not affect the main matter under discussion here, i.e. the conservation of this highly endangered taxon.

Distribution The species *Eurytides lysithous* occurs in Brazil and eastern Paraguay. The subspecies *E. l. harrisianus* formerly occurred in southern Espirito Santo and along the whole coast of the State of Rio de Janeiro, Brazil, but is now known only from Barra de São João, on the eastern coast of Rio de Janeiro. Potential habitat and strong colonies of the model species (*P. ascanius*) occur widely in lowland Rio de Janeiro, but no *E. l. harrisianus* have been seen in these sites since 1945 (4, 6). In the single known locality, four specimens were collected between 1977 and 1982 (4); the colony was strong and permitted extensive mark-recapture and biological work in 1984 (2).

Habitat and Ecology Much is now known of the biology of the taxon. It flies in habitats adjacent to the lowland swamps occupied by its model, rarely in the swamps themselves, and sometimes on adjacent hillsides. Males do not obviously frequent hilltops in search of mates, as many other *Eurytides* do, but are attracted to damp earth, where they were formerly often captured in Rio de Janeiro. Females are commonly encountered near their hostplants, which are various genera in the Annonaceae, especially *Xylopia* and *Rollinia* (2). The flight period is from September to February, with the pupal diapause lasting up to nine months (4, 6, 7). The adults fly rapidly and erratically along corridors in their dense, scrubby habitat, preferring areas with alternating large blocks of strong sun and deep shade. In the only known colony, two principal morphs are present, *platydesma* (mimicking *Parides ascanius*, about 60 per cent at present), and *oedippus* (mimicking *P. zacynthus* and *P. neophilus*, about 30 per cent at present) (2). The remaining 10 per cent are genetic recombinants, including typical *harrisianus*. These two morphs both fly from early morning (07.00 a.m.) throughout the day. However, the latter tend to be more active when their models are out, in the early morning and very late afternoon, while the former fly with *P. ascanius* in the warmer hours. Flight activity is strongly correlated with sun and high humidity. Flowers commonly visited include *Lantana*, *Inga* and *Eupatorium*. The colony numbered 50–200 individuals in late 1984 (2).

Threats The site of the only known colony is undergoing development as a recreational area. There is to be some emphasis on wildlife conservation, but without special care the butterfly is likely to be lost in the general alteration of habitat (4, 7). Unusual climatic conditions in 1983 greatly reduced the only known colony (1). At the current level of disruption, extinction of this colony is likely by 1990 (10). With the disappearance of *P. ascanius* from the site due to development of subcoastal marshes, the *platydesma* morph may be negatively selected and become quite reduced in numbers.

Conservation Measures In June 1982 *G. l. harrisianus* was proposed to the Brazilian Government agency responsible for inclusion on the official list of species threatened with extinction. Intensive searching is needed for any additional colonies which might occur in remote regions or in the Reserva Biológica Poço das Antas, Rio de Janeiro, where *P. ascanius* is still present to serve as a model. Officials responsible for the planning and management of the recreation area under construction at Barra de São João should be informed of the species' presence and petitioned to include measures for its conservation in plans for the area. If further populations can be found, urgent studies on the swallowtail's management ecology should be undertaken.

As the only clear Batesian mimic of the vulnerable *Parides ascanius*, this population occupies a unique position for mimicry research. Possible further studies of the model/mimic pair include abundance ratios, association in the field (especially in light of the unusual situation of the model and mimic occupying slightly displaced microhabitats) and synchrony of generations (10).

References

1. Brown, K.S., Jr. (1983). *In litt.* 6 April.
2. Brown, K.S., Fr. (1984). *In litt.* 20 November.
3. Burmeister, A.M. (1879). *Description Physique de la République Argentine* Vol. V. (Lépidoptères), Atlas p. 9, No. 23.
4. Callaghan, G., Laranja, J., Otero, L., Brown, K.S., Jr. (1982). Observations in 1940–1982 in Rio de Janeiro coastal lowlands.

5. D'Abrera, B. (1981). *Butterflies of the Neotropical Region. Part 1. Papilionidae and Pieridae.* Lansdowne Editions, Melbourne. xvi + 172 pp.
6. D'Almeida, R.F. (1922). *Mélanges Lépidoptérologiques, I. Etudes sur les Lépidoptères du Brésil.* R. Friedländer und Sohn, Berlin.
7. D'Almeida, R.F. (1966). *Catálogo dos Papilionidae Americanos.* Sociedade Brasileira de Entomologia. São Paulo, Brazil.
8. Hancock, D.L. (1983). Classification of the Papilionidae (Lepidoptera): a phylogenetic approach. *Smithersia* 2: 1–48.
9. Munroe, E. (1961). The classification of the Papilionidae (Lepidoptera). *Canadian Entomologist* 17: 1–51.
10. Otero, L.S. and Brown, K.S., Jr. (in press). Biology and ecology of *Parides ascanius* (Cramer, 1775) (Lep., Papilionidae), a primitive butterfly threatened with extinction. *Atala.*

This review has been extensively adapted from an entry in *The IUCN Invertebrate Red Data Book*, whose authors and contributors are gratefully acknowledged.

Eurytides (Eurytides) iphitas Hübner, 1821   VULNERABLE

Subfamily   PAPILIONINAE   Tribe   LEPTOCIRCINI

Summary *Eurytides iphitas*, the Yellow Kite, is a poorly known swallowtail from south–eastern and east–central Brazil. It has not been seen for many years and is believed to be in serious decline in the states of Rio de Janeiro and Espirito Santo.

Description *Eurytides iphitas* is a relatively large, tailed butterfly (Plate 2.3). The female is apparently unknown, but may be expected to resemble the male since the sexes of the close relative *E. dolicaon* are very similar. *E. iphitas* closely resembles *E. dolicaon*, but whereas in the former the long, narrow tails are only tipped with yellow, in the latter the tails are mainly yellow (2, 8, 9).

UFW/UHW creamy-buff with a black submarginal border, outer margin and apex, oblique black bands on the forewing and a row of small white submarginal spots on the hindwing (2).

LFW/LHW similar to the upperside, but with less black on the wing borders, more black scaling on the veins and more prominent white submarginal spots on the hindwing (2).

The immature stages of *E. iphitas* have not been described, but are probably similar to those of *E. dolicaon*, in which the final instar caterpillar is yellow–green with a black band and black spotting (3).

Distribution *E. iphitas* is endemic to south–eastern and east–central Brazil. It has not been seen for many years, even in a known colony locality at Bôca do Mato near Rio de Janeiro (1). It has previously been recorded from further north in the state of Espirito Santo, but no localities are known there at present (1, 2, 4).

Habitat and Ecology Very little information is available on this or other members of the *dolicaon* species-group (5, 7). According to Hancock (5) the larvae of the group are smooth and found on Annonaceae or Lauraceae; the pupa is slender with a long dorsal protuberance. *Eurytides iphitas* occurs at medium elevations (2) probably in seasonally moist, open deciduous woodland. *Eurytides dolicaon*, a close relative of *E. iphitas* which also occurs in this part of Brazil, lays its eggs on the underside edges of leaves of *Mespilodaphne indecora* (Lauraceae) (3). It is a multivoltine species, with generations renewed every month during summer. After April the development of caterpillars is slow and over-wintering in the pupal stage usually occurs until mid-August (3). *E. iphitas* might be expected to have similar habits.

Threats Habitat destruction in the south–eastern coastal states of Brazil is an increasingly serious threat to the wildlife. Urbanization, drainage and development will undoubtedly cause the decline of many species, as has been found with *Eurytides lysithous harrisianus*. There is insufficient information on the fauna of the several protected areas in the region (6) to know whether these measures have been effective. *Eurytides iphitas* is believed to be a relict species with a naturally small range being further reduced by destruction of suitable habitat. Its present rarity in regions close to human population centres is a matter for great concern. As the type-species of the genus, *E. iphitas* is of great importance to taxonomists.

Conservation Measures Searches for extant colonies are urgently needed in protected and unprotected areas, in order that data on habitat, ecology and present distribution may be gathered. Only then may its conservation status be properly determined and appropriate action taken. The situation of this butterfly could be extremely grave and no time should be lost in locating and protecting its present haunts.

References

1. Brown, K.S. Jr. (1983). *In litt.*, 6 April.
2. D'Abrera, B. (1981). *Butterflies of the Neotropical Region. Part I. Papilionidae and Pieridae.* Lansdowne Editions, Melbourne. xvi + 172pp.
3. D'Almeida, R.F. (1924). Les papilionides de Rio de Janeiro. Description de deux chenilles. *Annales de la Société Entomologique de France* 93: 23–30.
4. D'Almeida, R.F. (1966). *Catálogo dos Papilionidae Americanos.* Sociedade Brasileira de Entomologia. São Paulo, Brazil.
5. Hancock, D.L. (1983). Classification of the Papilionidae (Lepidoptera): a phylogenetic approach. *Smithersia* 2: 1–48.
6. IUCN (1982). *IUCN Directory of Neotropical Protected Areas.* Tycooly International, Dublin. 436 pp.
7. Munroe, E. (1961). The classification of the Papilionidae (Lepidoptera). *Canadian Entomologist* Supplement 17: 1–51.
8. Rothschild, W. and Jordan, K. (1906). A revision of the American Papilios. *Novitates Zoologicae* 13: 411–752. (Facsimile edition ed. P.H. Arnaud, 1967).
9. Tyler, H.A. (1983). *In litt.*, 30 September.

Graphium (Pathysa) epaminondas Oberthür, 1879 INSUFFICIENTLY KNOWN

Subfamily PAPILIONINAE   Tribe LEPTOCIRCINI

Summary *Graphium epaminondas* is one of three swallowtails confined to the Andaman Islands. Intensification and expansion of cultivation in former forest and clear felling for timber are severely reducing the area of natural habitat. A conservation plan for the islands is needed, coupled with a survey of important butterfly habitats.

Description *Graphium epaminondas* is a relatively large black, white and orange swordtail butterfly with a forewing length of 43 mm, and similar sexes (3).

UFW with a brown ground colour and broad transverse white bands; LFW may be lighter but is otherwise similar in appearance (2, 6).

UHW with a slightly scalloped outer edge and a long, tapering dark brown tail with a white tip. The discal area is white with brown shadows from the LHW pattern (2, 5). Postdiscal and submarginal areas are patterned with brown (2, 6). There is an orange and black tornal spot.

LHW with narrow black bands in the discal region and a postdiscal band of large, orange spots, small black lunules and a dark outer margin.

Distribution *Graphium epaminondas* is confined to the Andaman Islands (India) where it may only occur on South Andaman. The islands are situated between India and Burma in the south-east of the Bay of Bengal.

Habitat and Ecology Details of the life-history and foodplants of this species are unknown. *Graphium epaminondas* is in the *antiphates* species group, of which the larvae generally feed on Annonaceae or Lauraceae (4). *G. epaminondas* has sometimes been treated as a subspecies of *antiphates* (6), but is now generally accepted as a full species (2, 4). There are 13 species of Papilionidae recorded from the Andamans (1). In addition to *G. epaminondas*, *Papilio mayo* and *Atrophaneura rhodifer* are endemic to the islands. The natural vegetation of the islands is rain forest and all three butterflies are probably forest inhabitants.

Threats The forest flora of the Andaman Islands is of Malesian origin, but has its own unique character and a high level of endemicity. A number of articles and reports describe the increasing rate of forest destruction in the Andamans and, in the absence of protected areas, there is good reason for conservation concern. The main threats are the intensification and expansion of cultivation, clear-felling for timber and planting of forest monocultures. One cause is the growing human population of the Andamans, which is being boosted at a rate of up to 1000 per month (in 1980) by new settlers from Bihar in India and by Karens from Burma (7). The heavy monsoon rains are apparently causing severe erosion of the thin soils in deforested areas, ultimately leading to degradation and loss of natural habitats.

All three endemic papilionids are poorly known and very little has been published on their habitat and behaviour. A 1980 report states that *Graphium epaminondas* and *Atrophaneura rhodifer* are not rare, and that *Papilio mayo* males are common but females rare (1, 5). These assessments seem to be based mainly on numbers of museum specimens and an objective field assessment of conservation needs is required.

213

Conservation Measures At the moment there is too little information on *Graphium epaminondas* and the other endemic papilionids of the Andamans to propose conservation measures. The Botanical Survey of India has drafted proposals for a land use and conservation study of the Andamans, hopefully this will result in the development of protected areas. An entomological and zoological input into these studies might be encouraged.

References

1. Arori, G.S. and Nandi, D.N. (1980). On the butterfly fauna of Andaman and Nicobar Islands (India). I. Papilionidae. *Records of the Zoological Survey of India* 77: 141–151.
2. D'Abrera, B. (1982). *Butterflies of the Oriental Region. Part 1. Papilionidae and Pieridae.* Hill House, Victoria, Australia. xxxi + 244 pp.
3. Evans, W.H. (1932). *The identification of Indian butterflies.* Diocesan Press, Madras. 2nd ed., revised. 454 pp. + 31 plates.
4. Hancock, D.L. (1983). Classification of the Papilionidae (Lepidoptera): a phylogenetic approach. *Smithersia* 2: 1–48.
5. Haugum, J. (1983). *In litt.*, 2 June 1983.
6. Tsukada, E. and Nishiyama, Y. (1982). *Butterflies of the South East Asian Islands. Vol. 1 Papilionidae.* (transl. K. Morishita). Plapac Co. Ltd., Tokyo. 457 pp.
7. Wright, A. (1980). Destruction in the Andamans. *Oryx* 15: 315–316.

Graphium (Pathysa) idaeoides Hewitson, 1853   RARE

Subfamily PAPILIONINAE   Tribe LEPTOCIRCINI

Summary *Graphium idaeoides* is rare throughout its range in the Philippine islands of Luzon, Samar, Leyte and Mindanao. It is a remarkably realistic mimic of *Idea leuconoe* (Danainae), which has a much wider distribution in South East Asia. The main stronghold is eastern Mindanao, where rain forests are under increasingly severe threat from exploitation and recently from forest fires, even within national park boundaries.

Description *Graphium idaeoides* is a large white butterfly with reticulated black markings and a forewing length of 70–80 mm (Plate 2.4). It closely mimics the sympatric species *Idea leuconoe* (Nymphalidae: Danainae), which has a forewing length of around 73 mm (3). The sexes are alike (1, 3, 9).

UFW/LFW white with heavy black scaling over the veins, especially in the apical and submarginal regions and a row of large, black postdiscal spots. The basal part has a creamy or yellowish tinge to the ground colour, varying between island races (9).

UHW/LHW similar to upper surface.

Distribution *Graphium idaeoides* occurs in the eastern Philippines, from Luzon to Mindanao. It has been recorded from localities in the north-east, east and south of Luzon, from Leyte, Samar, and various localities throughout eastern Mindanao (3, 5). In addition, there is apparently a disjunct population in the Zamboanga Peninsula in the south-west of Mindanao (1, 3, 9).

Habitat and Ecology *Graphium idaeoides* is a rare and local species preferring the lowland rain forests (less than 330 m a.s.l.) on the eastern side of the Philippines, where rainfall is continuously high. Jumalon notes that *Graphium idaeoides* flies in forests 100–200 m above sea level in an area north-west of the Mt Diwaba range in Mindanao (3, 4). Adults are reported to fly along sheltered and shady streams at a height of 3–10 m, much in the manner of the model, *Idea leuconoe*, but with a slightly heavier flight (3). The main flight season is possibly February to April after the period of heavy rains, although specimens have also been taken in virtually every other month of the year (3, 8, 9). Males defend definite territories believed to extend up to 100 m along streams. They may adopt one or more perches, from which sorties are made to repel intruders of the same or different species. Neither the fast-flowing, upper reaches of the streams on hillsides nor the coastal 3 km or so of rivers are used by *G. idaeoides*, although its model may be seen in such places. Females are more elusive than males; Jumalon notes that of 45 specimens taken between 1958 and 1962, only three were females (3). Unlike some other papilionids, *G. idaeoides* does not fly through the tangled undergrowth of the forest (3). Many specimens have been taken along water-courses in secondary growth forest, indicating a certain tolerance of disturbance (3). Other species with which it shares its habitat include the birdwings *Troides magellanus* and *T. rhadamantus* (3). The species may have several broods per year, but the foodplants and young stages are completely unknown (9). *Graphium idaeoides* is in the *macareus* group of species (2, 7), some of which feed on Aquifoliaceae (2).

Threats Although widely distributed, *Graphium idaeoides* seems to be rare and local throughout its range (8, 9), often very difficult to locate (3). The hazards to this species are not unique, but common to many other Philippine endemics. The primary threat is habitat destruction caused by the accelerating rate of deforestation and soil degradation throughout the Philippines. Much of the habitat of *Graphium idaeoides* in eastern Luzon, Leyte and northern Mindanao has already been cleared for arable crops and coconut plantations, although there are still considerable areas of forest on Samar. Its main stronghold seems to be the rainforests in the Agusan Valley of north-eastern Mindanao and forested areas in the extreme south-east of the island (8), regions which are now under increasingly heavy threat from commercial logging and the 'kaingin' farmers (shifting cultivators) who follow in their wake. One of the finest national parks in the Philippines, Mt Apo National Park, certainly includes this butterfly in its lower reaches, but there have been reports that large areas of the Park have been leased for agricultural development. For further details of this, refer to the review of the Mt Apo endemic *Graphium sandawanum*. It is now feared that the drought of 1982–3 and the fire that followed in its wake have destroyed large areas of Mindanao's eastern forests. The full extent of defoliation will only be known when aerial or satellite surveys have been undertaken (6). General data on deforestation in Mindanao are given in the reviews of two Mindanao endemics, *Papilio osmana* and *Papilio carolinensis*. Within the next two decades the conservation status of all these butterflies will become increasingly serious.

Conservation Measures The reported threat to Mt Apo National Park is a serious matter not only for this butterfly, but for many other species of rain forest wildlife, including the Philippine national bird, the Monkey-eating Eagle (*Pithecophaga jefferyi*). Maintenance of the integrity of Mt Apo National Park is therefore a vital priority. The lowland forests of the Agusan watershed may also prove to be important for the conservation of this and many other lowland species since it is one of few extensive areas in Mindanao below 200 m a.s.l. The apparently small and very local nature of the populations of *Graphium idaeoides* is a matter for concern. Although the reason may be related to its status as a mimic of *Idea leuconoe*, further study is necessary in order to ascertain its foodplant and breeding ecology.

References

1. D'Abrera, B. (1982). *Butterflies of the Oriental Region. Part 1. Papilionidae and Pieridae.* Hill House, Victoria, Australia. xxxi + 244 pp.
2. Hancock, D.L. (1983). Classification of the Papilionidae (Lepidoptera): a phylogenetic approach. *Smithersia* 2: 1–48.
3. Jumalon, J.N. (1964). Haunt and habits of *Graphium idaeoides* (Hewitson) (Papilionidae: Lepidoptera). *The Philippine Journal of Science* 93(2): 207–216.
4. Jumalon, J.N. (1967). Two new papilionids. *Philippine Scientist* 1(4): 114–118.
5. Jumalon, J.N. (1969). Notes on the new range of some Asiatic papilionids in the Philippines. *The Philippine Entomologist* 1(3): 251–257.
6. Lewis, R.E. (1984). *In litt.*, 17 April.
7. Munroe, E. (1961). The classification of the Papilionidae (Lepidoptera). *Canadian Entomologist* Supplement 17: 1–51.
8. Treadaway, C.G. (1984). *In litt.*, 25 May.
9. Tsukada, E. and Nishiyama, Y. (1982). *Butterflies of the South East Asian Islands. Vol. 1 Papilionidae.* (transl. K. Morishita). Plapac Co. Ltd., Tokyo. 457 pp.

Graphium (Pathysa) megaera Staudinger, 1888   INDETERMINATE

Subfamily   PAPILIONINAE   Tribe   LEPTOCIRCINI

Summary   *Graphium megaera* is a small swallowtail confined to the forests of Palawan in the Philippines. The status of the species is uncertain but, along with many other endemic species of wildlife, it is likely to be declining in the face of rapidly accelerating deforestation. The national parks system in the Philippines is undergoing redevelopment. The Province of Palawan needs the protection of significant areas of wilderness to ensure the long term survival of its unique fauna and flora.

Description   This small tailless papilionid has a forewing length of about 35–40 mm. It has been variously stated that the female is unknown (4, 7, 11) and that the sexes are similar (3).
UFW with a dark brown ground colour, submarginal and postdiscal rows of white spots.
UHW similar but with a scalloped outer margin. White markings include horseshoe-shaped submarginal spots, postdiscal spots and small discal spots.
LFW/LHW. The lower surface has a lighter ground colour, but is otherwise similar.

Distribution   *Graphium megaera* is found only on the Philippine island of Palawan, where it is fairly widespread, particularly in central and northern regions (4, 10). *Atrophaneura atropos* is also endemic to Palawan and the review of that species later in this volume should also be consulted.

Habitat and Ecology   Palawan is a fairly large island (11 500 sq. km), 53 per cent of which is still covered with tropical lowland dipterocarp and montane forest (8). The greater part is in the southern and south-western regions (1). The western coast-line tends to have a more seasonal climate and the known areas of distribution of *Graphium megaera* are in the central highlands and eastern lowlands (11). The main habitat is lowland, but records indicate a potential distribution up to about 800 m (10). The butterflies are to be seen flying along the rivers, the males coming down to moist spots with a slow, gliding flight (11). Actively on the wing on fine mornings, the main flight season is March to April (11), sometimes going through into August (4, 10). Females are extremely shy and have been rarely, if ever, seen. The early stages and foodplants are unknown, but some other species in the *macareus* group feed on Aquifoliaceae (6, 7).

Threats   The main threat to the fauna of Palawan is deforestation by timber companies and agriculturalists. Much of the island is under concession to logging companies (44 per cent of total land area in 1976), and the rate of deforestation is increasing rapidly (1, 5). The same applies to the rest of the Philippines and nearby Sabah (5). Long-term plans include reforestation of logged sites but the experience in South East Asia is that this promise is rarely kept. Incursions by shifting cultivators and illegal squatters often follow in the wake of the logging road graders (5, 8). Mining for silica sand, copper, chrome, nickel and mercury may also have a local impact on forests, but these threats are more serious to freshwaters and along coastlines (2). The precise degree of threat to *Graphium megaera* is unlikely to be critical at this time, but with deforestation accelerating and the Philippine national parks system uncertain, there is concern for all Palawan's native and endemic wildlife.

Conservation Measures The whole of Palawan was at one time designated as a wildlife sanctuary and national park, but this was to no practical purpose (1, 2). Now it seems that the Philippine Government is restructuring its entire national park system in favour of creating fewer parks, but of a higher, international, standard (8, 9). It is not clear whether any new parks will be created on Palawan, although St Paul's Underground River National Park will presumably remain. There is certainly potential for siting more parks in Palawan, where large areas of wilderness are still available in the northern highlands around Cleopatra's Needle, in the central highlands around an area called The Teeth, and in the southern regions south of Mt Mantalingajan (1). It has been suggested that both Cleopatra's Needle and Mt Mantalingajan should be designated as Wildlife Sanctuaries, with access only for scientific research (9). The status of *Graphium megaera* and other butterflies needs urgent assessment in Palawan, together with rapid recommendations and action in siting well-protected national parks on the island.

References

1. Bruce, M.D. (Ed.) (1980). *The Palawan Expedition Stage I*. Traditional Explorations, Sydney. 47 pp.
2. Bruce, M.D. (Ed.) (1981). *The Palawan Expedition Stage II*. Associated Research, Exploration and Aid Pty Ltd., Sydney. 139 pp.
3. D'Abrera, B. (1982). *Butterflies of the Oriental Region. Part 1. Papilionidae and Pieridae*. Hill House, Victoria, Australia. xxxi + 244 pp.
4. Dacasin, G.A. (1984). *In litt.*, 25 April.
5. FAO/UNEP (1981). *Tropical Forest Resources Assessment Project. Forest Resources of Tropical Asia*. FAO, Rome. 475 pp.
6. Hancock, D.L. (1983). Classification of the Papilionidae (Lepidoptera): a phylogenetic approach. *Smithersia* 2: 1–48.
7. Munroe, E. (1961). The classification of the Papilionidae (Lepidoptera). *Canadian Entomologist* Supplement 17: 1–51.
8. National Environment Protection Council (1979). *Philippine Environment 1979. NEPC Third Annual Report*. NEPC, Ministry of Human Settlements, Republic of the Philippines. 158 pp.
9. Pollisco, F.S. (1982). An analysis of the national park system in the Philippines. *Likas–Yaman, Journal of the Natural Resources Management Forum* 3(12): 56 pp.
10. Treadaway, C.G. (1984). *In litt.*, 25 May.
11. Tsukada, E. and Nishiyama, Y. (1982). *Butterflies of the South East Asian Islands Vol. 1 Papilionidae*. (transl. K. Morishita). Plapac Co. Ltd., Tokyo. 457 pp.

Graphium (Arisbe) aurivilliusi Seeldrayers, 1896   INSUFFICIENTLY KNOWN

Subfamily PAPILIONINAE   Tribe LEPTOCIRCINI

Summary *Graphium aurivilliusi* is known only from the type series from present-day Zaire. More information on its status and distribution is needed.

Description The male *Graphium aurivilliusi* is a small blackish-brown tailless butterfly with green-tinged white markings and a wingspan of about 75 mm (4, 8). The female is unknown (8). This species is in the *ucalegon* species-groups of Berger (2) and Munroe (7), or the *adamastor* species-group of Hancock (6).

UFW/LFW blackish-brown with green-tinged white markings, these including three elongated cell spots, two large subapical spots and a broad discal band (8).

UHW/LHW with a broad, green-tinged white discal band, abdominal area, and many small postdiscal and submarginal spots (8).

Distribution This species is known only from the type series from the "Congo" in West Africa (4, 7, 8). There are presumably at least two specimens since the two illustrations given by Berger (1, 2) are different. The precise collecting locality is unknown (8) but is almost certainly in present-day Zaire (1, 2, 3).

Habitat and Ecology Unknown. Other members of the *adamastor* group feed on Anacardiaceae (6).

Threats This butterfly is so rare and poorly known that there are insufficient data to assess current threats. The forests of Zaire and Congo are better protected than in any other African countries.

Conservation Measures No proposals are possible without habitat information. Congo has 1300 sq. km of rain forest legally protected from logging, and Zaire has 56 900 sq. km (5). Logging in these countries is currently less than 0.2 per cent per year.

References

1. Berger, L.A. (1950). Catalogues raisonnées de la faune entomologique du Congo Belge. Lépidoptères–Rhopalocères, 1, Family Papilionidae. *Annales du Musée Royale du Congo Belge Sér. 4to (C.: Zool.)* 8: 1–104.
2. Berger, L.A. (1951). Systématique des Papilionidae de la faune éthiopienne. *3me Congrès National des Sciences, Bruxelles* 8: 47–50.
3. Carcasson, R.H. (1980). *Collins Handguide to the Butterflies of Africa*. Collins, London. Hardback edition, 188 pp.
4. D'Abrera, B. (1980). *Butterflies of the Afrotropical Region*. Lansdowne Press. xx + 593 pp.
5. FAO/UNEP (1981). *Tropical Forest Resources Assessment Project. Forest Resources of Tropical Africa. Part 1: Regional Synthesis*. FAO, Rome, 108 pp.
6. Hancock, D.L. (1983). Classification of the Papilionidae (Lepidoptera): a phylogenetic approach. *Smithersia* 2: 1–48.
7. Munroe, E. (1961). The classification of the Papilionidae (Lepidoptera). *Canadian Entomologist* Supplement 17: 1–51.
8. Williams, J.G. (1969). *A Field Guide to the Butterflies of Africa*. Collins, London. 238 pp., 24 col. pl.

***Graphium (Arisbe) weberi* Holland, 1917**   INSUFFICIENTLY KNOWN

Subfamily PAPILIONINAE   Tribe LEPTOCIRCINI

Summary *Graphium weberi* comes from Cameroon whence it is known only from a single specimen. Assessments of the taxonomic status and distribution of the species are required.

Description *Graphium weberi* is a small brown and ochreous tailless butterfly in the *ucalegon* group of Munroe (7), and the *adamastor* group of Hancock (4). It has a wingspan of about 80 mm (8). Since the species has been so rarely collected, doubts may be raised about its taxonomic status. The closely related and sympatric *G. ucalegonides* and *G. fulleri* are similar in appearance and known to be variable (5, 6).

UFW blackish-brown with a broad, pale ochreous median band separated into large spots, and a line of large subapical spots (8).

UHW also blackish-brown with a pale ochreous band and pale ochreous submarginal spots (8).

LFW/LHW pale brown with reddish wing bases (8).

Distribution This species from Cameroon, West Africa, is known only from the type specimen, held at the Carnegie Museum, Pittsburgh, Pennsylvania (2). It is not clear on what basis Williams noted that the sexes are alike (8).

Habitat and Ecology Believed to be confined to tropical rain forest in Cameroon (1, 5). Feeding requirements unknown, but other members of the group feed on Anacardiaceae (4).

Threats This butterfly is so rare and poorly known that there are insufficient data to assess current threats. However, it may be noted that Cameroon currently contains no forest which is legally protected from logging (3).

Conservation Measures Three forest sites in Cameroon are scheduled for full protection: Korup (c. 90 000 ha), Dja (c. 540 000 ha) and Pangar–Djerem (c. 480 000 ha). Legislation is expected during the current five-year plan. The World Wildlife Fund is assisting the newly-formed Department of Wildlife and National Parks in various ways. It is not known whether Weber's swallowtail occurs in these proposed parks.

References

1. Carcasson, R.H. (1980). *Collins Handguide to the Butterflies of Africa*. Collins, London. 106 pp.
2. D'Abrera, B. (1980). *Butterflies of the Afrotropical Region*. Lansdowne Press. xx + 593 pp.
3. FAO/UNEP (1981). *Tropical Forest Resources Assessment Project. Forest Resources of Tropical Africa. Part 1: Regional Synthesis*. FAO, Rome, 108 pp.
4. Hancock, D.L. (1983). Classification of the Papilionidae (Lepidoptera): a phylogenetic approach. *Smithersia* 2: 1–48.
5. Hancock, D.L. (1983). *In litt.*, 21 June.
6. Hancock, D.L. (1984). *In litt.*, 25 June.

7. Munroe, E. (1961). The classification of the Papilionidae (Lepidoptera). *Canadian Entomologist* Supplement 17: 1–51.
8. Williams, J.G. (1969). *A field guide to the butterflies of Africa.* Collins, London. 238 pp.

Note added in proof: In a very recent paper, Hancock (9) placed *Graphium weberi*: as a new synonym of *G. fulleri*, thereby invalidating the above assessment. *G. fulleri* is widespread in the forests of Cameroon, Gabon, Congo Republic, Ivory Coast, Ghana, Angola and Zaire.

9. Hancock, O.L. (1985). Systematic notes on some African species of *Graphium* Scopoli. *Papilio International* 2 (1-2): 97-103.

Graphium (Arisbe) levassori Oberthür, 1890   VULNERABLE

Subfamily PAPILIONINAE   Tribe LEPTOCIRCINI

Summary *Graphium levassori* is confined to Grande Comore in the Comoro Islands, where it is extremely scarce. There is no information on its biology or ecology and surveys and studies are urgently needed.

Description This pale yellow butterfly with black markings is in the *leonidas* group (4) and has similar sexes, though the females are slightly larger (Plate 2.5). The forewing length is 43–55 mm (3, 7).

UFW pale yellow with a black apex, outer and costal margins; wing base reddish with a small black cell spot, two yellow apical spots and a narrow yellow streak between the cell and the outer margin (3, 6).

UHW also pale yellow, with a black border to the scalloped outer margin. There is some brown scaling at the apex with two yellow lunules and a silvery-white costal margin (3, 6, 7).

LFW/LHW differs from the uppersides in having fawn instead of black scaling and brick-red wing base (3, 6, 7).

Distribution This species is endemic to Grande Comore in the Comoro Islands, between Mozambique, on the African mainland, and Madagascar in the Indian Ocean (3, 6).

Habitat and Ecology The four islands of the Comoros are volcanic in origin, Grande Comore being the most recent. On Grand Comore there are two uplifted areas, the volcanic massif of La Grille (1087 m) in the north, and the still active volcano of Karthala (2560 m) in the south. All the islands are characterized by moderate rainfall (1000–5000 mm per year) and soils with a very poor capacity for water retention. There are no streams or rivers on Grande Comoro, although subterranean hollows in the lava may hold water and encourage growth of forest (9). Patches of forest may still be found at low and middle altitudes e.g. near Nioumbadjou (550 m), but the main forests occur on the flanks of La Grille and Karthala up to a maximum of 1900 m (1). The forests on La Grille have been severely depleted by banana cultivation, but they are more extensive on Karthala, on whose western and south-western sides forest may grow at altitudes as low as 550–800 m (1).

Graphium levassori inhabits forested areas and is believed to be very local in its distribution (5). In 1980 it was reported from forests near M'Lima Manda, north of Karthala, where it was said to be not rare but difficult to capture (9). However, a further visit to search for the butterfly in 1983 was unsuccessful (8). Little is known of the habits or biology of this species, but other members of the *leonidas* group feed on Annonaceae, the Custard Apple family (4). *G. levassori* has been found at sea level in the north of the island, where the cultivated Custard Apple (*Annona reticulata*) is common (2), but this seems to be an exceptional record and has not been recently confirmed. The adults may be on the wing all year round since dry periods are not usually prolonged, but they are certainly on the wing at the beginning of the heaviest rains in April (9).

Threats *Graphium levassori* is one of many species of insects and other organisms, including *Papilio aristophontes* (see separate review), that are confined to the forests of Grande Comore and the other Comoro Islands. Even within these forests *G. levassori* seems to be restricted to medium altitudes and very patchily distributed. There is considerable evidence that the forests are under increasingly severe pressure and that if forest destruction continues to accelerate, extinctions are inevitable.

The estimated human population of Grande Comore in 1976 was 150 000, and growing rapidly. The island is mainly farmed at the subsistence level and income from the main cash crops such as vanilla, copra, cacao, sisal, coffee, cloves and essential oils is very low. The main subsistence crop is bananas, plantations and gardens of which clothe the island. Even forests on La Grille and Karthala, from a distance apparently dense and healthy, may be found on closer inspection to include few large trees and a dense understorey of banana trees (9). The wild trees are unable to reproduce and regenerate, and they are gradually retreating higher and higher (9). Bananas are grown at altitudes up to 1200 m; the destruction is particularly serious on La Grille, but also extensive on Karthala (1, 9). The gradual erosion of the forests by the planting of bananas has been continuing for at least the past 25 years (9).

Conservation Measures Even though *Graphium levassori* may be able to eke out an existence on cultivated Custard Apple, the only way to ensure its long-term survival is protection of its forest home from the constant incursion of banana growers. At present there are no protected areas on the islands, a situation which should be remedied as soon as possible. It is clearly necessary to create a series of reserves in representative biotopes. In the middle altitudes of Karthala it is almost too late to find undisturbed forest, but some areas should be protected and allowed to regenerate. At higher altitudes of 1200 m and above, the government could consider making the whole of Karthala a protected area. This would not affect many people and would ensure the survival of at least the higher altitude species for posterity.

Conservation measures specifically for *Graphium levassori* are not possible at this stage. More data are needed on its distribution and ecology in order that measures for its protection may be incorporated into efforts to conserve representative biotopes. Being large and spectacular as well as rare, *Graphium levassori* is prized by collectors (5). A farming or ranching programme on Custard Apple bushes could be developed into a useful local industry as well as a valuable measure for conservation purposes.

References

1. Benson, C.W. (1960). The birds of the Comoro Islands: results of the British Ornithologists' Union Centenary Expedition 1958. *Ibis* 103b: 5–106.
2. Collins, S. (1983). *In litt.*, 12 July.
3. D'Abrera, B. (1980). *Butterflies of the Afrotropical Region*. Lansdowne Press. xx + 593 pp.
4. Munroe, E. (1961). The classification of the Papilionidae (Lepidoptera). *Canadian Entomologist* Supplement 17: 1–51.
5. Paulian, R. (1983). *In litt.*, 10 May.
6. Paulian, R. and Viette, P. (1968). *Faune de Madagascar. XXVII Insectes Lépidoptères Papilionidae*. O.R.S.T.O.M. and C.N.R.S., Paris. 97 pp., 19 pl. (2 col.), 34 figs.
7. Smart, P. (1975). *The Illustrated Encyclopedia of the Butterfly World*. Hamlyn, London. 275 pp.
8. Turlin, B. (1983). *In litt.*, 15 September.
9. Viette, P. (1980). Mission lépidoptérologique à la Grande Comore (Ocean Indien occidental). *Bulletin de la Société Entomologique de France* 85: 226–235.

Graphium (Graphium) procles Grose–Smith, 1887   INDETERMINATE

Subfamily PAPILIONINAE   Tribe LEPTOCIRCINI

Summary *Graphium procles* is restricted to the Crocker Range in the East Malaysian state of Sabah in northern Borneo. It is best known from Mt Kinabalu, the highest mountain between the Himalayas and New Guinea, where it occurs in montane rain forest. Mt Kinabalu is designated as a national park, but economic interests such as copper mining, logging, farming and recreation are eroding the park boundaries.

Description *Graphium procles* is a tailless butterfly with a forewing length of about 39–42 mm. The female is very rarely seen, and has not been described in the literature.

UFW black with a broad median band tapering from the inner margin towards the apex, a row of small submarginal spots, and some irregular subcostal spots and streaks mainly in the subapical region. Markings are pale blue–green, becoming yellower towards the apex (7).

UHW black with a scalloped outer margin, and a continuation of the UFW median band which crosses the cell and tapers sharply towards the anal angle. The subcostal region is white and other markings are pale blue–green, including the submarginal spots (7).

LFW/LHW with paler markings and a slightly browner ground colour. The LHW has irregular dark orange or buff discal markings (6). The subcostal part of the LHW band is cream-coloured (6).

Distribution *Graphium procles* is endemic to Sabah (a state of Malaysia) in northern Borneo. It occurs in the west, only on Gunung (Mt) Kinabalu and the Crocker range (6, 7).

Habitat and Ecology *Graphium procles* is restricted to lower montane forest (4) at elevations over 1000 m (6). It is best known from Gunung Kinabalu, where the species as a whole is not uncommon, possibly the most frequently encountered species in the subgenus (1). Females, however, are extremely rare (7). The early stages and foodplants are unknown (7) but other members of the *eurypylus* group feed on Annonaceae, Lauraceae or Magnoliaceae (3). Adults have been recorded at all times of year, flying along streams or visiting moist spots (7). *Graphium procles* shares its habitat with the widespread *Graphium bathycles*, but their main flight seasons tend to be separate.

Threats The main threat to *Graphium procles* is destruction of its restricted montane habitat. Mt Kinabalu is the only part of the Crocker Range that is designated as a national park, and even this is under pressure from a number of quarters. Increased tourism in recent years has brought necessary developments such as access roads and a visitor centre, but plans for a golf course and other recreational facilities in the virgin forests of the Pinosuk plateau threaten to destroy habitats and are of questionable conservation value. Parts of the Pinosuk plateau have already been lost to forest clearance for cattle farming, timber processing and hydro-electric projects (2). Many of the lower slopes of western Kinabalu have long been cleared by shifting cultivators (2).

On eastern Kinabalu the discovery of porphyry copper has resulted in the lease of 2556 ha of national park land to a joint Japanese–Malaysian mining company. The lease runs from 1973 for 30 years and the Mamu Copper Mine has completely changed the eastern face of Kinabalu (2). These alterations to Mt Kinabalu are a serious threat to its butterflies and the great wealth of other endemic animals and plants that occur there. Two other threatened papilionids, *Papilio acheron* and *Troides andromache*, are also found on Mt Kinabalu (see separate reviews).

Conservation Measures Kinabalu is the highest mountain in South East Asia and the highest point between the Himalayas and New Guinea (5). Its value as an important biogeographical site and as a water catchment area led to its designation as a national park in 1964 (5). It now acts as a vital conservation area and a major tourist attraction. Short-term economic developments may threaten the long-term value of the region as a biological storehouse and source of tourist revenues. Sensitive planning is needed if further losses of these rare habitats and their fauna and flora are to be avoided.

References

1. Barlow, H.S., Banks, H.J. and Holloway, J.D. (1971). A collection of Rhopalocera (Lepidoptera) from Mt Kinabalu, Sabah, Malaysia. *Oriental Insects* 5: 269–296.
2. Davis, S. (1983). Draft review for IUCN Plant Sites Red Data Book.
3. Hancock, D.L. (1983). Classification of the Papilionidae (Lepidoptera): a phylogenetic approach. *Smithersia* 2: 1–48.
4. Holloway, J.D. (1978). Butterflies and Moths. In *Kinabalu Summit of Borneo*. Sabah Society Monograph, 25–278.
5. Luping, D.M., Wen., C. and Dingley, E.R. (Eds) (1978). *Kinabalu, Summit of Borneo*. Sabah Society Monograph, Malaysia.
6. Robinson, J.C. (1975–6). Swallowtail butterflies of Sabah. *Sabah Society Journal* 6: 5–22.
7. Tsukada, E. and Nishiyama, Y. (1982). *Butterflies of the South East Asian Islands. Vol. 1 Papilionidae*. (transl. K. Morishita). Plapac Co. Ltd., Tokyo. 457 pp.

Graphium (Graphium) meeki (Rothschild & Jordan, 1901)   RARE

Subfamily   PAPILIONINAE   Tribe   LEPTOCIRCINI

Summary *Graphium meeki* is a very rare and locally distributed butterfly from the mountainous districts of Bougainville, Choiseul and Santa Isabel. With extensive mining, logging and agricultural activities within its range, there are fears that local population extinctions may be occurring even before the distribution of the butterfly is properly known. With more data on its distribution and biology, it could be a good candidate for small-scale farming.

Description The authors have not seen a specimen or photograph of a male *Graphium meeki*, but the sexes are probably similar, as in the closely related *G. agamemnon*.

Female: UFW ground colour black with blue–green markings consisting of rows of spots in the subcostal, subapical, submarginal and median areas (1, 5). Forewing length about 53 mm (7).

UHW black with a slightly scalloped outer margin, no tail, some white spots and a row of large blue–green discal-submarginal spots.

LFW/LHW similar to the upper sides but scaled with white and with a sub-basal red bar (1, 7).

Distribution This butterfly is endemic to the Solomon Archipelago, east of Papua New Guinea. It has only been recorded from Bougainville (Papua New Guinea), Choiseul, and Santa Isabel (Solomon Is) (4, 5).

Habitat and Ecology *Graphium meeki* is confined to mountainous, inland areas (5). Much of its range remains little explored and although generally rare it may be locally common. The foodplants are unknown, but other species in the *agamemnon* group feed on Annonaceae or Magnoliaceae (2). In northern Bougainville *G. meeki* has been found at 600–800 m in cleared areas of rain forest (6). The egg is large, pale yellow and laid on young leaves of *Litsea* sp. (Lauraceae). Caterpillars of the second or third instar are completely black with no markings (6).

Threats No specific threats have been identified so far (3), but the species is so rarely collected and poorly known that populations could be disappearing before they are recorded. The Solomon Islands Government requires an export permit for all butterfly captures, but this is difficult to enforce. In any case, habitat destruction is likely to be a more significant threat. There are major logging interests throughout the Solomons archipelago. Habitat destruction is extensive on Bougainville, where open-cast copper mining is a major industry.

Conservation Measures Further study of the distribution and biology of *Graphium meeki* is urgently needed. Forest exploitation could already be reducing the range of *Graphium meeki*, possibly causing extinctions of some of its apparently highly localized populations. The status of this species should be reviewed as soon as more data are forthcoming. There is a possibility that it could be Vulnerable, if not Endangered. If a small farming operation could be initiated it would be a good conservation measure as well as providing local employment. There would certainly be some demand for such a little-known species.

References

1. D'Abrera, B. (1971). *Butterflies of the Australian Region*. Lansdowne Press, Melbourne. 415 pp.
2. Hancock, D.L. (1983). Classification of the Papilionidae (Lepidoptera): a phylogenetic approach. *Smithersia* 2: 1–48.
3. Macfarlane, R. (1983). *In litt.*, 15 March.
4. Parsons, M.J. (1980). *In litt.*, 7 February.
5. Racheli, T. (1980). A list of the Papilionidae (Lepidoptera) of the Solomon Islands, with notes on their geographical distribution. *Australian Entomologists' Magazine*. 7: 45–59.
6. Racheli, T. (1984). *In litt.*, 18 June, incorporating observations from R. Straatman.
7. Rothschild, W. and Jordan, K. (1901). On some Lepidoptera. *Novitates Zoologicae* 1901: 401–407.

Graphium (Graphium) stresemanni (Rothschild, 1916)   RARE

Subfamily   PAPILIONINAE   Tribe   LEPTOCIRCINI

Summary   *Graphium stresemanni* is a little-known butterfly from Seram in Maluku Province (Moluccas), Indonesia. Although Seram has not yet attracted large-scale logging operations comparable to other areas of the province, the threat remains. The Manusela Reserve on Seram promises to protect the island's wildlife but has yet to be designated as a national park. There is a substantial trade in butterflies on Seram, involving whole villages. Documentation and rationalization of this trade is needed.

Description   *Graphium stresemanni* is a brown, tailed butterfly with white, pale blue, pale green and yellow markings and a forewing length of about 40 mm (2, 8). The species was not listed by Munroe (4) because it was originally described as a subspecies of *G. weiskei*, but it is now generally accepted as a full species of some rarity (2, 6, 7, 8). The sexes are broadly similar.

UFW dark brown ground colour with pale blue patches below the cell and in the postdiscal area, yellow subapical spots and a row of small white, submarginal spots (2, 4).

UHW with a similar brown ground colour, pale green markings in the basal and discal regions, and a row of larger, white, submarginal spots.

LFW/LHW similar to the upperside (2).

Distribution   *Graphium stresemanni* is only found on Seram (otherwise known as Serang or Ceram), one of the larger and more southerly islands of Maluku Province (Moluccas), Indonesia. Several of the specimens listed by Okano (6) came from around the 1000 m level in the Manusela range of central Seram.

Habitat and Ecology   Very little is known about this butterfly and its ecological requirements have not been researched. Its young stages and foodplants have not been recorded, but the *sarpedon* group, of which *stresemanni* is a member, feed on Winteraceae, Monimiaceae, Lauraceae or Hernandiaceae, rarely other groups (3). The main range of the group is the eastern Palaearctic, but a number of species reach into Indo–Australia. *G. sarpedon* extends throughout South East Asia and New Guinea to Queensland, Australia, but other species are more restricted, in some cases as a result of the fluctuations in sea level resulting from the Ice Ages. *G. empedovana* is found from Palawan to Sumatra while *G. codrus* extends eastwards, from Sulawesi and the Philippines across to the Solomons. A few species occur only over a very small range, either because they have evolved there or because they are now restricted to a small fragment of a biome that may once have been more extensive. *G. sandawanum* (see next review) is restricted to montane forest on Mt Apo, Mindanao, and *G. monticolum* is only found in similar habitat on Sulawesi (3). Seram rises to over 3000 m in its central mountain range and *G. stresemanni* is probably another isolated montane relict species, perhaps with its main populations in the Manusela range.

Seram is the largest of the Maluku islands, very mountainous with central peaks rising to over 3000 m and with a rainfall of around 2000 mm falling mainly between May and October (1, 9). The natural vegetation includes lowland and montane forests. The lowland forests are mainly semi-deciduous and some of the common tree genera found there are *Canarium*, *Eucalyptus*, *Intsia*, *Agathis*, *Diospyros* and *Palaquium*.

Threats Maluku Province has a unique and scientifically extremely important fauna. At least 25 species of butterfly (1), 14 birds (5) and five mammals as well as many snails and reptiles are endemic to the islands. Seventy per cent of the islands' area is still untouched, but large-scale commercial logging is affecting the forests of the northern and central islands (9) and may in future spread south to Seram. Much of Seram consists of steep hills and deep valleys that are essential not only to the wildlife, but also as watersheds supplying the coastal towns and villages. These regions need to be permanently protected from logging operations.

Many of Seram's indigenous people traditionally trade in wildlife. Wild sago and damar resin are important forest products, and plants, cockatoos and lories have long been taken for barter. This formerly local trade has recently become lucrative and international, and butterflies have become increasingly in demand (9). According to a recent report, there are several whole villages on Seram whose inhabitants depend totally on butterfly catching (9). About 90 species of butterfly occur in Maluku, including two threatened birdwings (*Ornithoptera aesacus* (Obi only) and *O. croesus*) as well as *Graphium stresemanni* and a number of other endemic butterflies, including the swallowtails *Troides criton*, *Papilio deiphobus* and *Papilio tydeus*. *Graphium stresemanni* has been seen on sale in Hatumetan on the southern coast of Seram (10) as well as in a number of trade catalogues.

Conservation Measures Manusela Reserve (proposed as a national park) covers a large area of Seram (180 000 ha (5, 9)), and extends from sea level to the peak of Seram's (and the Province's) highest mountain, Gunung Pinaia (3027 m). All Seram's ecosystems are represented and the island's 14 endemic birds and five endemic mammals have also been recorded there (5, 9). *Graphium stresemanni* and many other endemic butterflies are certainly to be found in the area. There is a great need for more information from Manusela.

Feasibility studies and a management plan for the proposed Manusela National Park have been prepared (9) but as yet there is no legislation to give the reserve full national park status. At present the reserve has no headquarters, no facilities and presumably no guards (1). Steps to ensure the future of the reserve are urgently needed since the pressure of economic development in the forests of north and central Maluku must eventually spread southwards into the relatively untouched island of Seram. Part of the northern lowlands of the reserve have already been logged over, prior to reserve status being gazetted (5).

Seram is too large to be adequately served by one national park, even one as large as Manusela, and the Indonesian Conservation Plan recommends another in the mountainous interior of the Sahuai peninsula (5). Gunung Sahuai in west Seram is covered in primary rain forest and is an important water catchment area and resource for wildlife. Logging concessions have already been requested (5) and gazetting as a protected area is urgently needed.

With such a superb butterfly fauna, Maluku Province could do more to increase awareness of this important heritage. Local people involved in the butterfly trade are already aware of the international demand for specimens. Studies towards the development of butterfly farms or ranches would help to meet this demand and at the same time encourage local interest and conservation. The local tourist board could encourage appreciation of the butterfly fauna through its brochures and advertisements.

References

1. Anon. (undated). *Indonesia National Parks and Nature Reserves*. Directorate General of Tourism, Jakarta. 150 pp.
2. D'Abrera, B. (1971). *Butterflies of the Australian Region*. Lansdowne Press, Melbourne. 415 pp.
3. Hancock, D.L. (1983). Classification of the Papilionidae (Lepidoptera): a phylogenetic approach. *Smithersia* 2: 1–48.
4. Munroe, E. (1961). The classification of the Papilionidae (Lepidoptera). *Canadian Entomologist* Supplement 17: 1–51.
5. National Park Development Project, UNDP/FAO (1981). *National Conservation Plan for Indonesia Vol 7: Maluku and Irian Jaya*. FAO, Bogor.
6. Okano, K. (1983). On the data of the type-specimen *Graphium stresmanni* (sic) (Papilionidae) *Tokurana (Acta Rhopalocera)* 5: 88.
7. Saigusa, S., Nakanishi, A., Shima, H. and Yata, O. (1977). Phylogeny and biogeography of the subgenus *Graphium* Scopoli. *Acta Rhopalocera* 1: 2–32.
8. Smart, P. (1975). *The Illustrated Encyclopedia of the Butterfly World*. Hamlyn, London. 275 pp.
9. Smiet, F. (1982). Threats to the Spice Islands. *Oryx* 16(4): 323–328.
10. Walker, M.V. (1983). *In litt.*, 11 April.

Graphium (Graphium) sandawanum Yamamoto, 1977   VULNERABLE

Subfamily PAPILIONINAE   Tribe LEPTOCIRCINI

Summary *Graphium sandawanum* is confined to Mt Apo in Mindanao, the Philippines. The mountain has been a national park since 1936 but has suffered constant encroachment by squatters. In a recent proclamation the government has agreed to lease 56 per cent of the park's area to the squatters; much of this area is still under virgin forest. Drought followed by fires in 1983 caused a further loss of forest on Mt Apo. There is international concern for the fate of the park, which is a stronghold for the Endangered Philippine Monkey-eating Eagle as well as this rare and beautiful butterfly.

Description This recently discovered, black and pale yellow, short-tailed butterfly has a forewing length of about 38 mm in the male, and 41 mm in the female. The sexes are similar, the female with a slightly paler ground colour (3, 7, 11).

UFW with a dark brown or black ground colour and a very broad yellow–green discal band running from the anal margin and tapering into three subapical spots (3, 7, 11).

UHW similar, with the yellow–green discal band tapering towards the anal angle. There is also a row of pale green submarginal lunules (3, 7, 11).

LFW/LHW. The lower surface is blackish-brown with a paler distal border. The discal band is yellow green as on the upper surface, but there are a number of small red or yellow markings on the distal edge of the band on the hindwing. The ratio of the red-spotted form to the yellow-spotted form has been estimated at 7:3 in the male (11).

Distribution This remarkable species was discovered as recently as 1977 on Mt Apo (2954 m) in central Mindanao, Philippines (12). So far it has been found nowhere else. After some 50 years of collecting by lepidopterists in the Mt Apo region it is quite extraordinary to find that in its chosen locations both males and females of this species are relatively abundant (10). No-one knows why it was not discovered earlier.

Habitat and Ecology *Graphium sandawanum* is considered to be an Ice Age relict of Palaearctic continental stock (11). According to Tsukada and Nishiyama, it usually flies in the mossy, montane forest of Mt Apo (11), the broad green bands on the wings making it very conspicuous. The foodplants and young stages are unknown, but other members of the *sarpedon* group feed on Winteraceae, Monimiaceae, Lauraceae or Hernandiaceae, rarely other groups (4). *Graphium sarpedon* itself also flies on Mt Apo (12), but is less common at altitudes over 1000 m, where *sandawanum* flies (11). *Graphium sandawanum* has been captured from January to June, October and December, and is probably present all year round, possibly with reduced populations during the heavy rains of February to May (7, 10, 11). Mt Apo may be approached from Davao or Cotabato. The butterflies apparently fly in greater abundance on the western, Cotabato, side (11).

Threats Mt Apo is the highest mountain in the Philippines (2909 m) and was designated as a national park as long ago as 1936, with a total area of 76 900 ha (9). Since then it has undergone a number of amendments in terms of classification and land use (9), but none so alarming as the recent newspaper and other reports that 56 percent of the area, almost 42 000 ha, is to be reclassified as 'disposable land' and handed over to illegal squatters (2, 7). Squatters have settled in the park since 1945 and current records indicate that 4034 families are living within the park boundaries (2). Attempts have been made in the past to reforest the lower slopes of Mt Apo and restore the natural beauty and integrity of the park (1). The latest proclamation indicates that these measures have been largely unsuccessful. The 42 000 ha will be leased to the squatters for 25 years, but only 35 percent of this is currently suitable for commercial agriculture. The other 65 percent is still under forest but is likely to be exploited very quickly since the new proclamation will allow squatters to take up loans to expand their farms (2).

A further threat only recently being fully realized is the serious drought of 1982–3 that has affected eastern Mindanao, eastern Borneo and perhaps other islands. In the wake of the drought, huge fires have destroyed vast areas of forest. In Mindanao whole mountains have been burnt through and both Mt Apo and Mt Katanglad have suffered badly (8). The full extent of defoliation is not yet known and will only be discovered from aerial or satellite survey.

Public concern has been mainly for the Endangered Phillippine Monkey-eating Eagle (*Pithecophaga jefferyi*) (6) but there is undoubtedly a wide range of other species, including *Graphium sandawanum*, whose livelihood is threatened by these recent developments.

Recent reports confirm that quantities of *G. sandawanum* specimens are reaching butterfly dealers in Manila and are being sold on the international market (5). The most likely source of the material is local youths employed by dealers in remote cities (possibly even overseas) to collect the butterflies on Mt Apo. Such arrangements are not only irresponsible but are also quite illegal since the butterfly only occurs within national park boundaries. The Philippine authorities should take steps to prevent such blatant abuse of their national legislation.

Conservation Measures There is evidence that the proclamation to lease such a huge area to squatters on Mt Apo was not supported by a thorough survey of the forests and agricultural land within the boundaries adopted (2, 7). Those areas irretrievably occupied by squatters are of no value for the conservation of Mt Apo's wildlife, but to expand squatters' privileges into virgin forested land of a national park would appear to be a short-sighted move. Much basic scientific work remains to be done on Mt Apo. Undoubtedly many species await discovery and only thorough survey work will reveal the full international importance of the park. Meanwhile, the national park authorities should take immediate steps to prevent the constant erosion of the park's boundaries and wildlife. Conflicting land-use needs will not be solved by ignoring the problem.

References

1. Anon. (1980). Logging firms tapped for Mt Apo reforestation. *Evergreen* (Publ. Bureau of Forest Development) 5(4): 18.
2. Anon. (1983). BFD moves to save Philippine Eagle. *Bulletin Today*, 1 December: 8.
3. D'Abrera, B. (1982). *Butterflies of the Oriental Region. Part 1. Papilionidae and Pieridae*. Hill House, Victoria, Australia, xxxi + 244 pp.
4. Hancock, D.L. (1983). Classification of the Papilionidae (Lepidoptera): a phylogenetic approach. *Smithersia* 2: 1 48.
5. Jumalon, J.N. (1984). *In litt.*, 10 July.

6. King, W.B. (1981). *Endangered Birds of the World. The ICBP Bird Red Data Book.* Smithsonian Institution Press, Washington D.C.
7. Lewis, R.E. (1984). *In litt.*, 5 and 9 February.
8. Lewis, R.E. (1984). *In litt.*, 17 April.
9. Pollisco, F.S. (1982). An analysis of the national park system in the Phillippines. *Likas Yaman, Journal of the Natural Resources Management Forum* 3(12): 56 pp.
10. Treadaway, C.G. (1984). *In litt.*, 25 May.
11. Tsukada, E. and Nishiyama, Y. (1982). *Butterflies of the South East Asian Islands Vol. 1 Papilionidae.* (transl. K. Morishita). Plapac Co. Plapac Co. Ltd., Tokyo. 457 pp.
12. Yamamoto, A. (1977). A new species of Graphium from Mindanao (Lepidoptera: Papilionidae). *Tyô to Ga.* (*Transactions of the Lepidopterists' Society of Japan*). 28: 87–88.

Graphium (Graphium) mendana (Godman and Salvin, 1888)   RARE

Subfamily PAPILIONINAE   Tribe LEPTOCIRCINI

Summary *Graphium mendana* has spread throughout the Solomon archipelago with the exception of San Cristobal, but is apparently rare everywhere. Logging and agriculture are probably the main threats to the species, but considerably more biological and ecological data are needed before a rational conservation programme can be proposed.

Description *Graphium mendana* is a relatively large butterfly with a forewing length of up to 52 mm. The sexes are similar. Four subspecies have been described (1, 2, 4, 6).

UFW with a black upper surface and a median band of spots from the middle of the inner margin to the wing apex. The colour, originally described as 'cyan' (blue–green) (2) seems to vary between green and yellow (1, 2, 6, 7). Submarginal spots may be present in some subspecies (1, 2, 6, 7). In the original description *G. mendana* was believed to be closely related to *G. sarpedon* (2), but Munroe placed it in the *eurypylus* group (5). In a more recent revision, *G. mendana* has become the nominate species in a new group (3).

UHW with a scalloped outer margin and a long, broad tail with a rounded tip. The UHW is black with a large, pale blue–green basal patch extending along the abdominal part of the wing and a white anal margin (1, 2).

LFW brown–black with smaller round spots forming the median band; LHW dull black with no basal patch. Markings on the LHW vary according to subspecies; *G. m. acous* (Ribbe) has a green basal spot and two red spots before the anal angle, whereas *G. m. neyra* (Rothschild) only has one red spot (1, 2, 7).

Distribution *Graphium mendana* is endemic to the Solomon Archipelago. The four subspecies are distributed from Bougainville in the north-west to Malaita and Guadalcanal in the south-east. *G. m. acous* is only known from Bougainville Island (Papua New Guinea) whereas all of the other races occur in the Solomon Islands. *G. m. neyra* is found in the New Georgia group of islands where it has been recorded from New Georgia, Rendova and Vella Lavella, and the nominate subspecies *G. m. mendana* is known from Santa Isabel and the north-east of Guadalcanal. A fourth subspecies *G. m. aureofasciatum* Racheli was described in 1979 from Dala, north-west Malaita (1, 7).

Habitat and Ecology *Graphium mendana* seems to be very rare throughout its range, and very poorly known. The foodplant is believed to be in the Piperaceae (3), but no further details of breeding or habitat are given in the literature. Nevertheless, it has been successfully bred in captivity (7).

Threats *G. mendana* is to be found on most of the islands of the Solomon archipelago except San Cristobal. In the absence of habitat data it is difficult to catalogue threats in any detail, but forestry, agricultural activity and other forms of habitat destruction are certainly of most concern (4). *G. mendana neyra* may be vulnerable to logging operations in the New Georgia group (4).

Conservation Measures More data on the habitat and ecology of *Graphium mendana* are needed before suitable conservation measures can be put forward.

Breeding habits and details of habitat are essential prerequisites to a captive breeding programme that could benefit the butterfly and its breeders. Habitat data will also indicate where disruption of native vegetation is most likely to affect the species, and where protected areas would be most effective. Export of butterflies from the Solomon Islands requires a permit, but this is difficult to enforce, and collecting is in any case unlikely to constitute a major threat.

References

1. D'Abrera, B. (1971). *Butterflies of the Australian Region*. Lansdowne Press, Melbourne. 415 pp

2. Godman, F.D. and Salvin, O. (1888). New species of butterflies collected by Mr. C.M. Woodford in the Solomon Islands. *The Annals and Magazine of Natural History*. (6)1: 209–214.

3. Hancock, D.L. (1983). Classification of the Papilionidae (Lepidoptera): a phylogenetic approach. *Smithersia* 2: 1–48.

4. Macfarlane, R. (1983). *In litt.*, 15 March.

5. Munroe, E. (1961). The classification of the Papilionidae (Lepidoptera). *Canadian Entomologist* Supplement 17: 1–51.

6. Racheli, T. (1979). New subspecies of *Papilio* and *Graphium* from the Solomon Islands, with observations on *Graphium codrus* (Lepidoptera, Papilionidae). *Zoologische Mededelingen* 54(15): 237–240.

7. Racheli, T. (1980). A list of the Papilionidae (Lepidoptera) of the Solomon Islands, with notes on their geographical distribution. *Australian Entomologists' Magazine*. 7: 45–59.

Battus zetides Munroe, 1971   VULNERABLE

Subfamily PAPILIONINAE   Tribe TROIDINI

Summary *Battus zetides*, the Zetides Swallowtail, is recorded from both Haiti and the Dominican Republic on the Caribbean island of Hispaniola. It occurs only very locally in stands of deciduous forest at high altitudes. With severe human population pressure and habitat destruction in Haiti, moderated only slightly in the Dominican Republic, the status of *Battus zetides* is cause for serious concern. The species may already be extinct in Haiti.

Description The sexes are similar in general appearance, with a forewing length of 35–40 mm (males) or 39–44 mm (females) (1, 3) (Plates 2.6 and 2.7). *Battus zetides* was originally described as *zetes* (Westwood, 1847) (10), later found to be an invalid homonym (9).

UFW with a submarginal row of yellow ochre spots over a dark, chocolate brown ground colour (1, 9, 10). UHW with a medium length tail and a scalloped edge, a fringe of yellow crescents and a subdiscal band of large orange or yellow ochre spots (1, 9, 10), slightly darker than on the UFW (3). The tailless specimen figured in reference (9) is mutilated (see 1, 6).

LFW similar to the UFW, with orange or pale yellow submarginal spots and discal marks (1, 9, 10). LHW dark brown basally, with a boldly contrasting silvery-white discal band, a postdiscal row of red lunules and a row of white submarginal spots fringed with yellow (1).

The recently-described caterpillars have yellow–orange longitudinal stripes on a tan or darker brown ground colour (13). Dorso-lateral tubercles are only found on the second and third thoracic segments and the first, eighth and ninth abdominal segments (13). The pupa is dark apple green, yellowish dorsally, with a tapering thoracic process (13).

Distribution *Battus zetides* is found on the Caribbean island of Hispaniola in the Greater Antilles (9, 10). It has been recorded both from Haiti (9, 10) (although not since the 1950s) and the Dominican Republic (3), the two countries that share the island. The history of its capture and recent rediscovery in southern Dominican Republic is well described in recent papers by Gali and Schwartz (3) and Weintraub (13). The three presently known localities are in Pedernales Province, Independencia Province and Peravia Province (13).

Habitat and Ecology *Battus zetides*, for many years very poorly known, was recently found in some numbers on the southern slopes of the Massif de la Selle in Pedernales Province in the southern Dominican Republic (3). Two further localities with smaller populations have also been identified, one on the northern slopes of the Massif in Independencia Province, the other in Peravia Province in the Cordillera Central (13). The habitat of the larger population on the Massif was an enclave of upland broad-leaved deciduous forest at an altitude of about 1000–1300 m, completely surrounded by forests of *Pinus occidentalis* (3, 13). On the northern slopes the species flies slightly lower, at 900–1000 m, and in the Cordillera Central the locality is at 1450 m (13). These deciduous woodlands are generally dense but the butterflies prefer to fly in sunny glades and rides, soaring slowly and deliberately at a general level of about 3.5 m, but ranging as high as 10 m and as low as 2 m (3). The butterflies were not found in small patches of deciduous trees in a ravine on the Massif de la Selle, an observation that led to speculation that *Battus zetides* may require extensive stands of deciduous forest for its survival (3). However, both the colonies recently discovered were in medium-sized patches of remnant forest of less than 10 ha (13). Adults are on the wing between 9.00 a.m. and 4.00 p.m. except when the weather is overcast. Nectar plants include species from the Boraginaceae, Labiatae, Leguminosae and Verbenaceae. Adults have been recorded in March–April and July–October and are probably continuously brooded, although more common in the wet season (13). The foodplants are believed to be *Aristolochia bilabiata* and *Aristolochia montana* (13). The eggs are laid in clusters of up to 15 and are gregarious in the first two instars.

Deciduous broad-leaved forests were once extensive in Hispaniola, growing at altitudes above 800 m in Haiti (2, p. 175) and at variable altitudes in the Dominican Republic, beginning at 500–900 m and reaching 1000–2200 m (2, p. 299). Pines may also be dominant at these levels, interdigitating in a complex pattern of stratification which seems to vary considerably according to rainfall, aspect, slope and other physical features.

Threats   Hispaniola as a whole has a very high population density and this is the basic reason for the threats to this butterfly and other forms of wildlife. The total human population of the island is now over 10 million, living in an area of 76 150 sq. km, but the Haitian sector is more seriously overcrowded than the Dominican Republic, with over 4.8 million people (1978) living in 27 750 sq. km (2, 5, 8). This represents the highest population density in Latin America. Three-quarters of Haiti and a smaller, but nevertheless large, proportion of the Dominican Republic is mountainous, with a natural cover of forest. With a high proportion of rural people (85 per cent in Haiti (8)), demand for land and forest products is high. Many forests are degenerating through over-exploitation for charcoal, stakes and building timbers (2). Others are cleared and burnt for agriculture by subsistence farmers and for plantations of coffee and cocoa (2, 3).

In Haiti, degradation of soils is particularly serious. Without proper conservation efforts on the mountain slopes, slash-and-burn agriculture exposes the top-soil to erosion by the heavy rains of 1000–2000 mm per year. Annual fires exacerbate the problem, preventing forest regrowth and resulting in a bare landscape of impoverished grassland dotted with spiny legumes and cactus plants (2). Forest remains only in steep ravines. In Haiti total forest cover was reduced from 80 per cent to 9 per cent (including secondary fallow) between 1958 and 1978 (12, p. xi). An estimate in 1980 shows a further reduction to only 1.7 per cent (2). In 1980 the estimated cover of broad-leaved forest in Haiti was 36 million ha, but the projection for 1985 was only 30 million ha, a loss of nearly 17 per cent in five years (2). As stated above, *Battus zetides* has not been recorded in Haiti since the 1950s and may be extinct there.

In the Dominican Republic the human population is less dense and the pressures on natural habitats are somewhat ameliorated. Nevertheless, there is no doubt that the rich broad-leaved deciduous forests and pine forests of middle altitudes are under increasing threat. On the slopes of the Massif de la Selle, where *Battus zetides* was recently found, most of the deciduous forest has been cleared for coffee plantations and subsistence crops (3). In 1980 the area of broad-leaved forest in the Dominican Republic was estimated at 444 million ha, spread between the eastern region, the Sierra de Neiba, the Cordillera Central and the Sierra de Baoruco (2). Projections for 1985 were optimistic, with an expected reduction to 432 million ha, a loss of only 2.7 per cent in five years (2). Nevertheless, there is good reason to suppose that *Battus zetides* does not occur throughout these forests. It is very localized in relict patches of deciduous forests, none of which is protected and all of which are vulnerable to deforestation. The two smaller populations are particularly vulnerable to shifting agriculturalists, a proportion of which is comprised of illegal squatters crossing the border from Haiti (13). The larger population is situated on a bauxite mining concession and is currently relatively free of threat. However, the price of tin is depressed and if the concession is not worked it may be returned to the government. In this event the site would almost certainly be deforested and put to agricultural use. This site also contains populations of the Endangered Haitian Solenodon (*Solenodon paradoxus*) and the Indeterminate Hispaniolan Hutia (*Plagiodontia aedium*) (12, 13).

Conservation Measures Haiti has such a large and widely distributed human population that it can only afford minimal reserved areas (5, 12). In 1980 there was one small and badly damaged reserve (La Citadelle) and a further five proposed (5). There is no evidence that *Battus zetides* survives in any of these areas, or indeed in Haiti as a whole. If it does survive, there can be little doubt that it is Endangered. If suitable relict patches of deciduous forest can be found, they should be afforded all possible protective measures.

In the Dominican Republic there are five protected areas of which two, Armando Bermudez N.P. and José del Carmen Ramirez N.P., lie on the northern and southern slopes of the Cordillera Central (5). *Battus zetides* may occur in these parks, both of which include broadleaved hardwoods and *Pinus occidentalis* at higher altitudes (5). Confirmation of the presence of *Battus zetides* and other forest butterflies in these parks would be very valuable, not only for the butterfly but also for increasing the general appreciation of these important protected areas. Once found, studies of the foodplants and young stages of the butterfly would be required in order to assess possible management measures. At the same time, further data on the distribution of *Battus zetides* throughout the Dominican Republic are needed, together with an assessment of its status in Haiti.

The most important known site for *B. zetides* is the mining concession site on the Massif de la Selle. All possibilities for declaring this locality a reserve for the butterfly should be explored.

References

1. D'Abrera, B. (1981). *Butterflies of the Neotropical Region. Part 1. Papilionidae and Pieridae*. Lansdowne Editions, Melbourne, in association with E.W. Classey, Faringdon. 172 + xvi pp.
2. FAO/UNEP (1981). *Proyecto de Evaluacion de los Recursos Forestales Tropicales. Los Recursos Forestales de la America Tropical*. FAO, Rome. 343 pp.
3. Gali, F. and Schwartz, A. (1983). *Battus zetides* in the Republica Dominicana. *Journal of the Lepidopterists' Society* 37: 170–171.
4. Hancock, D.L. (1983). Classification of the Papilionidae (Lepidoptera): a phylogenetic approach. *Smithersia* 2: 1–48.
5. IUCN (1982). *IUCN Directory of Neotropical Protected Areas*. Tycooly, Dublin. 436 pp.
6. Lewis, H.L. (1973). *Butterflies of the World*. Harrap, London. 312 pp.
7. Munroe, E. (1961). The classification of the Papilionidae (Lepidoptera). *Canadian Entomologist* Supplement 17: 1–51.
8. Paxton, J. (1981). *The Statesman's Year-Book*. Macmillan Press, London. 1696 pp.
9. Riley, N.D. (1975). *A Field Guide to the Butterflies of the West Indies*. Collins, London. 244 pp., 24 col. pl.

10. Rothschild, W. and Jordan, K. (1906). A revision of the American Papilios. *Novitates Zoologicae* 13(3): 411–752. (Facsimile edition ed. P.H. Arnaud, 1967).
11. Schwartz, A. (1984). *In litt.*, 16 May.
12. Thornback, J. and Jenkins, M. (1982). *The IUCN Mammal Red Data Book Part 1*. IUCN, Gland. 516 pp.
13. Weintraub, J.D. (unpublished manuscript). Notes on the biology of *Battus zetides* Munroe (Lepidoptera: Papilionidae).

Parides ascanius (Cramer, 1775)   VULNERABLE

Subfamily   PAPILIONINAE   Tribe   TROIDINI

Summary *Parides ascanius*, the Fluminense or Ascanius Swallowtail, is a strikingly beautiful butterfly which occupies only a small fraction of its potential 'restinga' habitat (subcoastal swamps and thickets) in the state of Rio de Janeiro, south-eastern Brazil. It is a primitive species, lacking vigour, and most of its habitat is being drained and subdivided for residential, industrial, and recreational purposes. Many colonies active before 1970 are now extinct. Because of habitat pressure it is on the Brazilian list of animals threatened with extinction, the first insect so designated. Only about ten self-supporting colonies are now known, many in areas under pressure for development. Large colonies need to be located in regions suitable for establishment of conservation units.

Description The sexes of this relatively large, tailed butterfly are similar in appearance although the female lacks the androconial fold found on the hindwing of the male (2, 7) (Plate 2.2). *Parides ascanius* is the nominate species of the *ascanius* species-group, which includes 11 species in published lists (4, 5), but is expected to be reduced to nine in a forthcoming checklist of neotropical Lepidoptera (1).

UFW black with a broad white median band (2).

UHW also black, with a deeply scalloped outer margin, a relatively long tail, and an extension of the median band of the forewing to the inner margin. This wing is washed with rose coloured scales, especially anally, and there is also a row of red, hourglass-shaped submarginal spots (2).

Distribution *Parides ascanius* is presently known only from widely scattered points near the coast of the state of Rio de Janeiro, south-east Brazil, between the mouth of the Rio Paraíba do Sul (to the north) and Itaguai (to the south-west) (3). Formerly it may have occurred in favourable habitats on the coast of São Paulo state, but it is not known there today despite substantial searches (6).

Habitat and Ecology *Parides ascanius* inhabits only subcoastal and lowland swamps and thickets where its larval foodplant *Aristolochia macroura* (Aristolochiaceae) is abundant, and suitable flowers (mostly Compositae and Verbenaceae) are available to adults throughout the year (6). Very patchy distribution and wide-ranging male promenading indicate dispersed, possibly unstable, colonies with extensive competition from the sympatric *P. zacynthus* and *P. anchises nephalion*, the two most advanced members of the genus, both strong and aggressive species (6). *P. ascanius* occurs mostly on sandy soils in vegetation of low to medium height and flies only in the sun during the morning and late afternoon (6). It can be reared quite easily, and kept in captivity in proper conditions (6). The habitat is very rich in endemic plants and animals, almost all of which have far wider and less patchy distributions than *P. ascanius*, which seems to be a relict (6).

Threats Habitat destruction throughout its range, but particularly near the city of Rio de Janeiro, has rendered *P. ascanius* increasingly scarce (6). Many colonies known before 1970 no longer exist because of swamp drainage for development into recreational areas, banana plantations, pasture or buildings (6). In addition, there is some danger from competition with other *Parides* species (6). As a primitive and relict species with unusual affinities, *P. ascanius* has value for scientific study. Its beauty is also widely recognised and specimens sell for a high price when available.

Conservation Measures In an executive action (3481-DN) on 31 May, 1973, the Brazilian National Parks agency (through the IBDF, Brazilian Institute for Forest Development) placed *P. ascanius* on the official list of Brazilian animals threatened with extinction (as the synonym *Battus orophobus* D'Almeida). It is the only insect presently given this status, and is thereby protected from commerce (6). *Eurytides lysithous harrisianus* has also been proposed for listing (see separate review).

The recently-established (1974) Federal Biological Reserve of Poço das Antas, north-east of Rio de Janeiro, is a 5000 ha area which includes at least 1000 ha of ideal *P. ascanius* habitat. Suitable areas may be extended by river management. *P. ascanius* is known to live in the reserve in adequately large colonies (6).

A small but permanent colony of *P. ascanius* is also present in a swampy forest of less than 2 ha in the Parque Reserva Marapendi, Restinga de Jacarepaguá in Rio de Janeiro. However, the presumed source of this colony is a patch of the food plant with its own colony in an unprotected nearby swamp, which is due to be drained (6). The Parque Reserva Marapendi is too small and inadequately protected. Attempts to transplant the food plant into suitable parts of the reserve have failed. Key parts of the nearby swamp should be protected from draining and building in order to ensure the future of this small colony (6). Should this colony fail, however, the population left in the Reserve of Poço das Antas cannot be assigned full and exclusive responsibility for preservation of *P. ascanius*. The main hope for the species lies in 1000 sq. km of almost impenetrable swamps in lowland Rio de Janeiro, south and west of Itaguai and Campos. Colonies may well be flourishing there, but this is not certain (6). When possible, surveys in this region should be carried out.

Commerce in wild specimens should be discouraged internationally as well as nationally (6). Official captive breeding of the species in its habitat could be encouraged (6). Populations in reserves need to be extended and carefully monitored, to guarantee the well-being of this butterfly in coastal Rio de Janeiro (6).

References

1. Brown, K.S., Jr. (1984). *In litt.*, 4 January.
2. D'Abrera, B. (1981). *Butterflies of the Neotropical Region. Part I. Papilionidae and Pieridae.* Lansdowne Editions, Melbourne. xvi + 172 pp.
3. D'Almeida, R.F. (1966). *Catálogo dos Papilionidae Americanos.* Sociedade Brasileira de Entomologia, São Paulo, Brazil. 366 pp.
4. Hancock, D.L. (1983). Classification of the Papilionidae (Lepidoptera): a phylogenetic approach. *Smithersia* 2: 1–48.
5. Munroe, E. (1961). The classification of the Papilionidae (Lepidoptera). *Canadian Entomologist* Supplement 17: 1–51.
6. Otero, L.S. and K.S. Brown, Jr. (in press). Biology and ecology of *Parides ascanius* (Cramer, 1775) (Lepidoptera, Papilionidae), a primitive butterfly threatened with extinction. *Atala.*
7. Rothschild, W. and Jordan, K. (1906). A revision of the American Papilios. *Novitates Zoologicae* 13(3): 411–752. (Facsimile edition ed. P.H. Arnaud, 1967).

This review has been adapted from an entry in *The IUCN Invertebrate Red Data Book*, whose authors and contributors are gratefully acknowledged.

Parides hahneli (Staudinger, 1882)   RARE

Subfamily PAPILIONINAE   Tribe TROIDINI

Summary *Parides hahneli*, Hahnel's Amazonian Swallowtail, is an exquisitely beautiful butterfly that has been known for nearly a century, but only in the past decade have any reasonably dense colonies been discovered. It occupies a very specialized habitat in the lower middle Amazon Basin of Brazil and only three localities have ever been found for it. Until the most recent locality was found, just 20 collected specimens were known. A very primitive species, it may hold the key to the evolution of this swallowtail genus.

Description *Parides hahneli*, a beautiful black and yellow butterfly with a wingspan of 80 –100 mm, is the only tailed member of its genus in the Amazon basin of Brazil. It resembles sympatric ithomiine butterflies (*Methona*, *Thyridia*), which it apparently mimics, possibly to gain protection. The sexes are similar (5, 6, 7, 10). The larva is dark brown with yellow rings and resembles that of *P. pizarro steinbachi* and *P. vertumnus* (8, 9), other members of the same species-group. *P. hahneli* is the most primitive member of the *aeneas* species-group, which includes 25 species according to current lists (9, 10). However, the assemblage is probably artificial and will be divided up in the forthcoming checklist of neotropical Lepidoptera (3), leaving only 12 species in the *aeneas* group. Six species in the group are possibly threatened, and are reviewed in the following pages.

UFW/LFW black, long, narrow, and rounded, with three broad bands of yellow–grey patches (5).

UHW/LHW also black, with a scalloped outer margin and a long spatulate tail. There is a large patch, yellower than those of the forewing, covering all but the borders of the hindwing. The LHW has a red spot near the tail (5).

Distribution Until 1970 *Parides hahneli* was known only from the region of Maués, south of the middle Amazon Basin in Brazil, but it was then discovered in the Rio Arapiuns area more to the east. There are sight and unconfirmed capture records from the region of Manaus (and Manacapuru) to the north-west of Maués, but no other records have been located for this distinctive species in spite of extensive searching by lepidopterists and commercial collectors over many years (2, 5). The butterfly is very rare and has a patchy distribution.

Habitat and Ecology *Parides hahneli* is apparently restricted to ancient sandy beaches now covered by scrubby or dense forest vegetation (2). Like many other troidine swallowtails using *Aristolochia* as a host plant, *P. hahneli* is probably extremely localized in occurrence, limited by the distribution and density of the plant. In addition, the adults need to feed on a continuous supply of nectar plants. The particular host species of *Aristolochia* for *P. hahneli* is not identified but it is also used at least by the sympatric *P. chabrias ygdrasilla* (1, 9). Sympatric butterflies include *Heliconius egeria*, also partial to sandy areas and very local in conjunction with its foodplant, *Passiflora glandulosa*. Many plants are endemic to the Maués region. In general, the butterfly seems to occupy a poor but very specialized habitat which is not very diverse but has a high degree of endemism, demonstrated by this insect and other organisms (2). *P. hahneli* may suffer from food plant competition with sympatric troidine swallowtails (2).

P. hahneli retains a primitive pattern, morphology and tails, and is related to species of southern Brazil and Central America. It seems to represent a bridge between these archaic *Parides*, now pushed to the margins of the Neotropics, and the modern species in the *P. aeneas* and *P. anchises* groups. It persists in a marginal biotope in the central Amazon, where it is extremely rare. This situation offers many opportunities for research on tropical ecological and evolutionary processes, which can only be solved in the context of the living organism (2).

Threats Over-collecting for commercial purposes represents the only current threat. However, there is potential for deleterious habitat changes prior to its ecological requirements being fully understood (2). The species is beautiful (and rather bizarre), participates in mimicry rings not usual in other *Parides*, and currently sells for a very high price (2). It is possible to rear this species in captivity (11), but collecting sites are a closely guarded secret in view of its commercial value. Scientific work on the species has hardly begun.

Conservation Measures The butterfly may occur in the Amazon National Park between Itaituba and Maués, but this is not likely. A search should be made in the small Ducke reserve near Manaus (2). A general study of the habitat and colonies of *P. hahneli* should be undertaken, with a view towards proposing a reserve near Maués or Arapiuns. Such a reservation would permit the survival of this swallowtail as well as the many other endemic organisms of the region, and would be a significant component of the northern Brazilian land conservation system (2). Repeatedly scarred by rivercourse changes and affected by Pleistocene drying, this area is unique in its biological properties.

References

1. Brown Jr., K.S., Damman, A.J. and Feeny, P. (1981). Troidine swallowtails (Lepidoptera: Papilionidae) in southeastern Brazil: natural history and foodplant relationships. *Journal of Research on the Lepidoptera* 19: 199–226.
2. Brown, K.S., Jr. (1982). *In litt.*, 19 January.
3. Brown, K.S., Jr. (1983). *In litt.*, 23 October.
4. Brown, K.S., Jr. (1983). *In litt.*, 19 September.
5. D'Abrera, B. (1981). *Butterflies of the Neotropical Region. Part I. Papilionidae and Pieridae*. Lansdowne Editions, Melbourne, in association with E.W. Classey, Faringdon. 172 + xvi pp.
6. D'Almeida, R.F. (1966). *Catálogo dos Papilionidae Americanos*. Sociedade Brasileira de Entomologia. São Paulo, Brazil. 366 pp.
7. Hancock, D.L. (1983). Classification of the Papilionidae (Lepidoptera): a phylogenetic approach. *Smithersia* 2: 1–48.
8. Moss, A.M. (1919). The papilios of Para. *Novitates Zoologicae* 26: 295–319.
9. Moss, A.M. Manuscript material preserved with the Moss collection in the British Museum (Natural History).
10. Munroe, E. (1961). The classification of the Papilionidae (Lepidoptera). *Canadian Entomologist* Supplement 17: 1–51.
11. Rothschild, W. and Jordan, K. (1906). A revision of the American Papilios. *Novitates Zoologicae* 13(3): 411–752. (Facsimile edition ed. P.H. Arnaud, 1967).

This review has been adapted from an entry in *The IUCN Invertebrate Red Data Book*, whose authors and contributors are gratefully acknowledged.

Parides pizarro Staudinger, 1884   INSUFFICIENTLY KNOWN

Subfamily PAPILIONINAE   Tribe TROIDINI

Summary *Parides pizarro* is a little known swallowtail from Peru and Brazil. There are two subspecies (three if *P. steinbachi* is included), one fairly common but the other known from very few specimens. Further research is needed to confirm its conservation status.

Description *Parides pizarro* is a relatively large butterfly with sexes that are similar in general appearance (2, 7). The forewing is entirely black, and the hindwing, which has a scalloped outer margin, is black with a cream-coloured discal patch consisting of three to four discal spots in the male, and three to six in the female (2, 7).

Distribution Two subspecies of *P. pizarro*, *P. p. pizarro* and *P. p. kuhlmanni*, are currently recognized in the literature, although some authorities believe that *P. steinbachi* (see next review) represents a third (1, 5). Until new findings are published we adopt those of the most recent published work (4). The situation is confused by D'Abrera (2), who lists *kuhlmanni* as a subspecies of *P. steinbachi*, a view not accepted by other authorities (1, 5). *P.p. pizarro*, is from the upper Amazon in north-eastern Peru (formerly part of Ecuador) and has been recorded in the Departments of Loreto, San Martin and Ucayali (3, 7). *P.p. kuhlmanni* is known from two specimens from the Department of Madre de Dios in eastern Peru and was originally described from a single male specimen found in the neighbouring state of Acre in western Brazil. A further two Brazilian specimens are held in the National Museum in Rio de Janeiro (1). *P. pizarro* is in the *aeneas* species-group (4, 6).

Habitat and Ecology *Parides pizarro* occurs in lowland primary forest, probably below 500 m (5). The immature stages and foodplant are unknown, but the latter is likely to be *Aristolochia*. More information is needed on the life history, biology and ecology of *P. pizarro*.

Threats Cannot be assessed without further data.

Conservation Measures *P. p. pizarro* is said to be moderately common in the reaches of the upper Amazon of eastern Peru, but *P. p. kuhlmanni* has apparently been recorded only from two specimens from Peru and a small number from Brazil. The Peruvian specimens were both taken in the Tambopata Nature Reserve, a privately managed 5000 ha site on the Tambopata River, 25 km south-east of Puerto Maldonado (5). This extraordinarily rich reserve has so far yielded 893 species of butterflies (5). Although *P. pizarro* may be effectively protected by this Reserve, there is still insufficient information to be sure. Further research on its biology and habits are needed before conservation needs can be properly assessed.

References

1. Brown, K.S., Jr. (1983). *In litt.*, 23 October.
2. D'Abrera, B. (1981). *Butterflies of the Neotropical Region. Part I. Papilionidae and Pieridae.* Lansdowne Editions, Melbourne. xvi + 172 pp.

3. D'Almeida, R.F. (1966). *Catálogo dos Papilionidae Americanos*. Sociedade Brasileira de Entomologia. São Paulo, Brazil.
4. Hancock, D.L. (1983). The classification of the Papilionidae (Lepidoptera): a phylogenetic approach. *Smithersia* 2: 1–48.
5. Lamas, G. (1983). *In litt.*, 4 April.
6. Munroe, E. (1961). The classification of the Papilionidae (Lepidoptera). *Canadian Entomologist* Supplement 17: 1–51.
7. Rothschild, W. and Jordan, K. (1906). A revision of the American Papilios. *Novitates Zoologicae* 13(3): 411–752. (Facsimile edition ed. P.H. Arnaud, 1967).

Parides steinbachi Rothschild, 1905   INSUFFICIENTLY KNOWN

Subfamily PAPILIONINAE   Tribe TROIDINI

Summary *Parides steinbachi* is a very poorly known swallowtail from Bolivia. Its precise distribution is unclear and its taxonomic status questionable. Some authorities would place it as a subspecies of *P. pizarro*. More data are needed to assess the threats to this apparently very rare taxon.

Description *Parides steinbachi* is a relatively large butterfly, with a forewing length of 35–40 mm (8). The female is larger than the male, but otherwise similar in appearance, with minor differences in wing markings (2, 8). *P. steinbachi* is listed as a good species by Hancock (4) but is considered by a number of authorities to be a subspecies of *P. pizarro* (1, 6) (see previous review) and is expected to be published as such in a forthcoming checklist of neotropical Lepidoptera.

UFW/LFW ground colour black with a cream or white median patch (2, 8).

UHW/LHW black with a scalloped outer margin, white marginal spots and a discal patch of red spots with an innermost white spot. LHW has a red spot in both sexes, close to the anal angle (2, 8).

The caterpillars are likely to be fleshy with segmental tubercles, and the pupae with prominent abdominal ridges (7). Specimens of larvae and pupae are held in the Moss collection in the British Museum (Natural History) (1).

Distribution This butterfly is confined to Bolivia (8). The localities given in (8) are rather confusing, the name Mapiri being not uncommon in Bolivia. However, the balance of evidence is that *P. steinbachi* is mainly or entirely in eastern Bolivia (3, 8). Zischka (10) reports the species from Santa Cruz and Buena Vista up to the Brazilian border. There is also a specimen from Santa Cruz province in the Tyler collection (9).

Habitat and Ecology The original series was collected between February and June, apparently from altitudes between 200 and 1000 m in the eastern Andes. There is no information on the life history of the species, but in common with other troidine swallowtails, it probably feeds on *Aristolochia* (7).

Threats Threats to this very rare and poorly known butterfly cannot be assessed at present. The economic situation in Bolivia is presently so serious that the entire national park system is believed to be under threat. Lack of funds, equipment, management presence and enforcement combine to negate the intended protection of Bolivia's wildlife.

Conservation Measures No rational conservation measures can be proposed without more information on the biology, distribution and habitat of the species. Large areas of eastern Bolivia remain relatively undeveloped. The Huanchaca National Park is over 5000 sq. km in extent (5), but it remains to be seen whether *Parides steinbachi* occurs in the region.

References

1. Brown, K.S., Jr. (1983). *In litt.*, 23 October.
2. D'Abrera, B. (1981). *Butterflies of the Neotropical Region. Part I. Papilionidae and*

Pieridae. Lansdowne Editions, Melbourne. xvi + 172 pp.

3. D'Almeida, R.F. (1966). *Catálogo dos Papilionidae Americanos*. Sociedade Brasileira de Entomologia. São Paulo, Brazil.
4. Hancock, D.L. (1983). Classification of the Papilionidae (Lepidoptera): a phylogenetic approach. *Smithersia* 2: 1–48.
5. IUCN (1982). *IUCN Directory of Neotropical Protected Areas*. Tycooly International, Dublin. 436 pp.
6. Lamas, G. (1983). *In litt.*, 4 April.
7. Munroe, E. (1961). The classification of the Papilionidae (Lepidoptera). *Canadian Entomologist* Supplement 17: 1–51.
8. Rothschild, W. and Jordan, K. (1906). A revision of the American Papilios. *Novitates Zoologicae* 13(3): 411–752. (Facsimile edition ed. P.H. Arnaud, 1967).
9. Tyler, H.A. (1983). *In litt.*, 30 September.
10. Zischka, R. (1950). Catálogo de los insectos de Bolivia. *Folia Universidad Cochabamba* 4: 51–56.

Parides coelus Boisduval, 1836   INSUFFICIENTLY KNOWN

Subfamily   PAPILIONINAE   Tribe   TROIDINI

Summary *Parides coelus* is an extremely rare and poorly known troidine swallowtail from the state of Cayenne in French Guiana. Further research is needed before conservation measures can be proposed.

Description A large butterfly with a forewing length of about 49 mm in the male and 55 mm in the female (1). The sexes are similar but the male has prominent white androconial folds and smaller red spots on the hindwings (1, 4, 5, 6). *P. coelus* is a member of the *aeneas* species-group (5).

UFW/LFW ground colour blackish-brown, lighter in the apical and subapical regions and with a diffuse white spot at the apex of the cell, slightly larger on the LFW (1, 6).

UHW/LHW with a deeply scalloped outer margin, no distinct tail but a discal band of four bright red spots (six on the LHW) contrasting with the blackish-brown ground colour. A large whitish or grey androconial fold is present on the anal margin of the male, replaced by an extra pair of red–orange spots in the female (1, 6).

Distribution *Parides coelus* is only known from French Guiana in South America. The type female was caught in the area of Cayenne (2).

Habitat and Ecology Virtually nothing is known of this extremely rare species. The male and female were described as separate species, although they are fairly similar in appearance (1, 2, 6). In common with most other troidine swallowtails, the foodplant is likely to be *Aristolochia*, the larva fleshy with segmental tubercles, and the pupa with prominent edges or rows of tubercles (5).

Threats French Guiana has immense tracts of forests and it seems unlikely that forest destruction could be responsible for the apparent rarity of this species. Nevertheless, a new highway has been built from the coast to the Brazilian border, bringing settlement and forest destruction in its wake. Although commercial logging is still limited in extent, it is expanding rapidly. Further study is needed to assess what threats, if any, are affecting *Parides coelus*. Many troidine swallowtails feeding on *Aristolochia* have very patchy distributions.

Conservation Measures So little is known of this rare butterfly that conservation measures cannot be proposed. It is first necessary to discover its distribution, habitat and biology. There is no information on its presence or absence in the two protected areas on the mainland of French Guiana, namely the Sinnamari and Basse Mana National Parks (3). The former is mainly mangrove and the latter coastal wetland (3). French Guiana is a French Overseas Department subject to French laws. If necessary, *Parides coelus* could be placed on the French list of protected species, but research and protection of its habitat would be more useful objectives at this stage.

References

1. D'Abrera, B. (1981). *Butterflies of the Neotropical Region. Part I. Papilionidae and Pieridae.* Lansdowne Editions, Melbourne. xvi + 172 pp.

2. D'Almeida, R.F. (1966). *Catálogo dos Papilionidae Americanos*. Sociedade Brasileira de Entomologia. São Paulo, Brazil.
3. IUCN (1982). *IUCN Directory of Neotropical Protected Areas*. Tycooly International, Dublin. 436 pp.
4. Lewis, H.L. (1973). *Butterflies of the World*. Harrap, London. 312 pp.
5. Munroe, E. (1961). The classification of the Papilionidae (Lepidoptera). *Canadian Entomologist* Supplement 17: 1–51.
6. Rothschild, W. and Jordan, K. (1906). A revision of the American Papilios. *Novitates Zoologicae* 13(3): 411–752. (Facsimile edition ed. P.H. Arnaud, 1967).

Parides klagesi Ehrmann, 1904   INSUFFICIENTLY KNOWN

Subfamily   PAPILIONINAE   Tribe   TROIDINI

Summary *Parides klagesi* is a small, very rare and poorly known troidine swallowtail from the northern and eastern foothills of the Bolivar highlands in Venezuela. Further research is needed before conservation measures can be proposed.

Description *Parides klagesi* is in the *aeneas* species group (4, 6) and somewhat resembles *P. coelus*, but is a much smaller butterfly with a forewing length of only about 30 mm (2). The male and female are very similar in appearance, the former being only slightly darker and smaller (2).

UFW/LFW black with a white median patch and the distal part of the forewing somewhat paler than the rest.

UHW/LHW black with a scalloped outer edge, and a discal band of six pinkish spots, in the female or three in the male, which has a whitish anal hair-pouch.

Distribution *Parides klagesi* is only known from northern and eastern Venezuela, where it has been recorded from the region of the Caura, Suapure and Caroní rivers, which flow north into the Orinoco from the highlands of Bolivar (1, 3, 7).

Habitat and Ecology Little information on the habitat or ecology of this species has been recorded. As with most other troidine swallowtails, the foodplant is likely to be *Aristolochia*, the caterpillar fleshy and tuberculate, and the pupa with tubercles or prominent edges (6). Specimens have been taken in recent years in the region between the Pan American highway and the lower reaches of the River Caroní (1).

Threats No information on this region of Venezuela has been obtained and with so little known about the butterfly it is not possible to assess threats at present. Undoubtedly the building of the Pan American Highway through eastern Venezuela will attract development and settlement, possibly reducing the range of the butterfly.

Conservation Measures *Parides klagesi*, like its close relative in French Guyana, *P. coelus*, is too poorly known to permit rational conservation measures. Further research on its distribution, habitat and biology is needed. The only national park in the Bolivar highlands is Canaima, a 30 000 sq. km park 600 km to the east in La Gran Sabana (5). *P. klagesi* may occur in parts of the Park, but has not been recorded there so far. The Pan American Highway appears to bisect the Park; the effect of this on the wildlife remains to be assessed.

References

1. Brown, K.S., Jr. (1983). *In litt.*, 23 October.
2. D'Abrera, B. (1981). *Butterflies of the Neotropical Region. Part I. Papilionidae and Pieridae.* Lansdowne Editions, Melbourne. xvi + 172 pp.
3. D'Almeida, R.F. (1966). *Catálogo dos Papilionidae Americanos.* Sociedade Brasileira de Entomologia. São Paulo, Brazil.

4. Hancock, D.L. (1983). Classification of the Papilionidae (Lepidoptera): a phylogenetic approach. *Smithersia* 2: 1–48.
5. IUCN (1982). *IUCN Directory of Neotropical Protected Areas*. Tycooly International, Dublin. 436 pp.
6. Munroe, E. (1961). The classification of the Papilionidae (Lepidoptera). *Canadian Entomologist* Supplement 17: 1–51.
7. Rothschild, W. and Jordan, K. (1906). A revision of the American Papilios. *Novitates Zoologicae* 13(3): 411–752. (Facsimile edition ed. P.H. Arnaud, 1967).

Parides burchellanus Westwood, 1872   VULNERABLE

Subfamily PAPILIONINAE   Tribe TROIDINI

Summary *Parides burchellanus* is a large, velvet–black butterfly which inhabits riverine forest in central Brazil. It is very rare and its habitat is subject to flooding and forest clearance.

Description *Parides burchellanus* is a large, velvet–black butterfly with a forewing length of about 45 mm. The forewing is without markings except for a fringe of some small, white spots. The UHW has a row of small, red postdiscal spots which are smaller and paler on the lower surface. The hindwing has a scalloped outer margin, no tails, and a large hair-pouch on the anal margin of the male. Apart from this, the sexes are alike. The species is placed in the *aeneas* group by Munroe (5). It is believed to be very close to, and possibly even conspecific with, *P. aeneas* from Amazonas, with which a putative intermediate is known (1).

Distribution *Parides burchellanus* is found in central Brazil from northern Goiás state to western São Paulo (1, 3, 4, 6). The most recent specimens have come from the Rio Maranhão on the border of Goiás state and the Federal District of Brazilia (2). A specimen described by Schaus as being from Bolivia requires confirmation. It was sent to him from a correspondent who had been resident in Rio de Janeiro (6).

Habitat and Ecology *Parides burchellanus* is always associated with gallery (riparian) forest along rivers running through the cerrado landscape of central Brazil. Away from the rivers the land is dry for much of the year, but where ground-water occurs along ravines, streams and river-banks tall forest develops in a humid microclimate. *P. burchellanus* is rarely far from the river bank and may be seen flying a few centimetres above the water (1, 3). Colonies contain between 10 and 50 individuals in a favourable season; apparently six generations develop per year, as in other Brazilian *Parides*, with a low period in June–July (1). The foodplant remains unknown but, in common with other troidine swallowtails, is likely to be *Aristolochia*.

Threats Very few individuals or colonies have been captured or seen since the species was described over a century ago. It is associated with the even rarer *P. panthonus numa* which has the same pattern and habits, and has not been seen since 1920 (1). Colonies are subject to elimination by changes in water level. Natural floods are quite common in the cerrado landscape, but riverside clearing and agriculture increase their ferocity downstream. Changes in habitat and microclimate by cutting and opening up the gallery forest will also cause colonies to disappear. Colonies recently found along the Rio Maranhão may have since been destroyed by flooding (2).

Conservation Measures More information is needed on the precise localities of colonies and their breeding requirements. No conservation measures have been taken, but there are laws prohibiting the elimination of riparian vegetation. Once colonies have been located, these laws could be enforced. No captive breeding has been reported, but it should be feasible for river dwellers in the region (1).

However, the species is not particularly striking in appearance and may not be suitable for commercial development.

References

1. Brown, K.S., Jr. (1983). *In litt.*, 6 April.
2. Brown, K.S., Jr. (1983). *In litt.*, 23 October.
3. Brown, K.S., Jr. and Mielke, O.H.H. (1967). Lepidoptera of the central Brazil plateau. 1. Preliminary list of Rhopalocera (continued): Lycaenidae, Pieridae, Papilionidae, Hesperiidae. *Journal of the Lepidopterists' Society* 21: 145–68.
4. D'Almeida, R.F. (1966). *Catálogo dos Papilionidae Americanos*. Sociedade Brasileira de Entomologia. São Paulo, Brazil.
5. Munroe, E. (1961). The classification of the Papilionidae (Lepidoptera). *Canadian Entomologist* Supplement 17: 1–51.
6. Rothschild, W. and Jordan, K. (1906). A revision of the American Papilios. *Novitates Zoologicae* 13(3): 411–752. (Facsimile edition ed. P.H. Arnaud, 1967).

Atrophaneura (Atrophaneura) luchti Roepke, 1935 RARE

Subfamily PAPILIONINAE Tribe TROIDINI

Summary *Atrophaneura luchti* is a rare butterfly known only from a few specimens from the Ijen (Idjen) mountain range in eastern Java, Indonesia. Its biology and distribution are poorly known and more field information is needed as a basis for conservation assessment and planning. If the Indonesian Conservation Plan's recommendations are adopted, the species will be well protected in a large national park.

Description *Atrophaneura luchti* is a large, handsome butterfly with a forewing length of 60 mm (male) to 65 mm (female). The sexes have similar patterns but the male is smaller, with more pointed forewings. Adults have a red stripe along the sides of the head, thorax and abdomen (1, 3).

UFW/LFW elongated with a black ground colour fading to grey (3) or greenish grey (1) between the veins.

UHW/LHW tailless and scalloped with a black-spotted, creamy-white (3) or pale yellow (1) submarginal/postdiscal band on the black ground colour.

Atrophaneura luchti is sometimes treated as a subspecies of *A. priapus*, of which the subspecies *dilutus* flies in a separate part of eastern Java. *A. luchti* is clearly a close relative of *A. priapus*, and is believed to be an isolated relict of *priapus* stock (3). However, *A. luchti* is accepted as a good species by all recent authors (1, 2, 3) and a number of distinct characteristics are recognizable. The main difference from *A. priapus* is that both sexes have the sides of the abdomen covered by red hairs, rather than creamy yellow hairs (3).

Distribution This species is known only from the Ijen (Idjen) range of mountains in eastern Java, Indonesia.

Habitat and Ecology *Atrophaneura luchti* was described less than fifty years ago and is still very poorly known. The only specimens come from Gunung (Mt) Ijen and apart from the holotype, the species remained uncollected until 1979/1980, when more specimens were taken (1, 3). The larval foodplants remain unknown. *A. luchti* is a member of the *nox* group, some other members of which are known to lay their eggs on Aristolochiaceae (3) in the manner of other troidine swallowtails. Adults have been collected in November, January and March, and may fly throughout the year, taking nectar from flowers in the forest canopy (3).

The Ijen mountain range is volcanic in origin, with the highest points on Gunung Raung (3332 m) and Gunung Merapi (2800 m). This montane region has the lowest human population density in Java, most of the people being concentrated on the alluvial plains. Large areas of upland remain relatively undeveloped. The mountain ranges are mainly forested but with a high plateau, lakes and volcanic craters. The altitude at which *A. luchti* occurs is unknown, but is likely to be at medium or high elevations, most lowland habitat having been put to agricultural purposes.

Threats Threats to this butterfly and its habitat are presently unknown. Its very restricted range on an island with one of the world's highest human population densities is cause for concern and there is no information on its distribution within protected areas.

Conservation Measures There are six protected areas in the Ijen mountains of eastern Java (4). Three of these are very small: Pacur Ijen Reserve (9 ha) is contiguous with the Maeleng Reserve (see below); Ceding Reserve (2 ha) is a forest in the Ijen crater, probably already destroyed; and Cerakmanis Sempolan Reserve (16 ha) is on the slopes of Gunung Raung. The other three protected areas are Kawah Ijen Merapi Ungup–ungup Reserve (2560 ha) which surrounds the Kawah Ijen crater lake and includes mountains such as Gunung Merapi; Maelang Taman Buru Reserve (70 000 ha), which includes extensive undisturbed lowland forest; and the Gunung Raung proposed Reserve (about 60 000 ha), which is a marvellous area of montane forest well buffered by pine and teak plantations. In the recent Indonesian Conservation Plan it has been proposed that the three larger reserves should be combined into one protected area for future development into a national park. The three smaller reserves would automatically be included (31). It is likely, although unconfirmed, that *Atrophaneura luchti* occurs in at least the Maelang Reserve. Adoption of the Indonesian Conservation Plan's recommendation would ensure the future of the species and should be implemented at the earliest opportunity. There are many reasons why the Ijen range requires protection, not least of which is to maintain essential watersheds and thus ensure a constant water supply in the highly seasonal climate of eastern Java.

Atrophaneura luchti is clearly a rare, narrowly distributed, endemic species and further information on its biology and distribution is needed in order to ensure appropriate management measures.

References

1. D'Abrera, B. (1982). *Butterflies of the Oriental Region. Part 1. Papilionidae and Pieridae.* Hill House, Victoria, Australia. xxxi + 244 pp.
2. Hancock, D.L. (1983). Classification of the Papilionidae (Lepidoptera): a phylogenetic approach. *Smithersia* 2: 1–48.
3. Tsukada, E. and Nishiyama, Y. (1982). *Butterflies of the South East Asian Islands. Vol. 1 Papilionidae.* (transl. K. Morishita). Plapac Co. Ltd., Tokyo. 457 pp.
4. UNDP/FAO National Parks Development Project (1982). *National Conservation Plan for Indonesia. Vol 3. Java and Bali.* FAO, Bogor.

Atrophaneura (Losaria) palu (Martin, 1912)   INSUFFICIENTLY KNOWN

Subfamily   PAPILIONINAE   Tribe   TROIDINI

Summary *Atrophaneura palu* is a large and attractive black and white swallowtail known from very few specimens. It only lives in the area around Palu in Central Sulawesi, not far from the present site of the Lore Lindu Reserve. Field work is needed to assess the distribution, biology and conservation status of this apparently very rare species.

Description *Atrophaneura palu* is a large butterfly with a forewing length of up to 70 mm (2). As is usual in this genus the forewings of the female are a little broader and less angular than in the male, but otherwise the sexes ae similar in general appearance (2). *Atrophaneura palu* is a member of the *coon* species-group. Tsukada and Nishiyama consider *palu* to be a subspecies of *coon* (describing it as *Losaria coon palu*) (8) but Haugum *et al.* (4) follow Martin (5) in regarding it as specifically distinct. This judgement has been supported by Hancock, who recently examined the male genitalia (2).

UFW/LFW elongated with the black ground colour forming a border around the wing, but becoming transparent light grey over the inner part. The veins remain heavily scaled in black and other black lines extend from the outer margin and apex between the veins and across the cell (4, 5, 8).

UHW/LHW elongated with a long tail (broad and rounded at the tip, narrow at the base), and a scalloped outer margin. The black ground colour has a postdiscal band of seven white wedge-shaped spots. Only the larger, inner four of these are visible on the upper surface (4, 5, 8).

Distribution *Atrophaneura palu* is an isolated species only known from Sulawesi (Indonesia). Other members of the *coon* group occur much further west in Sumatra, western Java, other parts of western Indonesia and Peninsular Malaysia, but not on neighbouring Borneo. The type specimen was captured at an altitude of 2000 m at the village of Lewara, east of Palu on the west coast of Sulawesi. Further specimens, including the first male to be described, were taken at sea level west of Palu (2).

Habitat and Ecology No published information is available on the habitat, breeding biology or foodplants of *Atrophaneura palu*. The very wide altitudinal range of the butterfly is unusual; the type locality at 2000 m is presumably in a montane ecosystem quite different to the coastal habitat of the more recent discoveries. The foodplants of other members of the *coon* group are believed to be Aristolochiaceae (8). Palu, the provincial capital of Central Sulawesi, lies in a rain shadow with a rainfall of only about 700 mm per year, but the surrounding highlands have a normal rainy season from November to March, with a rainfall in excess of 3000 mm. The highlands are mainly forested, particularly on the slopes, but some valley bottoms are open grassland (1). According to unpublished information a breeding locality has recently been found by Japanese collectors in the region west of Palu. Unconfirmed reports suggest that the butterfly is confined to an arid and rather barren habitat (3). This would be most unusual for a troidine clubtail and might suggest a foodplant outside the Aristolochiaceae.

Sulawesi is one of the most distinctive and biologically interesting islands in

Indonesia. It has a fascinating swallowtail fauna recently reviewed by Haugum, Ebner and Racheli (4). Three species of *Atrophaneura* are known from the island; none are common but *A. dixoni* is fairly widespread and abundant in some localities (8). *Atrophaneura kuehni* lives on the north and east coast and is only slightly better known than *palu*. The northern subspecies, *mesolamprus*, is known only from the type specimens, and the eastern, nominate, form is known from very few collections (8). Other very rare species include *Papilio jordani* (see separate review) and *Graphium dorcus*, a rare but widespread species probably confined to high altitudes (8).

Threats The precise habitat of *Atrophaneura palu* has yet to be established, making it difficult to assess the status of the species. Like *Graphium dorcus*, *A. palu* may be mainly distributed in the poorly known and thinly populated central mountainous areas of Sulawesi. Although the locality of the species is remote, the degree of environmental disruption, particularly deforestation, should not be underestimated. Some of the presently restricted ranges of Sulawesi endemics may at one time have been much more widespread. Large areas stretching between Makassar and the Northern and Central Districts are covered with the unpalatable and useless grass *Imperata cylindrica* (lalang), a sure sign of over-exploitation and impoverishment of the soil (7). Natural habitats are increasingly being destroyed as a result of extensive, continuing and largely uncontrolled deforestation (7). Following the discovery of a new locality by Japanese collectors, *Atrophaneura palu* has recently appeared in the European trade at grossly inflated prices. This is a possible confirmation of the rarity and vulnerability of the species and is of particular concern when no details of habitat and ecology have been published.

Conservation Measures A UNDP/FAO development project has drawn up a conservation plan for Indonesia and made substantial recommendations for the conservation of the Sulawesi fauna (6). Lore Lindu Reserve, in Central Sulawesi, lies in the headwaters of the Palu River and is probably within 100 km of the type locality of *Atrophaneura palu*. Further study within the reserve and in other areas around Palu is needed to verify the presence of the butterfly and to ascertain its breeding biology, foodplant and conservation status.

References

1. Anon. (undated). *Indonesia, National Parks and Nature Reserves*. Directorate General of Tourism, Jakarta. 150 pp.
2. Hancock, D.L. (1984). A note on *Atrophaneura palu* (Martin) 1912. *Papilio International* 1(3): 71–72.
3. Haugum, J. (1984). *In litt.*, 14 April.
4. Haugum, J., Ebner, J. and Racheli, T. (1980). The Papilionidae of Celebes (Sulawesi). *Lepidoptera Group of 1968 Supplement* 9: 21 pp, 1 map, 2 pl.
5. Martin, Dr. (1912). Ein neuer *Papilio* aus Celebes. *Deutsche Entomologische Zeitschrift "Iris"* 26: 163–5. Plate given separately (1913) *Deutsche Entomologische Zeitschrift "Iris"* 27: Tafel 6.
6. National Parks Development Project, UNDP/FAO (1982). *National Conservation Plan for Indonesia Vol. VI: Sulawesi*. FAO, Bogor.
7. Straatman, R. (1968). On the biology of some species of Papilionidae from the island of Celebes (East-Indonesia). *Entomologische Berichten* 28: 229–233.
8. Tsukada, E. and Nishiyama, Y. (1982). *Butterflies of the South East Asian Islands Vol. 1 Papilionidae*. (transl. K. Morishita). Plapac Co. Ltd., Tokyo. 457 pp.

Atrophaneura (Pachliopta) jophon Gray, 1852   VULNERABLE

Subfamily PAPILIONINAE   Tribe TROIDINI

Summary *Atrophaneura jophon* is a rare species confined to the rain forests of south-western Sri Lanka. Over the past 30 years the human population has doubled and these forests have been devastated by the combined impact of timber extraction and agriculture. The 11 000 ha Sinharaja Forest Reserve is probably one of the last strongholds of the butterfly, but selective logging and illicit encroachment have already reduced the virgin forest area to 5000 ha. Although the entire forest has now been declared a UNESCO Man and Biosphere Reserve, exploitation of forest products continues to threaten the butterfly's habitat.

Description *Atrophaneura jophon*, the Ceylon Rose, is a large butterfly with a long tail to the hindwing, a forewing length of 60–64 mm, and similar sexes (Plate 3.1).

Forewing black with creamy-white markings in the discal and postdiscal regions, less well defined on the UFW which has heavier suffusion of black scales (2, 3, 10).

Hindwing deeply scalloped with a long spatulate tail. Black ground colour with creamy-white discal spots and pink submarginal spots, reduced to narrow lunules by heavy black suffusion on the UHW (2, 4, 10).

Distribution *Atrophaneura jophon* is endemic to Sri Lanka. It was formerly considered to be conspecific with *Atrophaneura pandiyana* from southern India (10), but has been recognized by Munroe (8), and Hancock (6) as a separate species.

Habitat and Ecology This butterfly flies at medium elevations (615–1230 m) in the wet, south-western zone of Sri Lanka (2, 10). Here the rainfall is over 2000 mm per year with no month completely dry and two main rainy seasons, from March to May and June to September (5). The natural vegetation is tropical evergreen rain forest, which is restricted to these south-western regions. The drier and more seasonal climate of the eastern and northern regions supports tropical savanna vegetation (5). Below 900 m the rain forest has a closed, dense canopy at 22–27 m with emergents to 45 m. Typical trees include species of the genera *Dipterocarpus, Shorea, Mesua, Doona, Hopea, Palaquium, Pygeum* and others (5). From 900 to 1500 m montane semi-evergreen forests grow to 18–24 m, with emergents to 30 m (5). The boundaries beween these two forests and the montane wet evergreen forests above 1500 m are indistinct.

The butterflies tend to fly low in the early morning and in the forest canopy later in the day (10). The eggs are yellow–brown, believed to be laid on the undersides of young leaves of Aristolochiaceae (10), as is the case in *Atrophaneura pandiyana* (2, 10). The caterpillars are purple–black with crimson tubercles and cream bands (10). Although usually rare and local, where it does occur it may be fairly numerous (10).

Threats Since the last comprehensive inventory of Sri Lankan forest resources in 1959–60, widespread deforestation has occurred. Of 2.9 million ha of closed broadleaved forest in 1956, only about 1.6 million ha now remains, excluding forest fallows (5). Of this total, only about 0.43 million ha is not exploited, over half of this for reasons of physical inaccessibility, the rest for legal reasons (5).

Deforestation in the country as a whole constitutes a very serious problem. The main cause is the pressure for agricultural land and forest products caused by high population levels and population growth rates, factors unlikely to be alleviated in the near future. The human population has doubled since the 1950s and now stands at almost 15 million.

The main threat to this narrowly endemic and local butterfly is the destruction of its habitat. Members of the subgenus all fly in dense forests and there is no evidence to suggest that they can survive serious disturbance.

Conservation Measures Numerous protected areas have been set aside in Sri Lanka and during the first half of this century the country boasted one of the best conservation programmes in Asia (7). However, the great majority of the reserves were in the dry northern and eastern regions (1), where this butterfly is not found. According to D'Abrera (2), the forests between Ratnapura and Deniyaya alluded to by Woodhouse and Henry (13) and Woodhouse (12) as the habitat of *jophon*, are now almost completely cleared for timber.

The last extensive area of primary lowland rain forest is the Sinharaja Forest Reserve, probably one of the last strongholds of *Atrophaneura jophon*. In the past the butterfly is known to have been well established here, but there is now evidence that the constant human disruption of the reserve is causing depletion of the butterfly populations (2).

Although the Sinharaja forest is 11 000 ha in extent, only about 5000 ha remains undisturbed, the rest having been seriously disrupted by a selective logging operation in 1972–75, illicit agricultural encroachment in many places along the reserve boundaries, and constant pressure for minor forest products (3). In April 1978 about 1822 ha of forest was declared a UNESCO Man and Biosphere Reserve and the rest of the area has since been included under this designation (9). A number of scientific studies have been carried out there. The reserve is the responsibility of the Forest Department of Sri Lanka, which has taken steps to ensure its protection, but policing the area is a continuing problem.

A survey of the butterfly fauna, with emphasis on *Atrophaneura jophon*, would add to the already evident value of the reserve. Sri Lanka has 15 species of swallowtails including another endemic, *Troides darsius*, two other species of *Atrophaneura*, *A. hector* and *A. aristolochiae* and two *Papilio* species restricted to Sri Lanka and southern India, *P. polymnestor* and *P. crino*. For many of these species the Sinharaja Forest Reserve will become an increasingly important habitat resource.

Sri Lanka's President has recently appointed a task force to examine the status of the island's natural resources in relation to their conservation and utilization for the future (11). Recommendations will be made for the development and implementation of a National Conservation Strategy (11). Unless the decline in Sri Lanka's forests is arrested quickly, *Atrophaneura jophon* and many other species will soon become very seriously endangered.

References

1. Crusz, H. (1973). Nature conservation in Sri Lanka (Ceylon). *Biological Conservation* 5: 199–208.
2. D'Abrera, B. (1982). *Butterflies of the Oriental Region. Part 1. Papilionidae and Pieridae*. Hill House, Victoria, Australia. xxxi + 244 pp.
3. Davis, S. (in prep.). Sinharaja Forest. Draft review for Plant Sites Red Data Book.
4. Evans, W.H. (1932). *The Identification of Indian Butterflies*. Bombay Natural History Society. 2nd ed. 454 pp.

5. FAO/UNEP (1981). *Tropical Forest Resources Assessment Project, Forest Resources of Tropical Asia*. FAO, Rome. 475 pp.

6. Hancock, D.L. (1983). Classification of the Papilionidae (Lepidoptera): a phylogenetic approach. *Smithersia* 2: 1–48.

7. Hoffmann, T.W. (1983). Wildlife conservation in Sri Lanka. Paper presented at Bombay Natural History Centenary Seminar, Bombay (in press).

8. Munroe, E. (1961). The classification of the Papilionidae (Lepidoptera). *Canadian Entomologist* Supplement 17: 1–51.

9. Sri Bharathi, K.P. (1979). Man and Biosphere reserves in Sri Lanka. *Sri Lanka Forester* 14(1–2): 37–38.

10. Talbot, G. (1939). *The fauna of British India, including Ceylon and Burma. Butterflies vol I*. Taylor and Francis Ltd., London, reprint New Delhi 1975. 600 pp.

11. Wijayadasa, K.H.J. (1983). Formulating a Sri Lankan strategy. *IUCN Bulletin Supplement* 4: 3.

12. Woodhouse, L.G.O. (1950). *The butterfly fauna of Ceylon*. Second Edition. Colombo Apothecaries' Co. Ltd., Colombo. 231 + xxxii pp.

13. Woodhouse, L.G.O. and Henry, G.M.R. (1942). *The Butterfly Fauna of Ceylon*. Colombo. 172 pp + 49 pls.

Atrophaneura (Pachliopta) schadenbergi (Semper, 1886)   VULNERABLE

Subfamily   PAPILIONINAE   Tribe   TROIDINI

Summary *Atrophaneura schadenbergi* is restricted to northern and central Luzon and the Babuyan Islands in the Philippines, where it occurs in grassland and wooded areas in foothills. It is threatened by loss of habitat caused by the demands of the rapidly expanding human population.

Description *Atrophaneura schadenbergi* is a relatively large butterfly (forewing length 50–60 mm) with narrow elongated wings and long spatulate tails (1, 3) (Plate 3.2). The sexes are similar in appearance, the female slightly larger than the male (3).

UFW/LFW white, heavily suffused with black scaling which covers the basal region, wing borders and veins, and occurs as stripes between all but the veins in the tornal region (1, 3). The white tornal flash is conspicuous in flight (3).

UHW/LHW scalloped and tailed, black with seven large submarginal spots, the lower three red and the remainder white. All these spots are white in the subspecies *micholitzi* (1, 3).

Distribution *Atrophaneura schadenbergi* is endemic to the northern and central parts of Luzon and the Babuyan Islands in the Philippines. The nominate subspecies occurs in the Cordillera Central and Sierra Madre of northern Luzon (2, 3), reaching as far south as the Bataan Peninsula and the Wawa Dam, north-east of Manila (3). The second subspecies, *A. s. micholitzi* is poorly known, but occurs at least in Camiguin, the Babuyan Islands, and possibly the northern tip of Luzon (1, 3).

Habitat and Ecology *Atrophaneura schadenbergi* is essentially an inhabitant of lowland foothills, although it is found at altitudes up to about 1200 m in the Sierra Madre of eastern Luzon, and on the slopes of the Cordillera Central in the west (3). It seems to have a variable habitat. At the Asin Hotspring near Baguio it flies in open wooded grassland, together with *Atrophaneura kotzebuea* (3). In the eastern and western foothills of the Sierra Madre (2) in eastern Luzon, and also in the limestone hills of Luzon's northern tip (3) it flies in heavily wooded secondary forests, while *A. kotzebuea* remains in the grassland (3). *A. schadenbergi* flies at heights of 1–4 m, particulary during the morning and late afternoon (2). The main flight season in the Sierra Madre seems to be February to August, with a peak in April (2). Other records, mainly from the Cordillera Central, indicate adults on the wing rather later in the year, from May to November (3). Whether it is continually breeding or producing one or two broods per year remains uncertain (3). The immature stages are unknown (3) and the foodplant, although unknown, is likely to be in the Aristolochiaceae.

Threats *Atrophaneura schadenbergi* has a limited distribution within which it occurs only locally (3). According to C.G. Treadaway, an authority on Philippine butterflies, it is in danger of disappearing as a result of the persistent degeneration and elimination of its habitat for agricultural purposes (2). A graphic description of the loss of suitable woodland is given by Tsukada and Nishiyama (3), who describe the difficulties encountered by S. Yamaguchi in finding forests in northern Luzon, where "the bald mountains stretched around him as far as the eye could see" (3).

Fortunately the foodplant of the butterfly can evidently survive in secondary forest, otherwise the plight of this butterfly might be even more serious.

Conservation Measures No measures have been taken to protect *Atrophaneura schadenbergi* and there are no known functional protected areas in northern Luzon. As a denizen of the foothills of northern Luzon the butterflies are probably surviving by migrating between forest patches, colonizing secondary growth as it matures. As human population pressures inevitably increase still further, the butterfly will become even more local and disjunct, finally surviving in isolated and highly vulnerable relict forests. The only way to ensure the continuing survival of this attractive species is the designation and effective management of suitable protected areas.

References

1. D'Abrera, B. (1982). *Butterflies of the Oriental Region. Part 1. Papilionidae and Pieridae.* Hill House, Victoria, Australia. xxxi + 244 pp.
2. Treadaway, C.G. (1984). *In litt.*, 25 May.
3. Tsukada, E. and Nishiyama, Y. (1982). *Butterflies of the South East Asian Islands. Vol. 1 Papilionidae.* (transl. K. Morishita). Plapac Co. Ltd., Tokyo. 457 pp.

Atrophaneura (Pachliopta) atropos Staudinger, 1888   INDETERMINATE

Subfamily PAPILIONINAE   Tribe TROIDINI

Summary *Atrophaneura atropos* is a handsome black-winged swallowtail confined to the western Philippine island of Palawan. Its status is uncertain, but it may be declining in the face of rapidly accelerating deforestation on the island. The national parks system of the Philippines is currently undergoing redevelopment. The Province of Palawan should protect large areas of its remaining wilderness as a resort for its unique flora and fauna, and also as resources for the growing tourist industry.

Description *Atrophaneura atropos* has entirely black and grey wings and red markings on the abdomen. The elongated forewing, which has a length of 45 – 55 mm, is grey with black margins, veins, parallel lines in the spaces between the veins, and lines in the cell. The hindwing is entirely black, and has a scalloped outer margin with a long spatulate tail. The sexes are alike, though the female is slightly larger. The lower surface is the same as the upper (3, 12).

This species is a member of the *polydorus* group of the genus *Atrophaneura* (6, 9), all of which have similar dark caterpillars with rows of dorsal and lateral tubercles, often red or red-tipped (6, 7). Some tubercles may be white and there is a white band on the abdomen (6, 7).

Distribution *Atrophaneura atropos* is endemic to the Philippine island of Palawan where it is found in the southern, central and northern lowlands (12). *Graphium megaera* is also restricted to Palawan and is reviewed separately in this volume.

Habitat and Ecology Palawan lies to the southwest of the main archipelago of the Philippines and just to the north of Sabah, East Malaysia. The island is dominated by a central mountain chain, with heavier and less seasonal rainfall on the eastern side. The lowland coastal plains in the east provide much of the island's arable land while in the west the mountains slope more directly to the coast. According to 1972 figures almost 70 per cent of the land area of the province was forested (1), but more recent studies indicate that only 53 per cent of Palawan is still under forest (10), a considerable reduction characteristic of all the Philippine islands (3). In the lowlands the forests are dominated by Dipterocarpaceae (notably *Dipterocarpus*, *Shorea*, *Parashorea* and *Pentacme*), which are highly prized timber trees (2, 3). In lower montane regions the dipterocarps gradually disappear to be replaced by oaks (*Quercus*) and a variety of other trees (*Tristania, Hopea, Eugenia, Agathis*, etc.) (3). On mountain tops upper montane mossy forests are found (3).

Palawan is on the eastern edge of the Asian continental shelf. Its fauna and flora bear many resemblances to that of nearby Sabah, with which Palawan was contiguous during the lowered sea levels of the last Ice Age. This is demonstrated by the distribution of *Atrophaneura neptunus*, which occurs from Peninsular Malaysia across Sumatra and Borneo to Palawan, where *dacasini*, a distinctive subspecies, flies (12).

Atrophaneura atropos flies in primary lowland rain forest throughout Palawan (10, 12). This species prefers humid forests and does not fly in the same forests as *Atrophaneura neptunus*, which prefers drier habitats (12). The foodplants of the two species are unknown, but likely to be in the Aristolochiaceae (5, 7, 9). *Atrophaneura atropos* flies slowly through the forest, quite often close to the ground and in open areas, or along the forest edge. They can be seen even on cloudy days or in light rain, sometimes stopping to take nectar from flowers (9). Adults are often counted as a rarity, but are in fact fairly widespread, can be common and fly all year round (9). There is evidence that the species survives well in secondary growth (4).

Threats By Asian standards Palawan is sparsely populated, with only 370 000 people (in 1979) spread over nearly 1.5 million ha (1, 2). Most of these people live in the eastern lowlands. In the less accessible central highlands, southern and western regions there are still large areas of forest and wilderness (1). *Atrophaneura atropos* probably still flies throughout these rain forest regions and is unlikely to be critically threatened at the moment. Nevertheless, the pace of deforestation is accelerating, particularly in the south (2), with 44 per cent of the land area under logging concessions in 1976, the area currently leased must be well over half of the total. The main practice is selective felling but with 'kaingin' farmers, the shifting cultivators, using logging roads for access to fresh areas, there is often no opportunity for the forest to recover. Mining of sand and minerals such as mercury and chrome have a local effect on forests, particularly when the operations close down and the employees remain and set up new villages and farms (2, 8). Although *A. atropos* and its foodplant apparently survive in secondary growth, there is evidence that clearance for upland rice after decreasing periods of fallow is having an adverse affect on populations (4).

Conservation Measures As has been stated in a number of other reviews of Philippine endemics, the national park system in the islands is undergoing a time of change. At one time as many as 63 parks were designated by the Philippine Government, 51 under the Bureau of Forest Development, but these are currently being restructured in favour of a reduced number of national parks to be set up by the Bureau and maintained to international standards (11). It is not clear whether any parks or wilderness areas will be sited in Palawan forests, although the St Paul's Underground River National Park will presumably remain. The regions around Cleopatra's Needle and Mt Mantalingajan have been suggested as wildlife sanctuaries with access only for scientific purposes (11). This is an important area for *A. atropos* and many other forms of wildlife (8). There is great potential for new and important parks and wilderness areas in Palawan, not only to protect the unique flora and fauna but also as a resource for the growing tourist industry. It is also essential to the well-being of the Philippine economy and of the Palawan environment that exploitation of non-protected forests should be done in a controlled and sustainable manner. In this way much of the wildlife will be retained.

References

1. Bruce, M. (1980). *The Palawan Expedition Stage 1*. Traditional Explorations, Sydney. 47 pp.
2. Bruce, M. (1981). *The Palawan Expedition Stage 2*. Associated Research, Exploration and Aid, Sydney. 139 pp.
3. D'Abrera, B. (1982). *Butterflies of the Oriental Region. Part 1. Papilionidae and Pieridae.* Hill House, Victoria, Australia. xxxi + 244 pp.
4. Dacasin, G.A. (1984). *In litt.*, 25 April.
5. FAO/UNEP (1981). *Tropical Forest Resources Assessment Project. Forest Resources of Tropical Asia.* FAO, Rome. 475 pp.

6. Hancock, D.L. (1980). The status of the genera *Atrophaneura* Reakirt and *Pachliopta* Reakirt (Lepidoptera: Papilionidae). *Australian Entomologists' Magazine* 7: 27–32.

7. Igarashi, S. (1979). *Papilionidae and Their Early Stages.* Vol 1: 219 pp., Vol 2: 102 pp. of plates. Kodansho, Tokyo. (In Japanese).

8. Jumalon, J.N. (1984). *In litt.*, 10 July.

9. Munroe, E. (1961). The classification of the Papilionidae (Lepidoptera). *Canadian Entomologist* Supplement 17: 1–51.

10. National Environment Protection Council (1979). *Philippine Environment 1979. NEPC Third Annual Report.* NEPC, Ministry of Human Settlements, Republic of the Philippines. 158 pp.

11. Pollisco, F.S. (1982). An analysis of the national park system in the Philippines. *Likas–Yaman (Journal of the National Resources Management Forum)* 3(12): 11–56.

12. Tsukada, E. and Nishiyama, Y. (1982). *Butterflies of the South East Asian Islands. Vol. 1 Papilionidae.* (transl. K. Morishita). Plapac Co. Ltd., Tokyo. 457 pp.

Troides (Troides) andromache (Staudinger, 1892)   INDETERMINATE

Subfamily PAPILIONINAE   Tribe TROIDINI

Summary *Troides andromache* is a birdwing which occurs at high elevations in a restricted part of Sabah and Sarawak in northern Borneo. Little is known of its ecology or precise range, but there is no doubt that threats to its habitat are multiplying. It occurs on Mt Kinabalu, an important national park for a variety of wildlife, but economic pressures are changing the park adversely. More information on *T. andromache* is required to assess its conservation status more precisely, and the planning of land-use in the area needs to be more effective.

Description *Troides andromache* is a large butterfly with a forewing length of about 65 mm (male) or 85 mm (female). The male is black, yellow, and grey in colour, while the female has additional brown and white scaling on the forewing (2, 3, 8, 9). Although two geographically separated subspecies, the nominate and *T. a. marapokensis* Fruhstorfer, have been distinguished by some authors, the latter is now regarded as a female form of the former (6).

Male: UFW entirely black with a violet sheen; LFW with a band of large greyish distal spots dusted with yellow.

Hindwing almost entirely yellow with black scaling narrowly over the veins, as a broad inner margin, and as large fringe spots producing an edentate submarginal band (2, 3, 8, 9).

Female: Forewing greyish-white lightly dusted with brown scales; brown apical area, outer margin and veins (2, 3).

Hindwing differs from male in the band of large, black discal/postdiscal spots. These join with the large fringe spots to leave very little of the yellow colour visible on the distal part of the hindwing (2, 3, 8, 9).

Distribution *Troides andromache* occurs in East Malaysia where it is found in Sabah and Sarawak in northern Borneo. The nominate subspecies is well known as occurring on Mt Kinabalu, particularly on the Pinosuk Plateau (1, 5, 6), but it has also been recorded from neighbouring mountains in the Crocker range. It has been suggested that it may also occur further south in Kalimantan, the Indonesian sector of Borneo, where there is plenty of suitable habitat but where little collecting has been done. *T. a. marapokensis* occurs further south in the area of Mt Marapok in Sarawak, a different mountain from the Mt Marapok south of Mt Kinabalu in Sabah, where the subspecies has not been found, despite incorrect statements to the contrary (2, 3, 8).

Habitat and Ecology *Troides andromache* occurs at high altitudes, from about 1000 to 2000 m above sea level. It has been described as an alpine butterfly (9) but this is misleading since the mountains are densely forested at such altitudes. It flies throughout the year, particularly in the morning and evening. Adults nectar on *Mussaenda* flowers and females fly lower than males (9).

Virtually nothing is known about the ecology of the early stages. The foodplant is an unidentified species of *Aristolochia* vine, on which oviposition has been observed (9).

Threats Destruction of the habitat of *Troides andromache* has occurred on a considerable scale (1, 4, 6). Parts of the Pinosuk Peninsula have been 'developed' for tourism, with construction of access roads, a golf course and a visitor centre, and with other facilities planned. Other areas have become a tea estate and a cattle ranch, or have been altered by timber processing and hydroelectric projects.

A large area of eastern Kinabalu, designated as a national park, has been leased (from 1973) for 30 years to a copper mining company. The Manu copper mine has changed the eastern face of Mt Kinabalu (4). For more details see the reviews of *Graphium (Graphium) procles* and *Papilio (Princeps) acheron*.

T. andromache is not a well-known species in trade and is only occasionally offered by dealers, usually as old rather than recently-collected specimens (2).

The province of East Kalimantan, where the species may possibly occur, has a low human population that is growing rapidly. Shifting agriculture and commercial logging are increasingly extensive in this region.

Conservation Measures Little is known about the ecology of *Troides andromache* and research and surveys are needed as a basis for practical conservation measures.

Although Mt Kinabalu was designated a national park in 1964 (7), economic developments in the short-term may reduce its value as an important wildlife area. Besides being the highest mountain in South East Asia, it has an exceptionally rich fauna and flora. *T. andromache* flies there with two other Rare swallowtails, *Graphium procles* and *Papilio acheron* (see separate reviews).

The undoubted reductions in the habitat of *T. andromache* do not give cause for immediate alarm (1), but if economic and tourist pressure increase, the butterfly may become more seriously threatened. It is clear that effective and sensitive planning and practical conservation are required to ensure the adequate survival of the fauna of which *T. andromache* is a part.

Some mention may be made of conservation in East Kalimantan, even though there are as yet no records of *T. andromache* from this province. East Kalimantan is the least densely populated of the five Kalimantan provinces and is of especially high conservation importance because of its richness and diversity (10). The very large Kayan–Menterang reserve of 1.6 million ha, in particular, should be thoroughly surveyed for *T. andromache*.

The species should be retained, until its status can be reviewed, under Appendix II of the 1973 Convention on International Trade in Endangered Species of Wild Fauna and Flora (CITES). Appendix II listing implies that commercial trade is allowed providing a permit from the country of export is obtained, this can provide a method of monitoring trade levels.

References

1. Barlow, H.S. (1983). *In litt.*, 29 June.
2. D'Abrera, B. (1975). *Birdwing Butterflies of the World.* Lansdowne Press, Melbourne. 260 pp.
3. D'Abrera, B. (1982). *Butterflies of the Oriental Region. Part 1. Papilionidae and Pieridae.* Hill House, Victoria, Australia. xxxi + 244 pp.
4. Davis, S. (in prep.) Mount Kinabalu. Draft review for IUCN Plant Sites Red Data Book.
5. Holloway, J.D. (1978). Butterflies and Moths. In *Kinabalu Summit of Borneo.* Sabah Society Monograph, 25–278.
6. Holloway, J.D. (1983). *In litt.*, 28 February.
7. Luping, D.M., Wen, C. and Dingley, E.R. (eds) (1978). *Kinabalu, Summit of Borneo.* Sabah Society Monograph, Malaysia.
8. Ohya, T. (1983). *Birdwing Butterflies.* Kodansha, Tokyo. 332 pp, 136 col. pls.
9. Tsukada, E. and Nishiyama, Y. (1982). *Butterflies of the South East Asian Islands. Vol. 1*

Papilionidae. (transl. K. Morishita). Plapac Co. Ltd., Tokyo. 457 pp.

10. UNDP/FAO National Parks Development Project (1981). *National Conservation Plan for Indonesia. Vol. 5: Kalimantan.* FAO, Bogor.

Troides (Troides) prattorum (Joicey and Talbot, 1922)   INDETERMINATE

Subfamily   PAPILIONINAE   Tribe   TROIDINI

Summary *Troides prattorum* is a large black and yellow birdwing butterfly with particular well-developed structural coloration producing a blue–green sheen on the angled hindwing. The species occurs only at high elevation in the Indonesian island of Buru, in the Moluccas. Its ecology is unknown. It occurs with the endemic *Delias apatela* and probably with two other endemic *Delias*. Research and survey are the first priority for conservation but a reserve for the unique butterfly fauna of Buru should also be established.

Description *Troides prattorum*, the Buru Opalescent Birdwing is a large, black butterfly with an unusual opalescent blue–green cast to the golden–yellow part of the hindwing. The female is larger with a forewing length of about 100 mm compared to about 85 mm in the male (1, 2).

Male: UFW black with narrow white borders to the veins on the upper part of the wing.

UHW slightly scalloped, almost entirely golden–yellow with black scaling narrowly over the veins, as a broad inner margin, and as large fringe spots producing an edentate submarginal band. The golden–yellow patch shows the dramatic characteristic of this species, the form of opalescent scaling. When observed fron various angles the entire patch changes to many different tones of opalescent blue–green.

LFW/LHW. The white scaling on the forewing is less extensive on the lower surface, which also has some black suffusion to the lower part of the hindwing golden–yellow patch (1, 2).

Female: UFW with more extensive white scaling in the female, especially in the top of the cell.

Hindwing with golden–yellow patch reduced by a broad discal band of black streaks to an irregular postdiscal line and a small patch around the cell apex. There is also more suffusion of dark scales distally but the opalescence is still present (1, 2).

Distribution *Troides prattorum* is known only from the Indonesian island of Buru in the Moluccas, which lies due east of Sulawesi. Its occurrence here is something of a zoogeographical mystery (1, 2).

Habitat and Ecology The habitat of *Troides prattorum* is the high plateau of the centre and west of the island. The type series of four specimens (one male, three female) was taken from 600 to 1600 m above sea level (3).

Nothing is known about the foodplants, early stages or ecology.

Threats *Troides prattorum* may be at risk because of its restricted area of occurrence, although as a high-altitude species it is perhaps less at risk than similar species at low elevations, which are frequently threatened by forestry and agricultural intensification. It is known, for example, that logging operations on steep limestone hills to the south of Bara are having a serious environmental impact.

Buru was apparently used from 1965 onwards as a prison island and access was particularly difficult (2). Conditions have eased recently and certain commercial collectors have been active on the island. Pairs of *T. prattorum* were very highly priced in 1980 and were commercially much more valuable than almost any other species of *Troides* (7). The effects of commercial exploitation are not known.

Conservation Measures Research on the ecology of *T. prattorum* is the first priority, together with an assessment of its precise distribution, current status and the effects of commercial collecting.

An effective reserve for this species, and others, should be established. The uncommon *Delias apatela* Joicey and Talbot (Pieridae) is endemic to Buru and flies in the same plateau biotope as *T. prattorum* (1, 5). *D. rothschildi rothschildi* Holland is more frequent (and occurs in Timor as a second subspecies) but is also an upland species. The only known specimen of *D. dumasi* Rothschild is also from the same or a similar biotope (1, 4).

A potential reserve of 145 000 ha has been proposed for Gunung Kelaput Muda in north-west Buru, and given a high priority in the National Conservation Plan for Indonesia (6). It is not yet known whether Buru's endemic butterflies are to be found in this region.

The species should be retained, until its status can be reviewed, under Appendix II of the 1973 Convention on International Trade in Endangered Species of Wild Fauna and Flora. Appendix II listing implies that commercial trade is allowed providing a permit from the country of export is obtained, this can provide a method of monitoring trade levels.

References

1. D'Abrera, B. (1971). *Butterflies of the Australian Region*. Lansdowne Press, Melbourne. 415 pp.
2. D'Abrera, B. (1975). *Birdwing Butterflies of the World*. Lansdowne Press, Melbourne. 260 pp.
3. Howarth, T.G. (1977). A list of the type-specimens of *Ornithoptera* (Lepidoptera: Papilionidae) in the British Museum (Natural History). *Bulletin of the British Museum of Natural History (Entomology)* 36: 153–169.
4. Talbot, G. (1929). *A Monograph of the Pierine Genus Delias*. Vol. 3. Bale & Co. Ltd., London. Pp. 117–172.
5. Talbot, G. (1937). *A Monograph of the Pierine Genus Delias*. Vol. 6. British Museum (Natural History), London. Pp. v + 261–656.
6. UNDP/FAO National Parks Development Project (1981). *National Conservation Plan for Indonesia. Vol. 7: Maluku and Irian Jaya*. FAO, Bogor.
7. Various trade catalogues (1980–1984).

Troides (Troides) dohertyi (Rippon, 1893)   INDETERMINATE

Subfamily PAPILIONINAE   Tribe TROIDINI

Summary *Troides dohertyi* is an unusually dark birdwing confined to the Talaud and Sangihe Islands north of Sulawesi, in Indonesia. The ecology of the species is very poorly known and research is needed. Human population pressure is very high on the islands and although the species may be protected in the Karakelang Hunting Reserve on the largest of the Talaud Islands, the proposed Gunung Sahendaruman Game Reserve on Great Sangihe Island should also be surveyed and gazetted.

Description *Troides dohertyi*, the Talaud Black Birdwing, is a large black or black–brown butterfly with a forewing length of about 73 mm (male) or 82 mm (female). The male has golden–yellow markings, but these may be absent in the female (1, 2, 5, 7) (Plates 3.3 and 3.4).

Male: Almost entirely black, rarely with coloration on the upper surface. Whitish-grey scaling along, but not over, the veins of the upper part of the LFW. The LHW has a golden–yellow, central discal band. There is usually suffusion of black scales from the base, which may obliterate the two costal spots of the discal band.

Female: Forewing black–brown ground colour with a variable amount of grey scaling. Grey scaling less diffuse on the LHW where it occurs around the veins.

Hindwing with a slightly scalloped outer edge and small yellow marginal spots. The golden–yellow central discal patch on the UHW varies from being large and well defined to entirely blackened. A similar whitish-grey patch on the LHW has a variable amount of dark suffusion.

Distribution *Troides dohertyi* is only known from northern Indonesia where it occurs on the Talaud Islands and Sangihe Island, between Sulawesi (Indonesia) and Mindanao (Philippines).

T. dohertyi has been variously regarded as a good species (1, 3, 7), even a distinctive species (2), or as a subspecies of *T. rhadamantus* (Lucas), together with *T. plateni* Staudinger (4, 5, 6, 7). *T. rhadamantus* is widely distributed in the Philippines and *T. plateni* is common, even in secondary forest, in Palawan.

Habitat and Ecology *Troides dohertyi* is a lowland species, which has been observed flying at sea level and taking nectar from *Mussaenda* flowers. It flies all the year round and throughout the daylight hours, though more particularly in early morning and late evening (7).

Little has been recorded on the ecology of the early stages. They are probably similar to those of *T. rhadamantus*, which have been well illustrated in colour (5). Both *T. plateni* and *T. rhadamantus* larvae are said to be common and easily reared or ranched (7). The foodplant is likely to be *Aristolochia tagala*.

Threats The precise degree of threat to *Troides dohertyi* has not been documented, hence the category Indeterminate is retained. That some degree of threat is present may be evidenced by two factors in particular. Firstly, human population pressure on the Sangihe and Talaud Islands is higher than in any other part of the province of Northern Sulawesi (8). The population growth rate in the province as a whole is 2.2 per cent per year (8). Pressure for land is likely to be most heavy in the coastal lowlands, where the birdwing is known to fly. There is no information on the capacity of *T. dohertyi* to adapt to secondary vegetation formations.

Secondly, the Sangihe and Talaud Islands are not particularly well served by protected areas. Two large hunting reserves, known jointly as the Karakelang Reserve, were recently established on the largest of the Talaud Islands (8). The main game animals are feral Balinese cattle (8). If these areas are kept under fairly natural vegetation and exotic species are not introduced, the forests could serve as useful refuges for the endemic Talaud fauna, including *T. dohertyi*. However, such circumstances are far from being assured. There are no protected areas on the Sangihe Islands and the only large piece of remaining forest is in the south of Great Sangihe, on Gunung Sahendaruman (8). *Troides dohertyi* is likely to be at risk because of its extremely restricted area of occurrence. Whether it is a good species or not, the taxon is one which should be assessed for any threats, particularly land-use changes in its island habitats.

Conservation Measures There is an immediate need for further data on the habitat requirements and conservation status of *Troides dohertyi*. As a preliminary measure, it should be sought in the Karakelang Hunting Reserve on the Talaud Islands, and in the proposed Gunung Sahendaruman Game Reserve on Great Sangihe.

The Indonesian Conservation Plan proposes to develop strict controls on hunting parties in the Karakelang Hunting Reserve or, failing this, to upgrade the site to a Game Reserve (8). Such attention to monitoring of the site would be essential to its long-term stability. The proposed Sahendaruman Game Reserve, being the last remnant of forest on Sangihe, should clearly be gazetted as quickly as possible (8). With the growing human populations in the area, encroachment may otherwise be inevitable. *Troides dohertyi* is not listed as a protected species in Indonesia, but this may be because the taxon is assumed to be included in *T. rhadamantus*. The species should be retained, until its status can be reviewed, under Appendix II of the 1973 Convention on International Trade in Endangered Species of Wild Fauna and Flora (CITES). Appendix II listing implies that commercial trade is allowed providing a permit from the country of export is obtained. This can provide a method of monitoring trade levels.

References

1. D'Abrera, B. (1975). *Birdwing Butterflies of the World*. Lansdowne Press, Melbourne. 260 pp.
2. D'Abrera, B. (1982). *Butterflies of the Oriental Region. Part 1. Papilionidae and Pieridae*. Hill House, Victoria, Australia. xxxi + 244 pp.
3. Hancock, D.L. (1983). Classification of the Papilionidae (Lepidoptera): a phylogenetic approach. *Smithersia* 2: 1–48.
4. Haugum, J. and Low, A.M. (1983). *A Monograph of the Birdwing Butterflies*. Vol. 2 (2): 105–240, col. pls. 5–12. Scandinavian Science Press, Klampenborg.
5. Igarashi, S. (1979). *Papilionidae and Their Early Stages*. Vol. 1: 219 pp., Vol. 2: 102 pp. of plates. Kodansha, Tokyo. (In Japanese).
6. Munroe, E. (1961). The classification of the Papilionidae (Lepidoptera). *Canadian Entomologist* Supplement 17: 1–51.
7. Tsukada, E. and Nishiyama, Y. (1982). *Butterflies of the South East Asian Islands. Vol. 1 Papilionidae*. (transl. K. Morishita). Plapac Co. Ltd., Tokyo. 457 pp.
8. UNDP/FAO National Parks Development Project (1982). *National Conservation Plan for Indonesia. Vol. 6. Sulawesi*. FAO, Bogor.

Ornithoptera tithonus De Haan, 1840 INSUFFICIENTLY KNOWN

Subfamily PAPILIONINAE Tribe TROIDINI

Summary *Ornithoptera tithonus* is a large, attractive birdwing which is restricted to western Irian Jaya and a few neighbouring islands. Its ecology and conservation needs are very poorly known. It may be best classified Rare because of its restricted distribution and possible 'relict' nature. On the other hand, there is concern because it has been offered in substantial numbers in commercial trade, the effects of which have not been monitored. A primary need is for research on ecology, distribution and possible conservation measures, including the effectiveness of existing and proposed reserves.

Description *Ornithoptera tithonus* is a relatively large birdwing, with a forewing length of 70–85 mm in the male and 105–110 mm in the female. The male is black, iridescent golden–green and golden–yellow, and the female is dark brown to black with yellow and white markings (Plates 3.5 and 3.6). The iridescent scaling of the male may vary from gold to green (2, 5). *O. t. waigeuensis* differs slightly from the nominate subspecies described below (4), and a second subspecies occurring on Misoöl has yet to be formally described (1, 7). *O. t. misresiana* has recently been synonymised with the nominate subspecies (7), but another subspecies, *cytherea*, has been described, which occurs in the Enarokei and Snow Mountains (Pegunungan Maoke) area (7).

Male: UFW black with iridescent golden–green radial, cubital and anal bands. UHW elongated and ovoid with a distinct anal notch, a long creamy brush border, a large, golden–yellow patch bordered outwardly with green and three black postdiscal spots. The wing margin and the broad anal area are black.

LFW iridescent greenish-yellow with black on the borders, apex, veins, postdiscal spots, as a large patch centred round the cell apex, and in a broad anal area (2, 5). LHW differs from UHW in narrower black margins and up to three extra postdiscal spots (2, 5).

Female: forewing black, becoming dark brown with age. White markings include a transverse cell spot, subapical streaks, discal spots and a series of submarginal spots.

Hindwing rounded with a slightly scalloped outer margin. Black with a broad, pale distal band containing large, black postdiscal spots (2, 5).

Distribution *Ornithoptera tithonus* is known to occur only in the western part of Irian Jaya on the mainland of New Guinea and on the neighbouring islands of Waigeo (Waigeu), Salawati (Salwatty) and Misoöl (Mysol). The nominate subspecies occurs on mainland Irian Jaya, with centres of distribution in the Arfak Mountains, the Onin Peninsula and at the southern end of Geelvinck Bay; it has also been taken far to the east, low down in the Snow Mountains area (4). Subspecies *waigeuensis* occurs in Waigeo and Salawati and the other subspecies is on Misoöl.

The distribution of *O. tithonus* is not well known but the butterfly appears to be absent where *O. chimaera* occurs and vice versa. It has been suggested that it is a relict species, the range of which has contracted in recent geological time (5).

Habitat and Ecology *Ornithoptera tithonus* inhabits hill forest areas. It does not usually ascend as high as *O. chimaera* and has been found at sea level. However,

273

it normally ranges up to about 1250 m above sea level and occasionally up to 1900 m or above (5).

Although virtually nothing has been recorded on the ecology, or even life history, of *O. tithonus*, the trade in apparently reared specimens implies that the biology is known locally in Irian Jaya. The field notes of C.B. Pratt indicate that both sexes nectar on particular trees, or that they assemble round a 'master tree' in the same way that some temperate butterflies do (5).

Threats Only generalised threats to this species can be currently recognised. However, many pressures on areas where it is likely to occur have been identified: for instance, logging, transmigration of people and prospecting for oil on Salawati Island (3), mining in Waigeo (9) and logging in the Arfak and Fakfak areas (9).

Ornithoptera tithonus has a restricted area of occurrence and is possibly a 'relict' species, with a range which may continue to contract (5). These factors and the possibility of further natural loss of populations may mean that *O. tithonus* should be classed as Rare.

However, there has also been a considerable trade in specimens, with one report that they have been collected in quantity (8). Prices have tended to be intermediate between those asked for *O. rothschildi* and those charged for much more widespread species such as *O. goliath* (8). The effects of commercial collecting on *O. tithonus* are not known.

Conservation Measures *Ornithoptera tithonus* is one of the least well known birdwings and its conservation depends on adequate research being done to determine at least the main features of its ecology.

Six nature reserves (Cagar Alam) in which *O. tithonus* has either been recorded or is likely to occur have either been established or proposed (9). Pulau Waigeo Barat Reserve (153 000 ha), Pulau Salawati Utara Reserve (62 000 ha) and Pulau Misoöl Reserve (105 000 ha) are all proposed nature reserves which should conserve both *O. t. waigeuensis* and the as yet undescribed Misoöl subspecies. A survey of Salawati to include groups other than birds, to make a more comprehensive biological inventory, and its wardening to combat the effects of hunting and human population pressure, have been recommended (3). On the mainland the most important areas proposed for protection are in Pegunungan Arfak (45 000 ha) and Pegunungan Fafak (51 000 ha). The nominate *O. t. tithonus* occurs in both areas, but is particulary numerous in the Arfak area (or perhaps has just been collected in greater numbers there). Overall priority for these four reserves is either one or two (9). It is possible that *O. t. tithonus* also occurs in the very large and important Lorentz Nature Reserve which has already been established (over 2 million ha, proposed reduction to 1 675 000 ha) and in the small Meriam Hill area which is proposed as a recreation park (Taman Wisata). *O. t. cytherea* occurs in the Enarotali Nature Reserve, although it is proposed that this reserve be replaced by a new Weyland Mountains Nature Reserve to the west (7, 9) (see review of *Ornithoptera paradisea*). It thus appears that if the Conservation Plan for the Irian Jaya Province is adopted, this birdwing will be well catered for in nature reserves. A detailed study of the biology of the species is necessary to effectively realise this potential.

O. tithonus is a protected species in Indonesia, and the effectiveness of this protection should be monitored. The species is listed under Appendix II of the 1973 Convention on International Trade in Endangered Species of Wild Fauna and Flora (CITES). Appendix II listing implies that commercial trade is allowed providing a permit from the country of export is obtained. This designation can provide a method of monitoring trade levels and should be retained.

References

1. Anon. (1984). Review of Ohya (1983). *Papilio International* 1: 84.
2. D'Abrera, B. (1975). *Birdwing Butterflies of the World*. Lansdowne Press, Melbourne. 260 pp.
3. Diamond, J.M. *et al.* (1983). Surveys of five proposed reserves in Irian Jaya, Indonesia: Kumawa Mts, Wandammen Mts, Yapen Island, Salawati Island and Batanta Island. Report to World Wildlife Fund and to the Directorate of Nature Conservation, Indonesia. 49 pp., 8 appendices.
4. Haugum, J. (1983). *In litt.*, 2 June.
5. Haugum, J. and Low, A.M. (1978, 1979). *A Monograph of the Birdwing Butterflies*. Vol. 1 (3): 193–308. Scandinavian Science Press, Klampenborg.
6. Munroe, E. (1961). The classification of the Papilionidae (Lepidoptera). *Canadian Entomologist* Supplement 17: 1–51.
7. Ohya, T. (1983). *Birdwing Butterflies*. Kodansha, Tokyo. 332 pp., 136 col. plates.
8. Various trade catalogues (1978–84).
9. UNDP/FAO National Parks Development Project (1981). *National Conservation Plan for Indonesia. Vol. VII. Maluku and Irian Jaya*. FAO, Bogor.

Ornithoptera rothschildi Kenrick, 1911   INDETERMINATE

Subfamily   PAPILIONINAE   Tribe   TROIDINI

Summary *Ornithoptera rothschildi* is a little-known birdwing with a restricted distribution at high elevations in the mountains of the Arfak area, north-western Irian Jaya, Indonesia. Virtually nothing is known about its ecology, but considerable numbers of specimens have entered the commercial trade. The first needs are for research, survey and assessment of status and a thorough review of the effects of commercial exploitation.

Description *Ornithoptera rothschildi*, Rothschild's Birdwing, is a relatively large birdwing with a forewing length of 63–80 mm in the male and 80–93 mm in the female. The male is black, iridescent pale green and golden–yellow, and the female is dark brown with creamy-white and yellow markings (1, 2, 4, 6) (Plates 4.1 and 4.2). *O. okakeae* Kobyashi and Koiwaya, is considered to be a natural hybrid between *O. rothschildi* and *O. priamus poseidon* (4).

Male: UFW black with iridescent pale green markings. UHW with black margins and anal area; inner part golden–yellow with a large iridescent green patch and a series of black spots.

LFW black with markings of iridescent yellow–green, including a broad median band, submarginal spots, subapical streaks and two streaks in the cell. LHW differs from the UHW in the narrow black margin (1, 2, 4).

Female: Forewing blackish-brown with creamy-white markings including small subapical spots, small submarginal spots, fringe spots and three discal spots. The LFW may have more extensive maculation (2, 4).

Hindwing elongated and rounded with a scalloped outer edge, an irregular black outer border, a very broad pale band which is cream–white discally becoming yellow distally and a complete series of black subdiscal spots (1, 2, 4, 6).

Distribution *Ornithoptera rothschildi* is endemic to the Arfak Mountains area of the Beran Peninsula in north-western Irian Jaya (Indonesia). It has the smallest range of any of the mainland New Guinea birdwings (4).

Habitat and Ecology *Ornithoptera rothschildi* is a montane species and is said to replace *O. tithonus* at high elevation (though the latter species is more widely distributed) (4). Most of the records of *O. rothschildi* are from about 1800–2450 m above sea level. The butterflies apparently prefer sheltered valleys and ravines which are sunny and protected from strong winds. The vegetation in this habitat is rich but shrubby, with some emergent larger trees (4).

Almost nothing is known about the ecology of *O. rothschildi*. Flight and behaviour seem to be typical for species of *Ornithoptera*. The species has been supplied in large numbers by commercial collectors recently and at least some of the specimens appear to have been reared from collected larvae or pupae. Nevertheless, there are no recorded descriptions of the early stages or foodplants, nor of larval behaviour and natural mortality.

Threats *Ornithoptera rothschildi* has a very restricted area of occurrence. Although no particular threats to its montane habitats have been identified, the area surrounding the Arfak Mountains is relatively densely populated and the cutting of wood in the forests is a general threat to the region (9). However, *O. rothschildi* would probably be classed as Rare if it were not for doubts about the effects of commercial collecting.

The very considerable trade in this species has caused comment and concern (3, 4, 7). High prices were originally quoted when specimens first became available, but by 1982 pairs were selling for £10 or less (8). There is no information on the effects of commercial exploitation on the butterfly and no means of knowing whether the resource is being managed or controlled in any way. The only indication that *O. rothschildi* is not threatened by commercial collecting is that it is not one of the species of birdwing which are protected in Indonesia, in contrast to *O. chimaera*, *O. goliath* and *O. paradisea*.

Conservation Measures The primary need for the conservation of *Ornithoptera rothschildi* is an adequate account of its habitat and ecology and an assessment of its status. Without at least an outline of its life history, and identification of its foodplants, natural enemies (if any) and actual threats to its abundance, no proposals for practical measures to conserve the species can be made.

However, proposals for a nature reserve (Cagar Alam) of 45 000 ha, the Pegunungan Arfak Reserve, which would include part of the range of *O. rothschildi*, have been given a high priority (9). *O. rothschildi* is only one of many rare and endemic species in need of protection by establishment of a reserve here. Included in the proposals is a small staff of wardens to check the spread of human settlement in the reserve (9).

The species should be retained, until its status can be reviewed, under Appendix II of the 1973 Convention on International Trade in Endangered Species of Wild Fauna and Flora (CITES). Appendix II listing implies that commercial trade is allowed providing a permit from the country of export is obtained; this can provide a method of monitoring trade levels.

References

1. D'Abrera, B. (1971). *Butterflies of the Australian Region*. Lansdowne Press, Melbourne. 415 pp.
2. D'Abrera, B. (1975). *Birdwing Butterflies of the World*. Lansdowne Press, Melbourne. 260 pp.
3. Haugum, J. (1983). *In litt.*, 2 June.
4. Haugum, J. and Low, A.M. (1979). *A Monograph of the Birdwing Butterflies*. Vol. 1 (3): 193–308. Scandinavian Science Press, Klampenborg.
5. Munroe, E. (1961). The classification of the Papilionidae (Lepidoptera). *Canadian Entomologist* Supplement 17: 1–51.
6. Ohya, T. (1983). *Birdwing Butterflies*. Kodansha, Toyko. 332 pp., 136 col. plates.
7. Pasternak, J. (1981). On the rediscovery of *Ornithoptera meridionalis tarunggarensis* Joicey and Talbot on a new locality in Kamrau Bay, south west Irian Jaya, Indonesia. *Transactions of the Himeji Natural History Association* 1981: 2–14.
8. Various trade catalogues (1977–84).
9. UNDP/FAO National Parks Development Project (1981). *National Conservation Plan for Indonesia. Vol. VII. Maluku and Irian Jaya*. FAO, Bogor.

Ornithoptera chimaera (Rothschild, 1904)   INDETERMINATE

Subfamily PAPILIONINAE   Tribe TROIDINI

Summary *Ornithoptera chimaera* is a specialised birdwing of montane biotopes and occurs throughout the mountainous areas of mainland Papua New Guinea; it is also found in a few mountain ranges of Irian Jaya. It is restricted to one species of foodplant and the density of larvae is low. Threats to the species have not been clearly defined, but its status should be effectively monitored. The butterfly is a good candidate for controlled utilisation to accompany its conservation.

Description The male of *Ornithoptera chimaera*, the Chimaera Birdwing, is a large black, iridescent yellow–green and golden butterfly with a wingspan of 120–155 mm. The female is larger, up to 190 mm, and is dark brown with pale markings (1, 2, 6, 11) (Plates 4.3 and 4.4). The description below is for the nominate subspecies. A second subspecies, *O. c. charybdis*, has males with more extensive, iridescent yellow–green scaling on the wings. A third subspecies, *O. c. flavidior*, described in recent monographs (2, 6), has now been synonymised with the nominate subspecies (9).

Male: UFW black ground colour with iridescent yellow–green radial, anal and cubital bands. UHW rounded with a noticeable anal notch, a large golden area, a yellow–green submarginal band, two or three subdiscal black spots and a narrow black margin (1, 2, 6, 11).

LFW lighter and more yellow, with black veins, wide black costal and outer margins and a row of black postdiscal spots (2, 6).

Female: forewing dark brown with white or grey markings including a small cell spot and small discal, subapical and submarginal spots.

Hindwing broad, dark brown with a scalloped outer margin and a broad light yellow–grey distal band containing large, brown discal spots (1, 2, 6, 11).

Distribution *Ornithoptera chimaera* is confined to mainland New Guinea. Subspecies *chimaera* is widely distributed along the central cordillera and has been recorded from the Finisterre Mountains and those of the Huon Peninsula, and from thirty-one 10 km squares in Papua New Guinea (10). There is an outlying record from eastern Irian Jaya (6).

O. chimaera charybdis is known from central Irian Jaya, from the Wandaman Mountains on the east coast of the Vogelkop (Berau Peninsula) through the Weyland Mountains (the main centre of occurrence) to an outlying locality in the Pegunungan Maoke (Snow Mountains) (6).

Habitat and Ecology *Ornithoptera chimaera* is a montane butterfly, occurring in areas of tall but fairly open primary forest, often in moderately to very steep-sided valleys along water courses. Adults frequent forest margins to collect nectar. They may be found in regions from 1200 to 2800 m above sea level, but normally occur between 1600 and 1800 m (10). Males congregate round special trees (7).

As far as is known, *O. chimaera* is monophagous on *Aristolochia momandul* throughout its range (10, 12). *A. pithecurus* (8) is thought to be synonymous with *A. momandul* (9). The eggs, which are 4 mm in diameter, are laid on the underside of leaves (12), and the larvae are usually solitary (10). The young larvae attack the tender leaves and shoots of the foodplant but mature caterpillars eat older leaves and may occasionally chew the bark of the main stem. Feeding occurs mainly in the early morning and late afternoon (10). In the field, the egg stage lasts 14 days, the larval stage probably about 2 months, and 49–70 days are spent as a pupa (10). Some larval cannibalism occurs and attack by the braconid wasp, *Apanteles* cf. *vitripennis*, may be heavy in some localities (10, 12). The early stages have been excellently figured in colour (8, 12).

Adults appear to range widely, particularly females in search of oviposition sites. Although the data come from a small sample, females which have been dissected were found to be carrying either no eggs or from 6–10 (10). There are no records of vertebrate predation on adult butterflies (10).

Threats *Ornithoptera chimaera* is restricted not just to the mainland of New Guinea but to areas of medium to high elevation, and to a single species of host plant. Its existing habitat in Papua New Guinea is limited but difficult to exploit commercially, e.g. by extraction of timber (10). Like many birdwings, *O. chimaera* is a K-selected species, producing few, well-protected offspring compared with many other butterflies. Although ecologically highly specialised, it can be quite common at times in some localities and cannot be considered as 'very rare' (10, cf. 1).

In Irian Jaya, *O. chimaera* is much less generally distributed and its habitat in the Wandamen Mountains may be under some threat from forest fires, depending on how high into the montane habitat these fires extend (3). Illegal collecting of vertebrates occurs in the Wandamen Mountains, but it is not known if protected birdwings, of which *O. chimaera* is one, are also collected (3).

O. chimaera is at risk in the Weyland Mountains area because if logging increases here, soil erosion would follow on the steep slopes of the mountains (13).

Certainly *O. chimaera* is in demand as specimens for study and display. No legal trade is possible in Papua New Guinea and until recently the trade was only in *O. c. charybdis*. There is no information on whether trade, legal or illegal, has had any adverse effect on the abundance of the butterfly.

Conservation Measures Although because of its high altitude habitat *O. chimaera* is not greatly threatened at the present time, its status needs continual monitoring in both Papua New Guinea and Irian Jaya. The effects of forest fires and timber extraction, where it occurs at such an elevation, need to be assessed.

In Papua New Guinea, the establishment of five nature reserve areas should be sufficient to conserve *O. chimaera*, without the need to undertake specific management of habitat and populations (10). The suggested locations for these reserves are Telefomin, Bundi, Naniwe Mission, Tapini–Woitape and Central Huon (10).

Two proposed nature reserves (Cagar Alam) in Irian Jaya almost certainly contain populations of *O. chimaera*. The Weyland Mountains Nature Reserve is proposed to replace the existing Enarotali Nature Reserve which lies to the east and is unsuitable because of its large human population (13) (see review of *Ornithoptera paradisea*). In elevation the proposed reserve varies from 900 to over 3800 m and so includes the entire altitudinal range of *O. chimaera* (10, 13).

The proposed Pegunungan Wandamen/Wondiwoi Nature Reserve includes most of the mountainous part of the Wandamen Peninsula, ranging from sea level to more than 2200 m and comprises 79 500 ha (13). This area contains a notable fauna of montane birds, and it has been suggested that a small local industry of natural history tours could be established, particularly to see bowerbirds (Ptilonorhynchidae) and birds of paradise (Paradisaeidae) (3). *Ornithoptera chimaera* could easily be included as a tourist attraction.

Suggestions have been made that *Ornithoptera chimaera* should be brought back into the area of legitimate trade (10). This is not the retrograde step in conservation terms that it superficially appears to be: "Ironically it is now becoming an accepted fact that the very demand for *Ornithoptera* is one of the main assets which will ensure their future survival if they can be exploited in the correct way." (10). A possible alternative, or first step, might be careful husbandry in a Wildlife Management Area in Papua New Guinea (7).

If *O. chimaera* were to be brought back into commerical trade, this should be under strict control and the effects on populations and abundance would need to be closely and effectively monitored. Information on the effects of trade on the numbers of any species of birdwing is urgently required. If controlled trade in *O. chimaera* were permitted, the protected status of the species would have to be rescinded, or at least suspended. If controlled utilisation of the resource were successful, the species could be removed from Appendix II of CITES (10).

References

1. D'Abrera, B. (1971). *Butterflies of the Australian Region*. Lansdowne Press, Melbourne. 415 pp.
2. D'Abrera, B. (1975). *Birdwing Butterflies of the World*. Lansdowne Press, Melbourne. 260 pp.
3. Diamond, J.M. *et al.* (1983). Surveys of five proposed reserves in Irian Jaya, Indonesia: Kumawa Mts, Wandammen Mts, Yapen Island, Salawati Island, and Batanta Island. Report to World Wildlife Fund and to the Directorate of Nature Conservation, Indonesia. 49 pp., 8 appendices.
4. Fenner, T.L. (1983). *In litt.*, 15 March.
5. Haugum, J. (1983). *In litt.*, 2 June.
6. Haugum, J. and Low, A.M. (1979). *A Monograph of the Birdwing Butterflies*. Vol. 1 (3): 193–308. Scandinavian Science Press, Klampenborg.
7. Hutton, A.F. (1984). *In litt.*, 1 February.
8. Igarashi, S. (1979). *Papilionidae and Their Early Stages*. Vol. 1: 219 pp., Vol. 2: 102 pp. of plates. Kodansha, Tokyo. (In Japanese).
9. Ohya, T. (1983). *Birdwing Butterflies*. Kodansha, Tokyo. 332 pp., 136 col. plates.
10. Parsons, M.J. (1983). A conservation study of the birdwing butterflies *Ornithoptera* and *Troides* (Lepidoptera: Papilionidae) in Papua New Guinea. Final report to the Department of Primary Industry, Papua New Guinea. 111 pp.
11. Smart, P. (1975). *The Illustrated Encyclopedia of the Butterfly World*. Hamlyn, London. 275 pp.
12. Straatman, R. and Schmid, F. (1975). Notes on the biology of *Ornithoptera goliath* and *O. chimaera* (Papilionidae). *Journal of the Lepidopterists' Society* 29: 85–88.
13. UNDP/FAO National Parks Development Project (1981). *National Conservation Plan for Indonesia. Vol. VII. Maluku and Irian Jaya*. FAO, Bogor.

Ornithoptera paradisea (Staudinger, 1893)   INDETERMINATE

Subfamily PAPILIONINAE   Tribe TROIDINI

Summary *Ornithoptera paradisea* is a particularly attractive, sexually-dimorphic butterfly which is widespread, though very localised, throughout northern mainland New Guinea as six described subspecies, several ill-defined. It is apparently extinct in one well-known area, but no obvious specific threats are known. It should be surveyed to establish its status, reserves should be established, and the future for ranched specimens as a marketable resource should be considered. For the present, it should remain protected in Papua New Guinea and Irian Jaya, and remain on Appendix II of CITES.

Description *Ornithoptera paradisea* is variously known as the Paradise Birdwing, Tailed Birdwing or Butterfly of Paradise. The male is a beautiful, black, iridescent yellow–green and golden, tailed butterfly with a wingspan of 100–130 mm (Plate 4.5). The dark brown female with yellow and white or creamish markings has an average wing span of 160 mm (2, 3, 6, 8, 10) (Plate 4.6). There is much variation both within and between subspecies. Six subspecies of *O. paradisea* have been described, including the nominate *O. p. paradisea*, but several are ill-defined, poorly-known, or both. The validity of the subspecies of this birdwing and the differences between them, must therefore be treated with caution, particularly as some populations have not yet been reliably referred to subspecies (6, 10). These include *O. p. arfakensis*, frequently encountered in the butterfly trade, and *O. p. tarunggarensis* formerly described as a subspecies of *O. meridionalis* (5).

Male: UFW of the nominate subspecies broad and elongated with a pointed apex and a rounded outer margin, black with broad, iridescent light yellow–green radial and cubital bands and a relatively short anal band. UHW reduced in size and almost triangular with a distinct apex, anal notch, and a tail at vein two which is a further 20–25 mm long (6). Black scaling limited to narrow costal and outer margins, and to a relatively narrow anal area. There is a broad yellow–green submarginal border and a variable, large, golden–yellow patch. Long white or cream hairs form a brush-border to the anal margin.

LFW predominantly covered with iridescent, light yellow–green scales, more golden than those on the upper surface. Slight black scaling around the veins, on narrow wing margins, and in a part of the wing apex (absent in eastern populations). LHW almost entirely light yellow–green and golden with a pale silver anal area (2, 3, 6, 8, 10).

Female: forewing pale to very dark brown with pale cream markings, including relatively large subapical streaks, submarginal spots and sometimes up to three discal spots. Hindwing elongated with a pale discal band, white towards the cell and yellow distally, containing a row of relatively small, black subdiscal spots (2, 3, 6, 10).

Distribution *Ornithoptera paradisea* is widely distributed on the mainland of New Guinea and has been reported more or less unreliably (sightings and dealers' records) from a number of adjacent islands (6). In Papua New Guinea, *O. paradisea* may be extinct in its well-known 19th century localities in Madang Province (9). There is a good sprinkling of localities in the north-west of the country in the Maprik and adjacent areas and recent records from the Lake Kutubu region (information about which post-dates the map in Haugum and Low (6, 9)). There are no recent

281

records from the eastern half of mainland Papua New Guinea, nor from any of the islands. In all, there are modern records from seventeen 10 km squares (9).

In Irian Jaya, *O. paradisea* is also very local, but the butterfly is well distributed, particularly in the west (though not the extreme west) and in the north-east (f. *borchi*); it has not been taken in the south-east of the country. There are scattered records from the Penunungan Maoke (Snow Mountains) westwards to Kamrau Bay, with a fairly well known area of occurrence in the Arfak region of the north-west (ssp. *arfakensis*) (6).

Habitat and Ecology *Ornithoptera paradisea* occurs in mature secondary and primary forest in hilly areas, where it frequents clearings, valleys, gulleys and gorges and also flies on slopes and ridges. It is normally found at altitudes between 200 and 800 m but there are records of occasional butterflies being found much higher, even once at 2000 m (1, 4, 6, 9).

The foodplant is a species of *Aristolochia* closely related to *A. momandul*, to which the manuscript name of *A.* "pseudo-momandul" has been given for ease of reference (9). The eggs are normally deposited singly on the underside of the leaves of the foodplant, sometimes, but rarely, on other objects nearby (1). The eggs are particularly large (4 mm diameter) and so probably few eggs are carried by a female, perhaps 8–10 at any one time. Eggs hatch in 10–12 days, the larval stage takes 36–40 days and the pupal stage 37 days (1). The immature stages therefore last about three months. The early stages and foodplant have all been well figured in colour (1, 8).

There is mortality in the egg stage from parasitic wasps. The larvae are taken by various vertebrate predators and attacked by braconid wasps. They are also destroyed by bad weather (1). Although mortality is said to be high, its significance has not been quantified.

Adults appear to be fairly localised in their movements and do not move long distances. They use valleys and gorges as flyways, and collect nectar frequently on various flowers. They probably live about three months.

Threats No clear threats to *Ornithoptera paradisea* have been identified. The possible extinction of the Madang populations gives rise to concern, but no plausible cause has been suggested. The birdwing may be Vulnerable but is best regarded as Indeterminate in status because of local extinction due to unknown causes, together with its patchy distribution. It should be classed as Rare if threats are not serious; although its area of distribution is not particularly narrow, it is very generally regarded as an uncommon butterfly.

In addition, it is much in demand as specimens. In recent years, considerable numbers of *O. p. arfakensis* have come into trade, but there is no evidence that this has been a threat to its continued existence or numbers (12).

Conservation Measures In Papua New Guinea, the conservation of this species falls into three parts. First, the distribution records and assessment of status of *Ornithoptera paradisea* appear to be incomplete. Survey of the butterfly is certainly necessary and should include a thorough field survey of the Madang area. Secondly, reserves need to be established under the authority of the Conservation Areas Act (1978), to prevent undue loss of habitat. Five National Reserve Areas have been suggested as the optimum number to conserve this and other *Ornithoptera* spp., at South Vanimo, Maprik, Frieda River, Lake Kutubu and East Erave (9). Thirdly, the utilisation of the resource, possibly as a tourist attraction, but particularly for farming of specimens, should be considered in the long term (7, 9). Even a limited, trial production of farmed specimens is clearly some years away.

In Irian Jaya, *O. paradisea* appears to be present in several reserves or proposed reserves. Perhaps the most important is the proposed Pegunungam Arfak Nature Reserve (Cagar Alam), which has overall priority one for establishment (see review of *O. tithonus*) (11). *O. p. arfakensis* is the subspecies which would be protected by this reserve. Other populations of *O. paradisea* are likely to occur in the established Lorentz Nature Reserve (see under review of *O. tithonus*) and in the proposed Weyland Mountains Nature Reserve. This area of 228 000 ha has been recommended to replace the Enarotali Nature Reserve of 300 000 ha which lies to its east (11). The Weyland Mountains reserve has an overall priority of one for establishment and almost certainly contains at least some of the area's populations of *O. paradisea* (11).

As in Papua New Guinea, an assessment of the birdwing's status and a survey of its ecological requirements are both necessary. Utilisation of the resource, both as a tourist attraction and in the form of carefully-controlled farming for the specimen trade, could also be considered.

For the present, however, *O. paradisea* should retain its status as a fully protected species in Papua New Guinea, as a protected species in Irian Jaya (Indonesia), and the butterfly should be retained in Appendix II of the 1973 Convention on International Trade in Endangered Species of Wild Fauna and Flora (CITES). Appendix II listing implies that commercial trade is allowed, providing a permit from the country of export is obtained; this can provide a method of monitoring trade levels. This situation should remain at least until effective conservation measures can be implemented.

References

1. Borch, H. and Schmid, F. (1975). The life cycle of *Ornithoptera paradisea* (Papilionidae). *Journal of the Lepidopterists' Society* 29: 1–9. 12 col figures.
2. D'Abrera, B. (1971). *Butterflies of the Australian Region*. Lansdowne Press, Melbourne. 415 pp.
3. D'Abrera, B. (1975). *Birdwing Butterflies of the World*. Lansdowne Press, Melbourne. 260 pp.
4. Fenner, T.L. (1983). *In litt.*, 15 March.
5. Hancock, D.L. (1983). A note on the status of *Ornithoptera paradisea tarunggarensis* (Joicey & Talbot) (Lepidoptera: Papilionidae). *Australia Entomologists' Magazina* 8: 93–95.
6. Haugum, J. and Low, A.M. (1979). *A Monograph of the Birdwing Butterflies*. Vol. 1 (3): 193–308. Scandinavian Science Press, Klampenborg.
7. Hutton, A.F. (1984). *In litt.*, 1 February.
8. Igarashi, S. (1979). *Papilionidae and Their Early Stages*. Vol. 1: 219 pp., Vol. 2: 102 pp. of plates. Kodansha, Tokyo. (In Japanese).
9. Parsons, M.J. (1983). A conservation study of the birdwing butterflies *Ornithoptera* and *Troides* (Lepidoptera: Papilionidae) in Papua New Guinea. Final report to the Department of Primary Industry, Papua New Guinea. 111 pp.
10. Smart, P. (1975). *The Illustrated Encyclopedia of the Butterfly World*. Hamlyn, London. 275 pp.
11. UNDP/FAO National Parks Development Project (1981). *National Conservation Plan for Indonesia. Vol. VII. Maluku and Irian Jaya*. FAO, Bogor.
12. Various trade catalogues (1978–84).

Ornithoptera meridionalis (Rothschild, 1897)   VULNERABLE

Subfamily   PAPILIONINAE   Tribe   TROIDINI

Summary *Ornithoptera meridionalis* is a birdwing butterfly with a high degree of sexual dimorphism. It is restricted to New Guinea, where it is a rare and very localized species of lowland forest. It occurs mainly in south-eastern Papua New Guinea but is also known from Irian Jaya, Indonesia. It is threatened by increased extraction of timber from its localities, many of which are very accessible. Reserves should be established for *O. meridionalis* and these could be managed in part as tourist showpieces. Protected status for the species should be maintained.

Description The male of *Ornithoptera meridionalis* has tailed hindwings, a wingspan of 80–115 mm (3) and is black, iridescent yellow–green and golden (Plate 5.1). The male is, on average, much smaller than any other male *Ornithoptera*, including that of the closely-related *O. paradisea*. The female is considerably larger than the male, black with white and yellow markings (1, 2, 4, 6, 10) (Plate 5.2).

Male: UFW similar to that of *O. paradisea*, black with broad, iridescent yellow–green radial and cubital bands and a short anal band. UHW curved inwards and much reduced in size, tailed, with a small, diamond-shaped tip, a reduced black anal area and an extensive yellow–gold patch.

LFW iridescent yellow–green, more golden than on the upper surface. Black scaling covers the veins, the narrow costal and outer margins, a large cubital patch and the wing apex. LHW similar to the upper surface with a pale grey brush border and no black scaling (1, 2, 4, 6, 10).

Female: Forewing black with white markings including a large cell spot, subapical streaks, submarginal spots and three large discal spots.

Hindwing rounded, black with a broad distal band, extending over the discal region and part of the cell. It is yellow and suffused with black scales on the distal side of a row of large, black subdiscal spots, and white on the discal side (1, 2, 4, 6, 10).

Distribution *Ornithoptera meridionalis* has been found only in the island of New Guinea. Until recently it was thought that the butterfly occurred only in Papua New Guinea. Females of a separate subspecies, *tarunggarensis*, occurring at Nomnangihé, 40 km south-west of Wanggar and highly disjunct with the Papua New Guinea populations, have been shown to be referable to *O. paradisea*, not *O. meridionalis* (3). However, *O. meridionalis* has been discovered recently to be truly present in Irian Jaya, in the region of Kamrau Bay about 200 km west of the Weyland Mountains, in the Weyland Mountains themselves and in the area around Lake Yamur (Jamur) to the east of the Weyland Mountains (3, 7, 9).

However, the name *tarunggarensis* has been most confusingly applied to these populations (7, 9). If, as seems likely, they represent a subspecies distinct from *O. m. meridionalis*, it currently lacks a valid name.

O. meridionalis is quite widely distributed in mainland Papua New Guinea, but very local. Its main area of occurrence is along the southern part of the south-eastern peninsula, but it has also been recorded from single localities in the Southern Highlands and East Sepik Provinces. It has not been reported from any of the islands and is known from only fourteen 10 km squares in the country (8).

Habitat and Ecology *Ornithoptera meridionalis* occurs mainly in lowland rain forest, both primary and mature secondary, usually between 20 and 200 m above sea level. However, it probably extends into hill forest if its foodplant is present. The specimen from the Southern Highlands Province was captured at 800 m (8). In the Kamrau Bay population in Irian Jaya, the butterfly is also a mainly lowland species, though one population was found to be established at about 700 m (9).

There is doubt as to the exact species of foodplants. Some accounts state that the larvae are monophagous on *Aristolochia dielsiana* (known in the literature as *A. schlechteri*), others that *A. pithecurus* (probably synonymous with *A. momandul*) is the main foodplant (5, 6, 11, 12). Larvae in the Kamrau Bay area feed on *A. dielsiana*, a species of *Aristolochia* indistinguishable from *A. schlechteri* (9).

The male butterflies fly poorly and are less often seen than the males of other *Ornithoptera* species; this is evidently correlated with the small size and unusual shape of the hindwing (9). Adult females seem to fly mainly in their home range, although they have also been seen flying along tracks in open secondary forest (8).

Females deposit eggs singly on the undersides of the foodplant leaves. The eggs are large (3 mm diameter) and females carry very few, usually between five and seven (8, 11, 12). Females of the Kamrau Bay populations apparently do not oviposit on foodplants growing on steeply-sloping terrain. In this population, 50–60 per cent of eggs were attacked by parasitic flies (9), probably a species of *Trichogramma* (Hymenoptera: *Chalcidoidea*). The parasite was not observed to attack eggs of *Ornithoptera priamus* on the same foodplant and may be specific to *O. meridionalis* (9). Feeding by mature larvae is characteristic: the central disc of the leaf is eaten from the apex, leaving a crescent-shaped area uneaten. Both larvae and pupae have been described and figured (6, 11, 12).

Natural enemies of the larvae include birds, tree frogs, a species of lizard and invertebrates such as large reduviid bugs and spiders (9).

Threats The Irian Jaya populations of *O. meridionalis* do not appear to be so much at risk as those of Papua New Guinea (8, 9). Indeed, it is probably the inaccessibility of the Irian Jaya localities, and paucity of human settlements, which has delayed their discovery.

However, the Papua New Guinea populations are seriously threatened by habitat destruction and change. These low-lying localities are particularly subject to commercial extraction of timber. Some of the main areas of occurrence are close to Port Moresby and are accessible by road, an unusual circumstance in Papua New Guinea. Logging is increasing in the region of the Brown and Vanapa Rivers, which lies in the centre of the range of *O. meridionalis* in south-eastern Papua New Guinea (5, 8).

Specimens of *O. meridionalis* are much in demand by collectors. The birdwing occurs only in New Guinea and is totally protected in Papua New Guinea. It is not protected in Irian Jaya, probably because the butterfly was so little known in that country until recently. No threats are known to be posed by illegal collecting and trade, nor by legitimate dealing. However, the butterfly is one of the most valuable known; a pair was offered in Britain for £650 in 1980 (13). It can be expected that commercial trade in specimens from Irian Jaya will be developed now that the localities are known there.

Conservation Measures Because of the main threat to *O. meridionalis* in Papua New Guinea, the most important measure to be taken for the butterfly's conservation is the establishment of National Reserve Areas under the Conservation Areas Act. The establishment of five reserves has been suggested, with priority given to the Brown and Vanapa Rivers areas (8). Because this area is easily accessible from

Port Moresby, the capital of Papua New Guinea, conservation could be combined with a tourist attraction by the establishment of a Wildlife Management Area. Enrichment of the habitat should be considered by planting the host of *O. meridionalis*, *Aristolochia dielsiana*, as has been done for the common birdwings by planting *A. tagala* (8).

The other four reserves in Papua New Guinea suggested for *O. meridionalis* are at Lake Kutubu, the Frieda River, Cape Rodney and Mamai Plantation. At Lake Kutubu, located in the Southern Highlands, three uncommon birdwing species (*Ornithoptera goliath*, *O. paradisea* and *O. meridionalis*) could be conserved (8).

The Irian Jaya populations of *O. meridionalis* appear to lie mainly outside the boundaries of existing or proposed reserves. This exception may be the proposed Weyland Mountains Nature Reserve (Cagar Alam), which has been put forward to replace the unsuitable Enarotali Nature Reserve (see review of *O. chimaera*) (14). Although the conservation of *O. meridionalis* in a national reserve in Irian Jaya is desirable, the recent discovery of the species in new areas suggests that survey and exploration are the first priorities.

In the long term, the future of *O. meridionalis* may be linked with its carefully controlled utilisation, both as a tourist showpiece, at least in Papua New Guinea, and as a sustainable resource of specimens (8).

O. meridionalis is a totally protected species in Papua New Guinea, so that any controlled utilisation would require amending legislation. The butterfly is listed under Appendix II of the 1973 Convention on International Trade in Endangered Species of Wild Fauna and Flora (CITES). This status should be retained. Appendix II listing implies that commercial trade is allowed, providing a permit from the country of export is obtained. If enforced, this can provide a method of monitoring the amount of trade.

References

1. D'Abrera, B. (1971). *Butterflies of the Australian Region*. Lansdowne Press, Melbourne. 415 pp.
2. D'Abrera, B. (1975). *Birdwing Butterflies of the World*. Lansdowne Press, Melbourne. 260 pp.
3. Hancock, D.L. (1982). A note on the status of *Ornithoptera meridionalis tarunggarensis* (Joicey & Talbot) (Lepidoptera: Papilionidae). *Australian Entomological Magazine* 93–95.
4. Haugum, J. and Low, A.M. (1978–9). *A Monograph of the Birdwing Butterflies*. Vol. 1 (3): 193–308. Scandinavian Science Press, Klampenborg.
5. Hutton, A.F. (1984). *In litt.*, 1 February.
6. Igarashi, S. (1979). *Papilionidae and Their Early Stages*. Vol. 1, 219 pp., Vol. 2, 102 pp. of plates. Kodansha, Tokyo. (In Japanese).
7. Ohya, T. (1983). *Birdwing Butterflies*. Kodansha, Tokyo. 332 pp., 136 col. plates.
8. Parsons, M.J. (1983). A conservation study of the birdwing butterflies, *Ornithoptera* and *Troides* (Lepidoptera: Papilionidae) in Papua New Guinea. Final report to the Department of Primary Industry, Papua New Guinea. 111 pp.
9. Pasternak, J. (1981). On the rediscovery of *Ornithoptera meridionalis tarunggarensis* Joicey & Talbot in a new locality in Kamrau Bay, south-west Irian Jaya, Indonesia. *Transactions of the Himeji Natural History Association* 1981: 2–14.
10. Smart, P. (1975). *The Illustrated Encyclopedia of the Butterfly World*. Hamlyn, London. 275 pp.
11. Straatman, R. (1967). Additional notes on the biology of *Ornithoptera meridionalis* (Rothschild). *Transactions of the Papua New Guinea Science Society* 8: 36–38.
12. Szent–Ivanny, J.J.H. and Carver, R.A. (1967). Notes on the biology of some Lepidoptera of the Territory of Papua New Guinea with descriptions of the early stages of *Ornithoptera*

meridionalis Rothschild. *Transactions of the Papua New Guinea Science Society* 8: 3–35.

13. Trade catalogue (1980).
14. UNDP/FAO National Parks Development Project (1981). *National Conservation Plan for Indonesia. Vol. VII. Maluku and Irian Jaya.* FAO, Bogor.

Ornithoptera alexandrae (Rothschild, 1907)   ENDANGERED

Subfamily PAPILIONINAE   Tribe TROIDINI

Summary *Ornithoptera alexandrae*, Queen Alexandra's Birdwing, is the world's largest butterfly. It is restricted to primary and advanced secondary lowland rain forest in or near the Popondetta Plain, a small area in the Northern Province of Papua New Guinea. Protected by law since 1966, the species is not often collected, but its habitat is now severely threatened by the expanding oil palm and logging industries. Conservation measures taken and proposed include the establishment of reserves and a wide range of ecological survey and research.

Description *Ornithoptera alexandrae* is the world's largest butterfly, the dark brown females having a wingspan of up to 250 mm (Plate 5.4). The male is smaller (wingspan 170–190 mm) and is light blue, yellow, green and black (3, 4, 9, 12) (Plate 5.3). The average head and body length is about 75 mm; the abdomen of both sexes is bright yellow and the ventral wingbases are bright red.

Male: UFW elongated, black with a long, broad, green radial band, a broad, blue–green anal band fused at both ends with a narrower cubital band to enclose a large, black sex brand. UHW elongated, black with a broad, blue–green submarginal band which is contined to the base along the costal and inner margins, and a broad, blue–green cell streak.

LFW blue–green with more blue towards the anal region; black veins, narrow margin and subapical streaks. LHW yellow becoming bluish towards the anal region, with black veins and narrow margins.

Female: forewing dark brown with relatively small, pale grey submarginal and discal spots which become smaller towards the apex.

Hindwing elongated, with a band of seven pale grey, yellow-powdered, wedge-shaped patches separated by broad bands over the veins. All but the innermost contain brown discal spots, and all are more yellow on the UHW.

Distribution The first specimen, a small, dull, atypical female, was collected in 1906 from the type locality high on the upper reaches of the Mambare River, well outside its present range (9, 13). To date, *Ornithoptera alexandrae* has only been recorded from nine 10 km grid squares on the Popondetta Plain in Northern Province, Papua New Guinea, and is known from only one other locality as a separate, high altitude population not far from the larger lowland population (18). It is reported that the 1951 eruption of Mt Lamington destroyed 250 sq. km of prime habitat, further fragmenting the already patchy distribution produced by agriculture and logging (10).

Habitat and Ecology *Ornithoptera alexandrae* occurs with its larval food-plant, *Aristolochia dielsiana* (formerly known as *Aristolochia schlechteri*), in secondary and primary lowland rain forest up to 400 m altitude on the volcanic ash soils of the Popondetta Plain, and in secondary hill forest on clay soils from 550 m to 800 m altitude in its other locality (8, 16, 17). It is strictly monophagous, although this is due to the oviposition specificity of the female as the larvae can mature equally well (and apparently even better) on the softer-leaved *Aristolochia tagala*, a vine which is common and far more widespread throughout Papua New Guinea (21, 22). The much commoner Papua New Guinea birdwing, *Ornithoptera priamus*, uses the same foodplant as *O. alexandrae*. Whether any competition occurs between the two species is uncertain. The duration of the early stages (from egg to adult emergence) exceeds four months and adults can live up to a further three months in the wild. Adults are subject to little predation but eggs are attacked by ants and heteropterous bugs. The larvae are preyed upon by toads, lizards and birds such as cuckoos, drongos and crow pheasants. Parasitism of larvae by unidentified tachinid flies, and of pupae by parasitic wasps, has been reported (21). Opinions vary as to whether parasitism occurs commonly or rarely (19, 21) and this is a topic which requires much more research. However, it is believed that the aposematic (warning) coloration of the larvae and adults is an indication that they can probably store the toxins that their foodplants are known to contain, using them for their own protection against more general predators (15). Adults are strong fliers but appear to remain in home ranges, ignoring other available habitat. It has been established recently that male butterflies often swarm around a large timber tree, *Intsia bijuga* (Leguminosae, known locally as Kwila), when it is in flower (10). Observations indicate that flying females will not accept males unless they have visited the flowers (10). Experimental confirmation of the behaviour pattern is needed, but the distribution of the tree may account for the absence of the butterfly from certain apparently suitable areas (10), although *I. bijuga* is a common and widely distributed species.

The eggs of *O. alexandrae* are extremely large (4 mm diameter) and it has been calculated that females, if their ovaries are continuously productive, have the potential to lay about 240 eggs during their lifetime (16). They possibly carry only 15–20 (maximum 30) at any one time (11).

Conventional mark-recapture methods cannot be used to estimate numbers of *O. alexandrae* as the species flies high and is too infrequently seen. Larval counts are also low (only one or two may be located during a day's survey) and the leaves of the foodplant vine are often 40 m high in the upper canopy, effectively precluding observation of larvae.

Threats The greatest current danger is the expanding oil palm industry in the Popondetta region, although cocoa and rubber plantations have also been a problem in the past. These have already claimed large tracts of forest known to have been habitat for *Ornithoptera alexandrae* (1). Negotiations to exploit the reserves of wood in the Kumusi Timber Area are also in progress. Localized extinctions are occurring due to the clearing of forest to make food gardens. During the Second World War, Popondetta was an important air base, and at one time contained 26 airstrips (7).

O. alexandrae is greatly prized by collectors and some illegal trade has undoubtedly occurred from time to time (see below). However, illegal collecting is not comparable with loss of habitat as a threat.

Conservation Measures In 1966, the Fauna Protection Ordinance gave *O. alexandrae* and six other birdwings legal protection from collection (5). The law has been stringently enforced on several occasions, resulting in fines for nationals and deportation of expatriates. Surveys by the Division of Wildlife are establishing the presence or absence of *O. alexandrae* in defined areas. A large Wildlife Management Area (WMA), comprising approximately 11 000 ha of grassland and forest, has been established north of Popondetta. Unfortunately, it is not at all clear how effective this WMA is in conserving wildlife in general, and *O. alexandrae* in particular. Reports suggest that the WMA has agriculture within its boundaries. Several thousand cuttings of *A. dielsiana* are being prepared and an area of 4 ha of government owned primary forest at the Lejo Agricultural Station is being planted as a future reserve and study area for *O. alexandrae*. The Wildlife Division has applied for a total of about 40 ha of government land that has been rejected for use as oil palm plantations because of the deeply dissected topography. The aim is to create reserves for *O. alexandrae* on government land, which can be protected by law in perpetuity. A trial planting of *A. dielsiana* cuttings under tall, shady, mature (c. 14 years old) oil palms has been undertaken at the Popondetta Agricultural Training Institute to study the growth of the vines in this artificial habitat and to see whether *O. alexandrae* will eventually utilize them. Provincial wildlife officers regularly hold educational meetings with people in the Northern Province, to explain why the butterfly needs to be conserved. Representations for conservation of the species have been made to the Government of Papua New Guinea by several international bodies, including the IUCN/SSC Butterfly (formerly Lepidoptera) Specialist Group.

There are well-defined plans for future conservation efforts. In particular, negotiations to establish new WMAs are in progress between the Wildlife Division and interested landowners. Proposals for three reserve areas within the Kumusi Timber area have been supported by the landowners and the timber company involved (Fletcher Forests, New Zealand). Implementation of the recent Conservation Areas Act (1978), which gives special protection to "sites and areas having particular biological, topographical, geographical, historic, scientific or social importance", is being considered for certain sites. The Act also provides for the active management of such areas (2). It is hoped that portions of prime *Ornithoptera alexandrae* habitat will be considered for inclusion under this Act.

The discovery of *Aristolochia dielsiana* on Siassi I. (= Umboi I.) in Morobe Province has prompted the suggestion that an establishment of *O. alexandrae* should be made there (19). Although *O. priamus* occurs on Siassi I. (and feeds on *A. dielsiana*), *Troides oblongomaculatus*, a potential competitor, does not. The suggestion of an establishment here is an important contribution to the positive conservation of *O. alexandrae*, but of course, a thorough survey of the island and the occurrence of *A. dielsiana* on it is necessary before an establishment can be planned.

O. alexandrae is the largest butterfly in the world and is aesthetically very attractive. The birdwings have long been held in high esteem by insect collectors and are in great demand worldwide. Species such as *O. alexandrae*, which are not only impressive but restricted in their range and hard to obtain, realise extremely high prices. Within the Division of Wildlife in Papua New Guinea, there is already a marketing agency which supplies insect dealers with the unprotected insects of the country (14). If the long term future of *O. alexandrae* is safeguarded, it could provide an extremely valuable income to the people of Papua New Guinea (6). Eventually the butterfly may become an added attraction to the growing tourist industry (20).

The resource provided by this butterfly could be utilised even now. It has been suggested that some of the 51 ex-pupa papered specimens of *O. alexandrae* (19 males, 32 females), confiscated as having been taken illegally, should be offered for sale, possibly by a 'sealed bid' method (19). These specimens are currently kept in papers at the Entomology Department, DPI, where they have no scientific purpose. Their sale could partially finance conservation programmes for this butterfly.

It may be possible to breed *O. alexandrae* in captivity so that its biology and the reasons for its monophagy can be more closely studied. However, extremely large flight cages are required if the species is to behave normally in captivity, and the cost is prohibitive. Experiments to breed selectively for a culture of *O. alexandrae* which oviposits on *Aristolochia tagala* may prove rewarding (20). Despite its attractions, *O. alexandrae* is poorly known because it is so rare, and further research on its life history, behaviour, natural enemies and population dynamics should be undertaken at the same time that conservation measures are put into effect.

O. alexandrae is currently included, together with all other birdwings, in Appendix II of the Convention on International Trade in Endangered Species (CITES). As the species is Endangered and totally protected in its only country of occurrence, it is more appropriate to include it on Appendix I if practical problems, such as its identification by custom officers, can be resolved (19). This change should not be formally proposed until consideration has been given to the partial financing of a practical conservation programme by the sale of confiscated specimens, mentioned above.

References

1. Anon. (1976). Appraisal of the Popondetta Smallholder Oil Palm Development. Report No. 1160, 25 September. Department of Primary Industry, Papua New Guinea.
2. Conservation Areas Act (1978). No. 52 Independent State of Papua New Guinea. 12 September.
3. D'Abrera, B. (1971). *Butterflies of the Australian Region*. Lansdowne Press, Melbourne. 415 pp
4. D'Abrera, B. (1975). *Birdwing Butterflies of the World*. Lansdowne Press, Melbourne. 260 pp.
5. Fauna Protection Ordinance (1966). No. 19 Independent State of Papua New Guinea.
6. Fenner, T.L. (1975). Proposal for experimental farming of protected birdwing butterflies with particular reference to *Ornithoptera alexandrae*. Unpublished manuscript, Department of Primary Industry. 5 pp.
7. Fenner, T.L. (1983). *In litt.*, 15 March.
8. Haatjens, H.A. (ed.) (1964). General report on the lands of the Buna–Kokoda Area, Territory of Papua and New Guinea. *C.S.I.R.O. Land Resources Series No. 10*, 113 pp.
9. Haugum, J. and Low, A.M. (1978). *A Monograph of the Birdwing Butterflies*. Vol. 1 (1): 1–84. Scandinavian Science Press, Klampenborg.
10. Hutton, A.F. (1982). *In litt.*, 20 June.
11. Hutton, A.F. (1984). *In litt.*, 1 February.
12. Igarashi, S. (1979). *Papilionidae and Their Early Stages*. Vol. 1: 219 pp., Vol. 2: 102 pp of plates. Kodansha, Tokyo. (In Japanese).
13. Meek, A.S. (1913). *A Naturalist in Cannibal Land*. T. Fisher Unwin, London. 238 pp.
14. National Research Council (1983). *Butterfly Farming in Papua New Guinea*. Managing tropical animal resources series. National Academy Press, Washington D.C. 34 pp.
15. Owen, D. (1971). *Tropical Butterflies*. Oxford University Press, Oxford. 214 pp.
16. Parsons, M.J. (1980). A conservation study of *Ornithoptera alexandrae* Rothschild (Lepidoptera: Papilionidae). First report, Wildlife Division, Papua New Guinea. 89 pp.
17. Parsons, M.J. (1980). A conservation study of *Ornithoptera alexandrae* Rothschild (Lepidoptera: Papilionidae). Second report, Wildlife Division, Papua New Guinea. 16 pp.
18. Parsons, M.J. (1980). A conservation study of *Ornithoptera alexandrae* Rothschild (Lepidoptera: Papilionidae). Third report, Wildlife Division, Papua New Guinea. 15 pp.
19. Parsons, M.J. (1983). A conservation study of the birdwing butterflies *Ornithoptera* and *Troides* (Lepidoptera: Papilionidae) in Papua New Guinea. Final report to Department of Primary Industry, Papua New Guinea. 111 pp.
20. Pyle, R.M. and Hughes, S.A. (1978). Conservation and utilisation of the insect resources of Papua New Guinea. Report of a consultancy to the Wildlife Branch, Dept. of Nature Resources, Independent State of Papua New Guinea. 157 pp.
21. Straatman, R. (1971). The life history of *Ornithoptera alexandrae* (Rothschild). *Journal of the Lepidopterists' Society* 25: 58–64.
22. Straatman, R. (1979). Summary of survey on ecology of *Ornithoptera alexandrae* Rothschild. Consultancy report to the Department of Agriculture, Stock and Fisheries, July 1970. 5 pp.

Ornithoptera aesacus (Ney, 1903)   INDETERMINATE

Subfamily PAPILIONINAE   Tribe TROIDINI

Summary *Ornithoptera aesacus* is a beautiful but little-known birdwing butterfly restricted to one small island, Obi, in the Moluccas (Maluku Province), Indonesia. Obi still has virgin forest on steep terrain around the highest peak in the centre of the island, but most of the lowland forest has been logged over, possibly a main cause of the apparent extreme rarity of the species. A national park has been proposed for Obi's central peak, an important conservation measure for this species. Surveys of the precise distribution of *O. aesacus* are needed in order to ascertain its ability to withstand habitat disturbance and its correct conservation status.

Description The male of *Ornithoptera aesacus* is iridescent turquoise and black with some yellow and green on the underside (Plate 5.5). The larger female (wingspan 150 mm) is black with pale markings (1, 2, 5) (Plate 5.6). Despite early doubts about the status of *O. aesacus* as a good species, modern research has firmly established its specific status (4, 5, 6, 7).

Male: UFW black with an iridescent light turquoise radial band and narrower anal and submarginal bands. The iridescent scaling has a violet–blue reflection. UHW iridescent light turquoise–blue with a narrow black margin, a golden tinge distally and up to four small golden–yellow spots.

LFW light blue becoming green towards the apex, with black wing margins, veins and markings. LHW iridescent turquoise–blue basally and in the cell, surrounded by green and yellow scaling. There is a narrow black margin and five large subdiscal spots and a golden–yellow anal patch, large subcostal spot and submarginal marks.

Female: UFW black with prominent white markings including large subapical streaks, large discal and submarginal spots and a large cell spot. UHW with a scalloped outer margin and a large pale grey distal band with grey suffusion distally and small black discal spots.

Despite earlier doubts about the status of *O. aesacus* as a good species, modern research has established its specific status without serious doubt (4, 5, 6, 7).

Distribution *Ornithoptera aesacus* is known to occur only in the small island of Obi (or Ombira) in the Moluccas (Maluku Province), Indonesia. Obi lies due south of Halmahera and due east of Sulawesi. The precise distribution of the butterfly within Obi is not known.

Habitat and Ecology Virtually nothing is known about the early stages and life history of *Ornithoptera aesacus*, despite the fact that the original series of three specimens (one male, two females) was bred (by the collector John Waterstredt) from larvae collected in May 1902 (5). The foodplant is stated to be "*Aristolochia* species" (1), but it is not clear whether this is based on recent observations or on an assumption. Nothing has been published on the habitat and behaviour of *O. aesacus*; it remains one of the least known birdwings.

Obi, the southernmost of the northern group of islands in the Moluccas is a mountainous island with broad coastal plains in the east and west, but steep coasts in the north and south. The native vegetation is seasonal monsoon forest and the island has remained forested with secondary formations, despite logging operations that have continued for many years (8).

292

Threats *Ornithoptera aesacus* is listed as Indeterminate mainly because of the undoubted extensive land-use changes that have taken place in Obi. Although the island is said to be still forested with secondary or disturbed growth (8) there are insufficient data to be confident that *O. aesacus*, can adapt to these conditions. The extreme rarity of specimens mitigates against this. Apart from the primary threat of habitat destruction, there appears to be some more or less clandestine trade in this species (5), but this is unlikely to pose a significant threat.

Conservation Measures No practical conservation measures for *Ornithoptera aesacus* have yet been attempted. The first priority is an assessment of the distribution and precise conservation status of the species on Obi, with particular attention to its ability to survive forest disturbance. At present there are no protected areas on Obi, but there is a proposal to gazette a new nature reserve of 45 000 ha centred on Obi's main peak (8). The proposed reserve embraces an altitudinal range of 500–1616 m; the mainly steep terrain is unsuitable for logging and would require little protection (8). The reserve would protect the endemic dove *Ptilinopus granulifrons* as well as *O. aesacus* and other fauna and flora unique to the island (8).

The species should be retained, until its status can be reviewed, under Appendix II of the 1973 Convention on International Trade in Endangered Species of Wild Fauna and Flora (CITES). Appendix II listing implies that commercial trade is allowed providing a permit from the country of export is obtained. This can provide a method of monitoring trade levels.

References

1. D'Abrera, B. (1971). *Butterflies of the Australian Region*. Lansdowne Press, Melbourne. 415 pp.
2. D'Abrera, B. (1975). *Birdwing Butterflies of the World*. Lansdowne Press, Melbourne. 260 pp.
3. D'Abrera, B. (1983). *In litt.*, 12 March.
4. Hancock, D.L. (1983). Classification of the Papilionidae (Lepidoptera): a phylogenetic approach. *Smithersia* 2: 1–48.
5. Haugum, J. and Low, A.M. (1978–9). *A Monograph of the Birdwing Butterflies*. Vol. 1 (1–3): 308 pp. Scandinavian Science Press, Klampenborg.
6. Munroe, E. (1961). The classification of the Papilionidae (Lepidoptera). *Canadian Entomologist* Supplement 17: 1–51.
7. Smart, P. (1975). *The Illustrated Encyclopedia of the Butterfly World*. Hamlyn, London. 275 pp.
8. UNDP/FAO National Parks Development Project (1981). *A National Conservation Plan for Indonesia. Vol. 7: Maluku and Irian Jaya*. FAO, Bogor.

***Ornithoptera croesus* Wallace, 1859**   VULNERABLE

Subfamily PAPILIONINAE   Tribe TROIDINI

Summary *Ornithoptera croesus* is a highly attractive birdwing butterfly with a unique golden–yellow coloration in the male. It is restricted, as two well-known subspecies and a little-known third, to a few islands in the Moluccas. It is at risk primarily because of deforestation but possibly also as a result of insecticidal spraying on a large scale. Conservation of *O. croesus* is inadequate and a thorough assessment of its status followed by protection of suitable habitat is needed.

Description *Ornithoptera croesus* has a wingspan of 130–150 mm in the male and 160–190 mm in the female (4). The male is dark brown, orange, and golden with a green and black lower surface. The female of the nominate subspecies is brown with white markings (1, 2, 3, 4, 6). The subspecies *O. c. lydius*, commonly seen in the butterfly trade, differs widely from the nominate subspecies described below and is illustrated in Plates 6.1 and 6.2. The female is unique among the birdwings in being a mimic of unpalatable Danainae species. A third subspecies *O. c. sananaensis* is only known from a single female and is stated to be intermediate between the first two.

Male: UFW ground colour very dark brown with a broad iridescent orange radial band and short anal streak (1, 2, 4, 8). UHW orange with a narrow black margin and a golden–yellow subcostal patch, discal and submarginal spots.

LFW black with iridescent green submarginal and discal spots, radial band and a patch in the cell. LHW yellow–green with black veins, subdiscal spots and a narrow margin, a yellow anal area and golden areas as on the upper surface (1, 2, 3, 4, 6).

Female: UFW dark brown ground colour with white markings including a cell spot, marginal fringe spots, submarginal and discal spots. UHW darker than forewing with yellow–brown distal patches and black subdiscal spots. LFW/LHW differs only in having paler markings (1, 2, 3, 4, 6).

Distribution The nominate subspecies, *Ornithoptera croesus croesus*, has been found only on the island of Bacan (Bachan, Batjan) in the Moluccas (Maluku). It is said to be very localized but less rare in its chosen sites than has been previously thought (3). *O. c. lydius* occurs in the neighbouring islands of Halmahera (also known as Jailolo Gilolo or Djailolo), Ternate, Tidore, and possibly Morotai (Morty) (3). Halmahera, by far the largest of these islands, is the most important locality, and is the source of most of the recent captures of this birdwing (9).

The only known specimen of *O. c. sananaensis* was taken on Sanana, the most southerly of the Sula Islands (8). Sanana lies about 240 km south-west of Bachan and this distribution requires confirmation.

Habitat and Ecology *Ornithoptera croesus* is a lowland butterfly in Bachan, where it occurs in swamps and other wet places (4). Its habitat has been regarded as difficult of access since the time of Alfred Russel Wallace, the discoverer of the butterfly (11). The larva and pupa of *O. c. croesus* have been briefly described but the larvae of *O. c. lydius* are stated to be unknown (3), a surprising fact since many of the recently-collected specimens have been reared in captivity (9). Excellent figures of the larvae and pupae have been published, but of which subspecies is not stated (4). The foodplants are not comprehensively known but include *Aristolochia gaudichaudii* (4, 8).

Threats In 1982 about 90 per cent of all forest in the northern and central Moluccas was under concession to large-scale commercial logging operations (7). In 1980 the entire production of 1.4 million cu. m of logs were exported and government control is reported to be slight (7). As a result, deforestation is being carried out in an irresponsible and unsustainable manner often on steep, easily eroded, slopes and with no reforestation—all in defiance of government regulations (7). Indonesia is one of the world's biggest timber producers but such exploitation without regard to conservation can only be a very short-term development strategy. *Ornithoptera croesus* is only one of many Moluccan endemics that are threatened by the devastating rate and nature of deforestation in the region (8).

Both *O. c. croesus* and *O. c. lydius* live in highly productive lowland forest, the most valuable timber concessions and the first areas to be deforested. There can be little doubt that *O. croesus* is declining in numbers and seriously at risk from further deforestation throughout its range. A further threat, the impact of which has not been assessed, is the reported use of large-scale insecticidal spraying in the swampy lowlands of Bachan (3), presumably as a mosquito control measure.

O. croesus has always been a much sought-after species (4, 9). From about 1979, large numbers of *O. c. lydius* appeared in trade. Prices were originally U.S. $90 a pair or more but by mid-1982 had fallen to $24 a pair in the U.S.A., perhaps indicating that the market had been satisfied (9). There have been no reports that *O. c. lydius* has been threatened by this increase in trade, but consideration might be given to the possibility of setting up a ranching programme on Halmahera.

Conservation Measures The Halmahera group of islands, including Morotai, Bacan and Obi, has by far the greatest number of endemic species, the widest range of land form types and the most varied climate in the whole of the Moluccas (10). Sadly, there is not a single reserve established or even approved on any of these islands. Seven reserves have been proposed in the National Conservation Plan for Indonesia, including one on Obi, one on Morotai, one on Bacan and four on Halmahera (10). There is no information on the likelihood of *Ornithoptera croesus* habitat being found within these proposed reserves and a survey is an essential preliminary step towards more specific conservation measures.

The species should be retained, until its status can be reviewed, under Appendix II of the 1973 Convention on International Trade in Endangered Species of Wild Fauna and Flora (CITES). Appendix II listing implies that commercial trade is allowed providing a permit from the country of export is obtained. This can provide a method of monitoring trade levels.

References

1. D'Abrera, B. (1971). *Butterflies of the Australian Region.* Lansdowne Press, Melbourne. 415 pp.
2. D'Abrera, B. (1975). *Birdwing Butterflies of the World.* Lansdowne Press, Melbourne. 260 pp.
3. Haugum, J. and Low, A.M. (1978–9). *A Monograph of the Birdwing Butterflies.* Vol. 1 (1–3): 308 pp. Scandinavian Science Press, Klampenborg.
4. Igarashi, S. (1979). *Papilionidae and Their Early Stages.* Vol. 1: 219 pp., Vol. 2: 102 pp. of plates. Kodansha, Tokyo. (In Japanese).
5. Munroe, E. (1961). The classification of the Papilionidae (Lepidoptera). *Canadian Entomologist* Supplement 17: 1–51.
6. Smart, P. (1975). *The Illustrated Encyclopedia of the Butterfly World.* Hamlyn, London. 275 pp.

7. Smiet, F. (1982). Threats to the Spice Islands. *Oryx* 16(4): 323–8.

8. Tsukada, E. and Nishiyama, Y. (1982). *Butterflies of the South East Asian Islands. Vol. 1 Papilionidae*. (transl. K. Morishita). Plapac Co. Ltd., Tokyo. 457 pp.

9. Various trade catalogues (1979–82).

10. UNDP/FAO National Park Development Project (1981). *National Conservation Plan for Indonesia. Vol. 7. Maluku and Irian Jaya*. FAO, Bogor.

11. Wallace, A.R. (1869). *The Malay Archipelago*. MacMillan, London. (Dover reprint edition, 1962, pp. 257–8).

Papilio (Pterourus) homerus Fabricius, 1793   ENDANGERED

Subfamily PAPILIONINAE   Tribe PAPILIONINI

Summary *Papilio homerus*, the Homerus Swallowtail, is a superb butterfly, rivalling in size some of the birdwings (Troidini) of South East Asia. This and the fact that it is uncommon and restricted to the island of Jamaica have made it particularly prized by collectors. However, the main threat seems to be destruction of habitat. The once continuous population is already divided into two isolated pockets and conservation measures are urgently needed.

Description *Papilio homerus* is one of the largest species in the genus (comparable with *Papilio antimachus*) and the largest American swallowtail, with a forewing length of about 75 mm (Plate 6.3). It is easily recognizable by the large size and generally slow flight (8). The sexes are alike (1–6).

UFW black or very dark brown with a broad yellow discal band extending across the wing, including a yellow bar across the cell, two subapical spots and two or three submarginal spots.

UHW broad, elongated, with a relatively long spatulate tail. The wing is black or dark brown with a broad yellow discal band, powdery blue postdiscal spots and brick-red submarginal lunules.

LFW/LHW similar to the upper surface with a dark brown ground colour. The hindwing has a much narrower yellow–brown discal band which is dusted with blue scales, and there are dark marks between the red and the blue spots (1–6).

The eggs are spherical and light green in colour (8). The first three larval instars are essentially black and white; the last two instars are dark brown with extensive saddles of green and a pair of conspicuous lateral eyespots on the thorax (8). The pupa is shades of brown with eight small white, dorsal spots; the shape is blunt with only slight dorso-ventral flattening (8).

Distribution *Papilio homerus* is confined to the Caribbean island of Jamaica, where it has been recorded from 7 of the 13 parishes (1). It is now restricted to two strongholds, the eastern population in St Thomas and Portland (where the Blue Mountains meet the John Crow range) and the western population in the 'Cockpit Country' of Trelawny Parish (1, 4, 6). Each locality is only a few square kilometres of forest (8). A third population in central Jamaica probably became extinct around the turn of the century (8). There is no record on any other Caribbean island, despite suggestions to the contrary (2). The nearest relatives are apparently *P. garamus* and *P. abderus* in Mexico.

Habitat and Ecology *P. homerus* is restricted to virgin forest on mountain slopes and in gullies at fairly low elevations (150–600 m, occasionally higher) (8). Flight is slow but powerful, and occurs throughout the day (0900–1800 hrs). The adults bask on high trees and bushes, visiting flowers daily, but only for short periods of time. Preferred nectar sources include species of *Blechum*, *Bidens*, *Asclepias*, *Lantana*, a malvaceous plant, and possibly a local *Ipomoea* (Morning Glory) (7). The larval foodplants are two endemic *Hernandia* species *H. catalpaefolia* (eastern population) and *H. troyiana* (western population) (7). Oviposition has also been observed on an endemic species of the Camphorwood tree (*Ocotea* sp.) in both locations (7). *Thespesia populnea* (Seaside Mahoe) has been recorded as a foodplant

(4) but does not occur in the known distribution of the insect (7). It may have been confused with *Hernandia* which is locally known as Water Mahoe (7). Seasonality is not marked, but the adults fly during February to April, September and October (4).

Little information is available on populations. *P. homerus* is said to be quite common in a few favoured localities, but the total population is certainly small. The population range is rapidly shrinking and the two populations are probably genetically isolated. Predation of larval instars by birds is high but the adult populations seem to be free of the marked fluctuations in numbers characteristic of some other Caribbean swallowtails.

Threats Records for the past 20 years together with former and present collecting or confirmed visual records clearly establish that both populations are declining (8). The main reasons for this are the destruction and alteration of habitat, first for timber extraction and then for coffee gardens and pine plantations. The rate of habitat destruction in the east has slowed since 1952, but that population is still threatened and on the decline (7, 8). The western population, although living in more mountainous country, is also suffering a reduction in range, and thorough surveys are needed. Collecting is difficult in such mountainous country, but commercial collecting by expatriates may be a minor threat since prices are high. The extent of collecting by Jamaicans is unknown.

Conservation Measures There have been no conservation measures specifically for this species. A detailed survey of the distribution of the species and its ecological requirements is needed, in order that consideration may be given to conservation. Some form of habitat protection seems essential, probably in the form of a patrolled nature reserve or national park (8). The Natural Resources Conservation Department of Jamaica should take some responsiblity for the protection of this beautiful butterfly, and should take steps to ensure the permanent protection of its breeding grounds. *P. homerus* is of great aesthetic value and a rational farming programme would help conservation efforts, and be an appealing commercial venture. The species has been reared with some difficulty in captivity (7) and further research is needed. Commercial farming would lessen pressure on wild populations, and funds should be made available to provide suitable facilities (7, 8). Collecting of wild specimens should be limited to those used for scientific purposes; permits should be issued only by the Natural History Division of the Institute of Jamaica (8).

References

1. Brown, F.M. and Heinemann, B. (1972). *Jamaica and its Butterflies*, E.W. Classey Ltd., Faringdon, U.K. 478 pp.
2. D'Abrera, B. (1981). *Butterflies of the Neotropical Region Part. 1, Papilionidae and Pieridae*. Landsdowne Editions, Melbourne. xvi + 172 pp.
3. Munroe, E. (1961). The classification of the Papilionidae (Lepidoptera). *Canadian Entomologist* Supplement 17, 51 pp.
4. Riley, N.D. (1975). *A Field Guide to the Butterflies of the West Indies*. Collins, London.
5. Rothschild, W. and Jordan, K. (1906). A revision of the American Papilios. *Novitates Zoologicae* 13: 411–752.
6. Smart, P. (1975). *The Illustrated Encyclopedia of the Butterfly World*. Hamlyn, London. 275 pp.
7. Turner, T.W. (1982). *In litt.*, 12 August.
8. Turner, T.W. (1983). The status of the Papilionidae, Lepidoptera of Jamaica with evidence to support the need for conservation of *Papilio (Pterourus) homerus* Fabricius and *Eurytides marcellinus* Doubleday. Unpublished report, 14 pp.

Papilio (Heraclides) esperanza Beutelspacher, 1975   VULNERABLE

Subfamily PAPILIONINAE   Tribe PAPILIONINI

Summary Little information is available on this extremely rare and recently described species from Mexico. The breeding locations are kept secret as a protection from commercial collectors, but a conservation plan is needed to ensure long-term protection and survival of the species.

Description *Papilio esperanza* is an attractive, medium-sized black, yellow and orange swallowtail (Plate 6.4). The female, known only from one specimen, has a similar pattern to the male, but is slightly larger (7). The male forewing length is 51–52 mm (1) whereas in the female it is 65 mm (7). The affinities of the species are in dispute. It was originally placed in the *glaucus* group (1, 9), but was transferred to the *thoas* group in the recent revision by Hancock (3), who found the male genitalia to be very similar to those of *P. androgeus* (4). However, a number of authorities familiar with the insect in the wild consider the nearest relative of *P. esperanza* to be *P. palamedes*, in the *troilus* species group, thus placing it in the subgenus *Pterourus* (6, 8).

UFW black with a broad yellow median band and a row of small, yellow submarginal spots.

UHW with a slightly scalloped outer margin with yellow fringe spots and a long tail. The broad UFW median band extends across the UHW, tapering slightly to the inner margin. There is an orange anal lunule, a further series of narrow, yellow submarginal lunules, and postdiscal groups of blue scales.

LFW differs from the upper surface in having broad yellow lines in the cell, an orange submarginal region and orange scaling over the wing apex.

LHW predominantly orange with yellow-bordered black veins, a concave black subbasal-abdominal line, and an irregular black postdiscal line with silvery-white scales over the middle of it (1, 7, 2).

Distribution *Papilio esperanza* is only known from 1700 m above sea level at La Esperanza in the Sierra de Juárez, Oaxaca State, Mexico (1, 7).

Habitat and Ecology *Papilio esperanza* flies in montane cloud forest with an annual rainfall of over 4000 mm (6). The dominant tree species in the forest is *Engelhardtia (Oreomunnea) mexicana* (Juglandaceae), a relict species of the Cenozoic era (10). The adult butterflies are only active on sunny days from 11.00–15.00 hours (6) and take nectar from the purple, aromatic flowers of *Eupatorium sordidum* var. *atrorubens*, a herb that grows along streams and gullies (5). Two broods are produced each year, one in March and the other in August (6, 7). Adults from the two broods show slight differences in the depth of colour and width of the yellow wing-bands (7). The foodplant of the caterpillars is unknown. Associated species include *Eurytides calliste*, *Papilio abderus*, *Anetia thirza*, *Paramacera chinanteca*, *Piscina zelys* and *Polygonia g-argenteum* (6). In nine years only 50 specimens have been collected, and the total population is believed to be declining (6).

Threats The distribution of the species is very limited. The precise breeding locations are kept secret in order to protect it from commercial collectors (6, 11). Threats to the habitat are unknown, but the cloud forest is restricted in extent and vulnerable to human interference (6).

Conservation Measures The Sociedad Mexicana de Lepidopterología has kept vigilant watch over the type locality of this exciting new discovery since 1975, preventing any commercial collecting (6). The Mexican Government plans to establish a reserve in the Sierra de Juárez, Oaxaca, which will include the habitat of *P. esperanza* (6). National and international conservation bodies are encouraged to assist in this worthy proposal

References

1. Beutelspacher, C.R. (1975). Una especie nueva de *Papilio* L. (Papilionidae) *Revista Sociedad Mexicana de Lepidopterología* 1(1): 3–6.
2. D'Abrera, B. (1981). *Butterflies of the Neotropical Region. Part I. Papilionidae and Pieridae.* Lansdowne Editions, Melbourne. xvi + 172 pp.
3. Hancock, D.L. (1983). Classification of the Papilionidae (Lepidoptera): a phylogenetic approach. *Smithersia* 2: 1–48.
4. Hancock, D.L. (1984). *In litt.*, 15 February.
5. Maza E., J. de la (1984). *In litt.*, 23 March.
6. Maza, E., J. de la (1983). *In litt.*, 4 October.
7. Maza, E., J. de la, and Díaz Francés, A. (1979). Notas y descripciones sobre la familia Papilionidae en México. *Revista de la Sociedad Mexicana de Lepidopterología* 4(2): 51–56.
8. Miller, L.D. (1984). *In litt.*, 6 January
9. Munroe, E. (1961). The classification of the Papilionidae (Lepidoptera). *Canadian Entomologist* Supplement 17: 1–51.
10. Rzedowski and Palacios (1977). El bosque de *Engelhardtia* (*Oreomunnea*) *mexicana*, en la región de la Chinantla (Oaxaca, México), una reliquia del Cenozóico. *Boletín de la Sociedad Botánica de México* 36: 93–119.
11. Tyler, H.A. (1983). *In litt.*, 13 March.

Papilio (Heraclides) aristodemus ponceanus Schaus, 1911   ENDANGERED

Subfamily PAPILIONINAE   Tribe PAPILIONINI

Summary *Papilio aristodemus ponceanus*, usually known as Schaus' Swallowtail, was formerly locally common in Dade and Monroe Counties, Florida, U.S.A., but is now restricted to North Key Largo, Elliott Key and several smaller keys between them. The Key Largo population is severely threatened by development and insecticide spraying, despite being federally protected in the U.S.A. since 1976. The Elliott Key population is protected by the Biscayne National Park, but suffers dangerously wide fluctuations in numbers.

Description An attractive medium-sized, tailed swallowtail with a wingspan of 86–95 mm (Plate 6.5). The sexes are similar in general appearance, the female slightly larger than the male (10).

UFW/UHW with brown ground colour, a bold, median yellow band, yellow submarginal spots, long spatulate tails and a red and blue anal spot (16).

LFW/LHW duller with a wide orange–brown band on the LHW followed by a distinctive postdiscal row of blue lunules (16).

The subspecies *ponceanus* is quite different from the nominate subspecies, being distinguished by the reduction in width of the dorsal oblique yellow band and more extensive yellow beneath. The cryptic larvae are brown mottled with tan, white and yellow, resembling bird droppings.

Distribution This subspecies is only recorded from Florida, U.S.A. It was formerly locally common in areas of Dade County (including Miami) and Monroe County, including Elliott, Sands, Largo, Old Rhodes, Totten, Porgy, Adams, Upper and Lower Matecumbe and possibly Lignumvitae Keys (23). It has been destroyed in most of its range and is now restricted to localized colonies on North Key Largo in the south, Elliott Key to the north, and probably Old Rhodes, Totten, Adams, and Porgy Keys in between (21). Other subspecies are found on Cuba and the Cayman Islands (*P. a. temenes*), Hispaniola (*P. a. aristodemus*), the Bahamas (*P. a. bjorndalae*) and possibly Puerto Rico (7, 25)).

The first colony of *P. a. ponceanus* was found in the Brickell Hammock of Miami (19), but was apparently destroyed by the city's growth (16). It was subsequently rediscovered in the Keys, but after a hurricane hit Lower Matecumbe in September 1935 *P. a. ponceanus* was feared extinct (11). Surveys after World War II showed that this was not so (13, 14, 15) but the most recent report of distribution only lists breeding populations on North Key Largo and the larger keys of the Biscayne National Park, such as Elliott Key (10, 21).

Habitat and Ecology Only tropical hardwood hammocks (patches of forest) containing the host plant Torchwood (*Amyris elemifera*) will support populations of Schaus' Swallowtail, although another rutaceous plant, the Wild Lime (*Zanthoxylum fagara*) has been observed as an oviposition site (19, 21). Torchwood and Wild Lime are pioneering shrubs or small trees in whose shade sprout the hardwood seedlings which eventually form the hammock. They are therefore abundant at the edge of hammocks, but scarcer within the understorey of the mature trees (18). Schaus' Swallowtail probably continually colonizes regrowth areas partially destroyed by storm or fire (18). Females oviposit single green eggs on the underside of young leaves at the tips of branches (18). Caterpillars hatch in four to seven days and moult four times at intervals of approximately 12 days. The final instar fastens itself vertically to a twig with silk and moults into a rusty brown or grey pupa (5). Emergence is normally slightly less than one year later, but some pupae in captivity have remained in diapause for two years (5, 18). Annual population fluctuations suggest that this behaviour occurs in the wild and may be an adaptation to avoid unfavourable conditions while making best use of good rainfall years (21). Reproduction is correlated with the beginning of the rainy season (April to June) which, perhaps with light intensity or day length, seems to trigger a synchronous emergence of adults. There is some evidence to suggest a partial second brood in some years (7), for example, specimens were discovered during September in 1969 (2). Adult life span does not seem to exceed one month, but during that time the adults are quite capable of flying across open water to adjacent islands (5). Adults take nectar from blossoms of Guava (*Psidium guajava*) and Wild Tamarind (*Lysiloma latisiliqua*) in the hammocks, or Cheese Scrub (*Morinda roioc*) on their edge (18).

The Schaus' Swallowtail's normal population size seems to be small at all stages, although numbers may follow a cyclical pattern (17). Large numbers of adults (up to 100 per day) were recorded in the keys of the Biscayne National Park in 1972 (3, 4, 6, 9) but populations were small during surveys there from 1973 to 1981 (6, 17). However, a survey in May 1982 on Elliott Key recorded about 15 individuals, a cause for guarded optimism (8). Possible reasons for the population decline are given below. Recorded numbers on Key Largo have always been low (9), or perhaps locally common (20) and most authorities consider that population to be doomed to extinction.

Threats The Schaus' Swallowtail is a rare example of an essentially tropical butterfly resident in a peripheral habitat in the only suitable area of the U.S.A. However, this is only one example of a number of less conspicuous tropical invertebrate species threatened in Florida by environmental degradation (12). From the 1940s until today the range of Schaus' Swallowtail has been progressively and irrevocably eroded by the destruction of hardwood hammocks by private develop-ment and the leisure industry. Schaus' Swallowtail is no longer present on Upper or Lower Matecumbe (21), and it is nearly lost from North Key Largo. In the latter locality prime habitat has been subjected to fires, development, and aerial spraying against mosquitoes (6). Despite its protection under federal law (23, 26, 27), the extinction of Schaus' Swallowtail on North Key Largo seems inevitable. Incidental injury to, or destruction of, deposited eggs, larvae or pupae of Schaus' Swallowtail was not an offence when the species was classified as Threatened(23), but the recent reclassification to Endangered status (26, 27) provides stricter protection (see below). Such incidental injuries are inevitable on building sites and the law has so far done little to protect the species' range. The extent of threat to Elliott Key, and to other smaller keys to the north of Key Largo, is presumably lessened by their inclusion in the Biscayne National Park. However, restriction to such a small range would inevitably increase the threat of extinction from natural disasters such as hurricanes, frost or disease (22). Natural disasters such as the 1935 hurricane, which struck only four months after the species was rediscovered on Lower Matecumbe Key, have threatened the swallowtail populations in the past (6). Successive droughts since the early 1970s, combined with hard winter conditions in 1977/78, have also been detrimental (21). Throughout its chequered history Schaus' Swallow-tail has suffered at the hands of collectors (16). Such a rare species fetches high prices (16).

Conservation Measures On 17 April 1975 the U.S. Fish and Wildlife Service published a proposed ruling of Threatened status for Schaus' Swallowtail (22). It was recognised that the range of the butterfly was greatly depleted, but its status was limited to Threatened because the current protection of the population on Elliott and other keys in the Biscayne National Park was considered substantial (22). In the Final Rulemaking of 28 April 1976, the eggs and immature stages were protected from collectors, although not from incidental or inadvertent damage, and collection of adults was permitted outside protected areas (23). However, this legislation was reinforced by Florida state law, which banned collection of adults or immature stages of any federally listed species, except by special permit (1, 26). In 1978 an authority considered the Biscayne National Park large enough to maintain Schaus' Swallowtail, unless natural catastrophe critically reduced the population (7). The staff of the National Park are aware of the butterfly, and co-operative in its conservation (7). Nevertheless, since the Final Rulemaking in 1976, the populations there have declined inexplicably. Recently, the Fish and Wildlife Service proposed a new Crocodile Lake National Wildlife Refuge on Key Largo (25). The report for the proposal notes the presence of Schaus' Swallowtail on North Key Largo, but it is not clear whether populations have been located within the proposed Refuge (25). Schaus' Swallowtail is classified as Endangered in the most recent assessment by authorities in Florida (12, 21) and a recovery plan by the Florida Game and Fresh Water Fish Commission recommended that Schaus' Swallowtail be federally listed as Endangered (25). In a notice of proposed reclassification in 1983 the status of Schaus' Swallowtail was to be raised from Threatened to Endangered, giving it full protection (26). This proposal was accepted in 1984 (27) and it is to be hoped that the other recommendations of the recovery plan will be adopted as expeditiously as possible.

The recovery plan for Schaus' Swallowtail gives details of the objectives which must be achieved to protect the butterfly from extinction (25). The priorities include protection of extant colonies and re-establishment of colonies within the species' historic range. The Biscayne National Park and Crocodile Lake National Wildlife Refuge will partially fulfil these aims, but should be expanded to include various important hammocks (25). Increased public awareness could result in voluntary conservation by informed developers and landowners in North Key Largo. Local authorities should inform planners of the butterfly's vulnerability. Scientific research should be encouraged, to facilitate large-scale artificial rearing and re-introductions (25). Suitable sites include Lignumvitae Key, a preserve of the Nature Conservancy (8), or the proposed Crocodile Lake Refuge on Key Largo (24). There is a great need for detailed scientific study of the population dynamics of Schaus' Swallowtail. Although habitat destruction and excessive collecting will harm the species, very little is known of the effects of parasites on young stages, or of fire and hurricane on maintenance of suitable habitat. Although only given subspecific rank, Schaus' Swallowtail is widely separated spatially from its conspecifics and natural reintroduction from conspecific stock is an unlikely event.

References

1. Baggett, H.D. (1982). Schaus' Swallowtail. In Franz, R. (Ed.), *The Rare and Endangered Biota of Florida, Vol. 6, Invertebrates*. University Presses of Florida, Gainesville. Pp. 73–74.
2. Baggett, H.D. (1982). Statement from 39–27.02 of the Rules of the Florida Game and Fresh Water Fish Commission. *In litt.*, 24 October.
3. Brown, C.H. (1976). A colony of *Papilio aristodemus ponceanus* (Lepidoptera: Papilionidae) in the upper Florida Keys. *Journal of the Georgia Entomological Society* 11: 117–118.

4. Brown, L.N. (1973). Populations of a new swallowtail butterfly found in the Florida Keys. *Florida Naturalist* April 1973: 25.
5. Brown, L.N. (1973). Populations of *Papilio andraemon bonhotei* Sharpe and *Papilio aristodemus ponceanus* Schaus (Papilionidae) in Biscayne National Monument, Florida. *Journal of the Lepidopterists' Society* 27: 136–140.
6. Brown, L.N. (1974). Haven for rare butterflies. *National Parks and Conservation Magazine* July 1974: 10–13.
7. Covell, C.V., Jr. (1976). The Schaus Swallowtail: a threatened subspecies? *Insect World Digest* 3: 21–26.
8. Covell, C.V., Jr. (1978). Project Ponceanus and the status of the Schaus Swallowtail (*Papilio aristodemus ponceanus*) in the Florida Keys. *Atala* 5 (1977): 4–6.
9. Covell, C.V., Jr. (1982). *In litt.*, 18 May.
10. Covell, C.V., Jr. and Rawson, G.W. (1973). Project Ponceanus: a report on first efforts to survey and preserve the Schaus Swallowtail (Papilionidae) in southern Florida. *Journal of the Lepidopterists' Society* 27: 206–210.
11. D'Abrera, B. (1981). *Butterflies of the Neotropical Region. Part I. Papilionidae and Pieridae*. Lansdowne Editions, Melbourne. xvi + 172 pp.
12. Franz, R. (Ed.), (1982). *The Rare and Endangered Biota of Florida, Vol. 6, Invertebrates*. University Presses of Florida, Gainesville. 131 pp.
13. Grimshawe, F.M. (1940). Place of sorrow: the world's rarest butterfly and Matecumbe Key. *Nature Magazine* 33: 565–567, 611.
14. Henderson, W.F. (1945). *Papilio aristodemus ponceana* Schaus (Lepidoptera: Papilionidae) *Entomological News* 56: 29–32.
15. Henderson, W.F. (1945). Additional notes on *Papilio aristodemus ponceana* Schaus (Lepidoptera: Papilionidae). *Entomological News* 56: 187–188.
16. Henderson, W.F. (1946). *Papilio aristodemus ponceana* Schaus (Lepidoptera: Papilionidae). *Entomological News* 57: 100–101.
17. Klots, A.B. (1951). *Field Guide to the Butterflies*. Houghton Mifflin, Boston. 349 pp.
18. Loftus, W.F. and Kushlan, J.A. (1982). The status of Schaus' Swallowtail and the Bahama Swallowtail butterflies in Biscayne National Park. Report M-649, National Park Service, Everglades N.P., Homestead, Florida.
19. Rutkowski, F. (1971). Observations on *Papilio aristodemus ponceanus* (Papilionidae). *Journal of the Lepidopterists' Society* 25: 126–136.
20. Schaus, W. (1911). A new *Papilio* from Florida, and one from Mexico (Lepid.) *Entomological News* 22: 438–439.
21. Spencer Smith, D. (1983). *In litt.*, 25 February.
22. U.S.D.I. Fish and Wildlife Service (1975). Proposed threatened status for two species of butterflies. *Federal Register* 40(78): 17757.
23. U.S.D.I. Fish and Wildlife Service (1976). Determination that two species of butterflies are Threatened species and two species of mammals are Endangered species. *Federal Register* 41(83): 17736–17740.
24. U.S.D.I. Fish and Wildlife Service (1980). Ascertainment report: Crocodile Lake National Wildlife Refuge, Monroe County, Florida. 25 pp.
25. U.S.D.I. Fish and Wildlife Service. (1982). Schaus' Swallowtail butterfly recovery plan. U.S.D.I. Fish and Wildlife Service, Atlanta, Georgia. 57pp.
26. U.S.D.I. Fish and Wildlife Service (1983). Proposed delisting of Bahama Swallowtail butterfly and reclassification of Schaus Swallowtail butterfly from Threatened to Endangered. *Federal Register* 48(168): 39096–7.
27. U.S.D.I. Fish and Wildlife Service (1983). Endangered and Threatened wildlife and plants; final rule to deregulate the Bahama Swallowtail Butterfly and to reclassify the Schaus Swallowtail Butterfly from Threatened to Endangered. *Federal Register* 49(171): 34501–4.

This review has been adapted from an entry in *The IUCN Invertebrate Red Data Book*, whose authors and contributors are gratefully acknowledged.

Papilio (Heraclides) caiguanabus Poey, 1851   INDETERMINATE

Subfamily PAPILIONINAE   Tribe PAPILIONINI

Summary *Papilio caiguanabus*, sometimes known as Poey's Black Swallowtail, is endemic to Cuba, the largest island in the Caribbean. It is rare throughout its range but there is insufficient information to assess the extent of threats to the survival of the species.

Description *Papilio caiguanabus* is a tailed butterfly whose sexes are similar but with slight colour differences (1, 6, 7). The forewing length is 48–50 mm (6).

UFW with a submarginal row of equal-sized yellow spots, (pale yellow in the female) over the black ground colour.

UHW markings include a postdiscal row of larger spots, deep yellow in the male, white in the female, and a red eye-spot.

LFW/LHW similar to the upper surface but with two red spots below the hindwing cell and a row of small blue discal lunules (3, 5, 6)).

P. caiguanabus is a member of the *thoas* species group and the mature caterpillar is black dorsally, dark brown along the side, with bold white markings (6).

Distribution *Papilio caiguanabus* is endemic to the Caribbean island of Cuba, where it is found up to an altitude of about 300 m (2). It is fairly widespread, occurring in Matanzas and Oriente provinces (6), but is rare everywhere. Records and sightings have been more frequent in the eastern provinces (6).

Habitat and Ecology The food plant has been reported as *Securinega acidothannus* (6), synonymized in 1919 with *Securinega acidoton* (L.) Fawcett & Rendle. This Caribbean species is known, at least in Jamaica, as Green Ebony. In the family Euphorbiaceae, *S. acidoton* may be a shrub of 2–3 m height or a tree of up to 6 m.

Feeding on Euphorbiaceae is very unusual for a species in Munroe's section IV (5) (Hancock's *Heraclides* species-group (3)), which generally use Rutaceae or Piperaceae, rarely Umbelliferae (3). Further research is needed to confirm this observation and to elucidate details of the habitat and breeding pattern of the butterfly.

Threats There is insufficient information to assess the extent of threat to *Papilio caiguanabus*, but its rarity throughout Cuba is certainly a cause for concern.

Conservation Measures There is little information available on the Cuban national parks legislation (4), and no information on the presence of *Papilio caiguanabus* within park boundaries. Cuba is the largest island in the Caribbean, with adequate area and resources for conservation programmes. Surveys of the distribution of this species are needed, followed by biological studies and an assessment of the degree of threat.

References

1. D'Abrera, B. (1981). *Butterflies of the Neotropical Region. Part I. Papilionidae and Pieridae.* Lansdowne Editions, Melbourne. xvi + 172 pp.
2. D'Almeida, R.F. (1966). *Catálogo dos Papilionidae Americanos.* Sociedade Brasileira de Entomologia. São Paulo, Brazil.

3. Hancock, D.L. (1983). Classification of the Papilionidae (Lepidoptera): a phylogenetic approach. *Smithersia* 2: 1–48.
4. IUCN/CNPPA (1982). *IUCN Directory of Neotropical Protected Areas*. Tycooly, Dublin. 436 pp.
5. Munroe, E. (1961). The classification of the Papilionidae (Lepidoptera). *Canadian Entomologist* Supplement 17: 1–51.
6. Riley, N.D. (1975). *A Field Guide to the Butterflies of the West Indies*. Collins, London. 244 pp., 24 col. pl.
7. Rothschild, W. and Jordan, K. (1906). A revision of the American Papilios. *Novitates Zoologicae* 13(3): 411–752. (Facsimile edition ed. P.H. Arnaud, 1967).

Papilio (Heraclides) aristor Godart, 1819 INDETERMINATE

Subfamily PAPILIONINAE Tribe PAPILIONINI

Summary *Papilio aristor*, the Scarce Haitian Swallowtail, was until recently known only from a few specimens collected near Port-au-Prince, Haiti, and Monte Christi, north-western Dominican Republic. Surveys are urgently needed in order to define its present distribution more precisely, particularly in the xeric areas which it is now known to inhabit.

Description This black swallowtail is similar to *P. caiguanabus* from Cuba (see previous review) but with additional discal spots on the UFW and rather smaller postdiscal spots of colour on the UHW. The female is larger (forewing length 50 mm) than the male (forewing length 40 mm), and the yellow spots on both wings are somewhat paler, but the sexes are otherwise similar (1, 7, 8).

UFW with a diagonal row of five small discal yellow spots from the costal margin, converging with a submarginal row of larger yellow spots.

UHW tailed with a postdiscal row of large yellow spots, yellow marginal crescents and a red anal spot.

LFW/LHW differs from the upper surface in having a diffuse yellow spot in the forewing cell, and paler postdiscal spots on the hindwing. The hindwing also has two red spots below the cell and a discal row of blue lunules. The immature stages are unknown (7) but, like *Papilio machaonides*, *P. caiguanabus*, and *P. esperanza*, *P. aristor* is a member of the *thoas* species group (3, 6). This implies a mottled brown mature larva with pale yellowish or white patches, an ornate metathoracic band and a raised band on the first abdominal segment (3).

Distribution This species is endemic to the Caribbean island of Hispaniola and is known from very few specimens. Most have their origin in the area around Port-au-Prince in Haiti (1, 2, 7, 8) but there is also a record from Monte Christi in north-western Dominican Republic (5). Recent studies in the Dominican Republic have discovered a number of new localities, but these have not yet been revealed (9).

Habitat and Ecology *Papilio aristor* is known to inhabit xeric areas of lowland country in the Dominican Republic, and probably in Haiti also (9). Studies in recent years by A. Schwartz and his colleagues indicate that the species is not as rare as has been implied by its common name (7). However, only one female and few males have been recorded in the literature so far (7, 8). The foodplants are unknown, but as a member of the *thoas* species-group *P. aristor* is likely to feed on Rutaceae, Piperaceae or Umbelliferae (3). Nothing is known of the young stages or breeding cycle, but adults are known to be on the wing in July. Details of the habitat of *P. aristor*, although known to lepidopterists who visit the region, have not yet been published.

Threats There are no data specifically addressing the threats to *Papilio aristor*. Nevertheless the threats to all forms of wildlife on Hispaniola, and particularly in Haiti, are well documented (10, and see review of *Battus zetides*). Land development for agriculture, timber extraction and cash crop plantations by a large and rapidly expanding population has taken a heavy toll on wildlife. Degradation and erosion of soils through poor agricultural practices is exacerbating the problem, causing a constant search for new land. *P. aristor* is said to inhabit xeric, lowland areas (9), possibly unsuitable for agricultural expansion. Nevertheless, natural habitats on Hispaniola are threatened to such a degree that Indeterminate category is retained for this species until more substantial data on its distribution and ecology are available. Xeric biomes are by no means immune from environmental damage; over-grazing or lowering of the water table may have a catastrophic effect.

Conservation Measures An assessment of the present distribution of *Papilio aristor* is required, together with a careful study of its ecological requirements. The only national park in Haiti is La Citadelle and this is poorly protected (4). The del Este and Isla Cabritos National Parks in the Dominican Republic contain subtropical dry forest characterized by mesquite, giant milkweed and several cactus species (4), and may possibly include suitable habitat for the butterfly.

References

1. D'Abrera, B. (1981). *Butterflies of the Neotropical Region. Part I. Papilionidae and Pieridae*. Lansdowne Editions, Melbourne. xvi + 172 pp.
2. D'Almeida, R.F. (1966). *Catálogo dos Papilionidae Americanos*. Sociedade Brasileira de Entomologia. São Paulo, Brazil.
3. Hancock, D.L. (1983). Classification of the Papilionidae (Lepidoptera): a phylogenetic approach. *Smithersia* 2: 1–48.
4. IUCN/CNPPA (1982). *IUCN Directory of Neotropical Protected Areas*. Tycooly, Dublin. 436 pp.
5. Miller, L.D. (1984). *In litt.*, 6 January.
6. Munroe, E. (1961). The classification of the Papilionidae (Lepidoptera). *Canadian Entomologist* Supplement 17: 1–51.
7. Riley, N.D. (1975). *A Field Guide to the Butterflies of the West Indies*. Collins, London. 244 pp., 24 col. pl.
8. Rothschild, W. and Jordan, K. (1906). A revision of the American Papilios. *Novitates Zoologicae* 13(3): 411–752. (Facsimile edition ed. P.H. Arnaud, 1967).
9. Schwartz, A. (1984). *In litt.*, 16 May.
10. Thornback, J. and Jenkins, M. (1982). *The IUCN Mammal Red Data Book. Part 1*. IUCN, Gland. 516 pp.

Papilio (Heraclides) garleppi Staudinger, 1892   INSUFFICIENTLY KNOWN

Subfamily PAPILIONINAE   Tribe PAPILIONINI

Summary *Papilio garleppi* is a very rare species known from only a few sites in Peru, Brazil, Bolivia and Guyana. Nothing is known of its biology or the reasons for its rarity and disjunct distribution.

Description *Papilio garleppi* is in the *torquatus* species-group (4). The male has a forewing length of 45–50 mm. The female is not described in the literature.

Male: UFW black–brown with a broad yellow postdiscal band which may or may not be interrupted by a black band from outer margin to the costal margin.

UHW with yellow band broadening across the discal zone; red anal spot preceded by two or three yellowish-buff spots.

LFW/LHW lighter in colour, with a row of rufous red discal spots, one small bluish dot and a row of yellow submarginal spots on the hindwing (2, 5).

Distribution Three subspecies have been described (2). *P. g. garleppi* is known from Mateo, Rio Juntas and Rio Chapare in Bolivia (3). *P. g. interruptus* occurs in eastern Peru as far south as the Rio and Cordillera de Carabaya, and in the Upper Amazon region of Brazil around São Paulo d'Olivença and Benjamin Constant, near the Peruvian and Colombian borders (3). The third subspecies, *P. g. insidiosus*, is recorded from the Maroni River in French Guiana, Surinam and Guyana (2, 3). No explanation for this disjunct distribution has been published. However, it has been suggested by K.S. Brown Jr. that *P. garleppi* could possibly be a hybrid between *P. torquatus* and *P. astyalus*, two sympatric species (1). This would explain the disjunct distribution of *P. garleppi* and is worthy of further research.

Habitat and Ecology Very few specimens of this species have ever been collected and no data on its biology or habitat have been published. In common with other members of the *torquatus* species-group, the caterpillars probably feed on Rutaceae (4). The type specimen came from the Rio Chapare, which lies between 200 and 500 m altitude in the eastern foothills of the Bolivian Andes. Similarly, specimens from western Brazil and eastern Peru seem to be from low-lying areas with an altitude of less than 100 m. One southern Peruvian locality however, Carabaya, is near Lake Titicaca at an altitudinal range of 3000–5000 m. It is unusual for a species to embrace such a diversity of habitats, but the range might be possible for an ancient species in this group (7). A mislabelling of the specimen might also be suspected.

Threats No specimens of *Papilio garleppi* are held in museums in Peru (6) and very few are held elsewhere, suggesting that the species is very rare. However, Zischka (8) claims that it is not rare in Chapare district, taken to be the region just to the north of Cochabamba and Santa Cruz. This observation certainly requires confirmation. Without details of its habitat or the precise limits of its distribution it is impossible to assess threats.

Conservation Measures Details of the biology, habitat and distribution of *Papilio garleppi* are needed before any conservation measures can be proposed. There is no information on the occurrence of the species within the national parks and protected areas of Peru, Brazil, Bolivia or Guyana.

References

1. Brown, K.S., Jr. (1984). *In litt.*, 4 January.
2. D'Abrera, B. (1981). *Butterflies of the Neotropical Region. Part I. Papilionidae and Pieridae.* Lansdowne Editions, Melbourne. xvi + 172 pp.
3. D'Almeida, R.F. (1966). *Catálogo dos Papilionidae Americanos.* Sociedade Brasileira de Entomologia. São Paulo, Brazil.
4. Munroe, E. (1961). The classification of the Papilionidae (Lepidoptera). *Canadian Entomologist* Supplement 17: 1–51.
5. Rothschild, W. and Jordan, K. (1906). A revision of the American Papilios. *Novitates Zoologicae* 13: 411–752. (Facsimile edition ed. P.H. Arnaud, 1967).
6. Tyler, H. (1983). *In litt.*, 13 March.
7. Tyler, H. (1983). *In litt.*, 30 September.
8. Zischka, R. (1950). Catálogo de los insectos de Bolivia. *Folia Univ. Cochabamba* 4: 51–56.

Papilio (Heraclides) himeros Hopffer, 1866   VULNERABLE

Subfamily PAPILIONINAE   Tribe PAPILIONINI

Summary *Papilio himeros* occurs in the south-eastern states of Brazil and possibly Argentina, and is known to be disappearing from its former haunts. Although healthy populations are known, the localities are restricted. There is insufficient information on its distribution and biology to propose conservation measures.

Description *Papilio himeros* is a member of the *torquatus* species-group (4, 7). It is a relatively large butterfly with a forewing length of 44–50 mm, the female being larger than the male (Plates 6.6 and 6.7).

The wings of the male have a blackish-brown ground colour with a broad yellow median band from the forewing subapical region which continues to cover the entire discal and basal part of the hindwing. The hindwing also has a red and blue eye-spot and a row of yellow submarginal spots. The outer margin is scalloped and the long spatulate tail has a yellow spot at the apex (2, 8).

The female differs in having a narrower median band which does not cover the base of the hindwing and submarginal lunules which are yellow near the apex but otherwise red (2, 4, 7, 8).

Distribution Brazil and, according to D'Almeida, also from Corrientes in Argentina (3). There are two subspecies: *P. h. baia* from Bahia State, and *P. h. himeros* from Minas Gerais, Rio de Janeiro (3) and Espírito Santo (1). This is the subspecies which is also reported from Corrientes in Argentina, although the locality is a long way south of the Brazilian sites (5). There is no published explanation for this apparently disjunct distribution.

Habitat and Ecology Generally found in open areas at low altitudes, particularly in February, August and September (3). The female is very rare (3). The species flies quickly and with an intense, tremulous flight (3). Its habits are similar to the other members of the *thoas*, *torquatus* and *anchisiades* groups in Munroe's (7) section IV (called *Heraclides* by Hancock (4)). The larvae presumably feed on Rutaceae and, as in other members of the *torquatus* group, are probably tuberculate (7) and solitary (9). Adults seek nectar at flowers, but apparently have some highly specialized requirements which prohibit them from persisting in most areas (1).

Threats The species has disappeared from the Cavalao area near Rio de Janeiro city, where many were captured earlier this century (1). *P. himeros* persists in some numbers around Linhares in Espírito Santo (1), but coastal development is probably gradually reducing the available habitat.

Conservation Measures Although some localities are known, the full range of this species has not been properly determined. There is no information on its occurrence in Brazilian national parks, of which there are several in the south-eastern region (6). Studies of its biology are needed in order to assess its precise feeding requirements. Only then can proper conservation measures be put forward. No information on captive breeding is available.

311

References

1. Brown, K.S., Jr. (1983). *In litt.*, 6 April.
2. D'Abrera, B. (1981). *Butterflies of the Neotropical Region. Part I. Papilionidae and Pieridae.* Lansdowne Editions, Melbourne. xvi + 172 pp.
3. D'Almeida, R.F. (1966). *Catálogo dos Papilionidae Americanos.* Sociedade Brasileira de Entomologia. São Paulo, Brazil.
4. Hancock, D.L. (1983). Classification of the Papilionidae (Lepidoptera): a phylogenetic approach. *Smithersia* 2: 1–48.
5. Hayward, K.J. (1967). *Genera et Species Animalium Argentinorum. 4. Insecta. Lepidoptera (Rhopalocera). Familiae Papilionidarum et Satyridarum.* Guillermo Kraft Lbla, S.G.; Bonariae (General editors Rossi, J.A.H. and Lillo, M.). 447 pp.
6. IUCN (1982). *IUCN Directory of Neotropical Protected Areas.* Tycooly International, Dublin. 436 pp.
7. Munroe, E. (1961). The classification of the Papilionidae (Lepidoptera). *Canadian Entomologist* Supplement 17: 1–51.
8. Rothschild, W. and Jordan, K. (1906). A revision of the American Papilios. *Novitates Zoologicae* 13: 411–752. (Facsimile edition ed. P.H. Arnaud, 1967).
9. Tyler, H.A. (1975). *The Swallowtail Butterflies of North America.* Naturegraph Publishers. viii + 192 pp.

Papilio (Heraclides) maroni Moreau, 1923   INSUFFICIENTLY KNOWN

Subfamily PAPILIONINAE   Tribe PAPILIONINI

Summary *Papilio maroni* is only known from the region around St Laurent on the River Maroni in western French Guiana. More information is required on the taxonomic status and biology of this apparently very rare species.

Description *Papilio maroni* is a relatively large butterfly with a forewing length of 52–55 mm. The sexes are similar though the female tends to be slightly larger (1, 4, 5). There is some doubt concerning the taxonomic status of this species (2). One authority considers it to be a form of *Papilio anchisiades*, a very variable species (9); others believe it may be a subspecies of *P. chiansiades* (7) or *P. isidorus* (8).

UFW black with a white central-median spot which is smaller in the male and is positioned closer to the inner margin (1, 4, 5).

UHW black with a deeply scalloped outer margin and no distinct tail. There are three long, red discal-postdiscal spots forming a short band from the inner margin. These may be divided, by black suffusion, into separate discal and postdiscal spots, and in the female this band may be extended towards the costal margin by additional submarginal spots, and another discal spot (1, 4, 5).

Distribution *Papilio maroni* occurs only in western French Guiana. It has been recorded from the area around St Laurent, on the River Maroni, which constitutes French Guiana's western border (3). The species is known from very few specimens.

Habitat and Ecology *Papilio maroni* is placed in the *P. anchisiades* group by Munroe (6) and Moreau (4). The larvae are probably tuberculate and gregarious when mature, feeding on Rutaceae (5).

Threats These cannot be assessed without further biological and distributional data. French Guiana still contains vast tracts of undisturbed forest and it would appear unlikely that forest destruction has contributed to the rarity of *Papilio maroni*.

Conservation Measures Verification of the taxonomic status of this species is needed. As yet there are insufficient data on its habitat, biology and distribution to pursue conservation measures. The Réserve Naturelle de Basse Mana is on the north-west coast. It runs from above the lower Mana River as far as the river's mouth at Maroni (3). The reserve is mainly coastal wetland, but it includes a variety of other habitats and could contain *Papilio maroni*. French Guiana is a French Overseas Department subject to French laws. *Papilio maroni* could be placed on the French list of protected species if it proved to be seriously threatened, but research and protection of habitat are more useful objectives at this early stage.

References

1. D'Abrera, B. (1981). *Butterflies of the Neotropical Region. Part I. Papilionidae and Pieridae.* Lansdowne Editions, Melbourne. xvi + 172 pp.
2. Hancock, D.L. (1983). Classification of the Papilionidae (Lepidoptera): a phylogenetic approach. *Smithersia* 2: 1–48.

313

3. IUCN (1982). *IUCN Directory of Neotropical Protected Areas*. Tycooly International, Dublin. 436 pp.

4. Moreau, E. (1923). Un *Papilio* nouveau de la Guyane Française. *Bulletin de la Société Entomologique de France* 1923: 144, 215.

5. Moreau, E. (1924). Description de *Papilio maroni* male de Guyane Française. *Bulletin de la Société Entomologique de France*. 1924: 93–94.

6. Munroe, E. (1961). The classification of the Papilionidae (Lepidoptera). *Canadian Entomologist* Supplement 17: 1–51.

7. Racheli, T. (1984). *In litt.*, 18 June.

8. Smart, P. (1975). *The Illustrated Encyclopedia of the Butterfly World*. Hamlyn, London. 275 pp.

9. Tyler, H.A. (1983). *In litt.*, 30 September.

Papilio (Chilasa) maraho Shiraki & Sonan, 1934   VULNERABLE

Subfamily PAPILIONINAE   Tribe PAPILIONINI

Summary *Papilio maraho* is a beautiful tailed butterfly endemic to Taiwan. Its habitat in the mountains is under severe pressure from growing human populations needing land for agriculture and forestry. Specimens are much in demand by collectors and sell for high prices. A combination of the destruction of critical habitats and excessive, intensive collecting gives cause for grave fears about the survival of the species.

Description *Papilio maraho* is an attractive black, red and cream butterfly with slender forewings and deeply indented edges to the hindwings giving the impression of large spatulate tails. The sexes are similar with a forewing length of about 60 mm (1). The species is believed to be in a mimetic association with troidine swallowtails of the genus *Atrophaneura*, possibly *A. polyeuctes* or *A. aristolochiae*, both of which are similar in general appearance.

UFW brown–black, darker along the veins.

UHW brown–black ground colour with a creamy-yellow cell and a row of bold, red submarginal lunules.

P. maraho was originally described as a subspecies of the Chinese species *P. elwesi*, but in a recent reassessment it was afforded full species status (2).

Distribution *Papilio maraho* is endemic to the island of Taiwan (Republic of China), lying about 145 km from mainland China and straddling the Tropic of Cancer. *P. maraho* is a replacement species for *P. elwesi*, a close relative that is found on the mainland and is very similar in appearance.

Habitat and Ecology The island of Taiwan is 385 km long and 145 km wide, almost 36 000 sq. km in area (5). The eastern half is very mountainous, with over 60 peaks exceeding 3000 m. The highest point is Yu Shan (Mt Morrison), 3950 m. The climate and vegetation of the island vary widely, from tropical monsoon forests to rocky snow-capped mountains (5). Although small in size, Taiwan has a diverse lepidopteran fauna, reflecting the diversity of habitat type. Shirozu (6) listed 361 species of butterflies in 161 genera. About 10 per cent of these species are endemic to Taiwan (3).

Very little is known about the habitat and ecology of *P. maraho* except that it favours montane habitats.

Threats Taiwan is an island with a large and growing population. Most of the 17 million or more people have traditionally concentrated in the lowlands, but increasing numbers have been moving to the mountains. As a result, the virgin hardwood and coniferous forests are now being cleared for agriculture, forestry and other developments, following the already almost total destruction of forests in the lowlands (5). The precise altitude preferred by *P. maraho* is unknown, but likely habitat in the hardwood forests below 2000–2300 m is now under severe pressure of replacement by conifer plantations (R.D. Schultz, in ref. 5). Destruction of critical habitat is probably the single most important factor in causing the decline of *P. maraho* and other butterflies in Taiwan today (5).

Taiwan has the world's largest and most intensive industry in butterflies (7). Details of the extent of the trade have been presented in section 5 of this book. *P. maraho* is a rarity that is never exploited in the low quality, high volume curio trade. Specimens sell to specialist collectors at high prices, generally £20–£50 per individual. As has been emphasized throughout this book, small-scale butterfly collecting alone is unlikely to have a serious impact on populations. However, when destruction of natural habitat is so extensive that only small areas of critical habitat remain, intensive and commercial collecting within these critical habitats can be disastrous. Because of soaring prices and decreasing habitat, *P. maraho* is very much in demand. Some dealers have expressed fears that it may already by seriously endangered and close to extinction (4).

Conservation Measures The conservation movement in Taiwan has gathered considerable pace in recent years (4), despite the lack of international support. The future of *Papilio maraho* and the many other endemic species of Taiwan depends upon the designation of protected areas. Four national parks are already established, but there is uncertainty about the inclusion of *P. maraho* range within their boundaries. A survey of the habitat of *P. maraho* must be a high priority. Biological studies would be of immense value in planning management strategies and developing farming or ranching.

For the time being, capture and trade of *Papilio maraho* in Taiwan should be discouraged by whatever means seem most appropriate. In international trading circles, dealers should consider a voluntary ban on trading in this species. The British Association of Entomological Suppliers has already done this for *Troides aeacus kaguya*, another Taiwanese endemic.

Taiwan has a "critical fauna" for international conservation of the Papilionidae (see section 4). Of the 32 swallowtail species known from the island, five are endemic (15 per cent). These are *Atrophaneura febanus*, *A. horishanus*, *Papilio thaiwanus*, *P. maraho* and *P. hoppo*. Reports on the status of all these species are required.

References

1. D'Abrera, B. (1982). *Butterflies of the Oriental Region. Part 1. Papilionidae and Pieridae.* Hill House, Victoria, Australia. xxxi + 244 pp.
2. Hancock, D.L. (1983). Classification of the Papilionidae (Lepidoptera): a phylogenetic approach. *Smithersia* 2: 1–48.
3. Marshall, A.G. (1982). The butterfly industry of Taiwan. *Antenna* 6(2): 203–4.
4. Morris, M.G. (1985). *In litt.*, 11 January.
5. Severinghaus, S.R. (1977). The butterfly industry and butterfly conservation in Taiwan. *Atala* 5(2): 20–23.
6. Shirozu, T. (1960). *Butterflies of Formosa.* T. Shoten, Tokyo. 65 pp.
7. Unno, K. (1974). Taiwan's butterfly industry. *Wildlife* 16: 356–359.

Papilio (Chilasa) osmana Jumalon, 1967   VULNERABLE

Subfamily PAPILIONINAE   Tribe PAPILIONINI

Summary *Papilio osmana* is an extremely rare swallowtail from southern Leyte and north-eastern Mindanao in the southern Philippines. Very little is known about this butterfly, but its habitat is being rapidly exploited and degraded by human activities. Further surveys and biological studies are needed in order that conservation methods and protected areas may be recommended.

Description *Papilio osmana* is a relatively large tailless butterfly belonging to the *veiovis* species-group. The sexes are similar with a forewing length of 47–52 mm (7). The species is believed to be in a mimetic association with *Salatura melanippus* (Nymphalidae: Danainae), and some individual variation is to be expected (7).

Male: UFW/UHW bluish grey ground colour scattered with blue scales over the basal half of the forewing. The veins are covered by brownish-black scales and there are also black wing margins, forewing apex and a broad, black distal border with elongated bluish-white submarginal-postdiscal spots. The hindwing has a small, orange anal spot (3, 7).

LFW/LHW differs in being lighter than above due to a reduction in the dark scaling. The anal spot is absent (3, 7).

Female: More robust and darker appearance with slightly more diffuse markings and the anal spot yellower (3, 7).

Distribution *P. osmana* was discovered relatively recently in southern Leyte and north-eastern Mindanao in the south-eastern Philippines (3, 7). New localities are to be expected from future work, particularly in Samar, Mindanao and possibly other southern Philippine islands, but *Papilio osmana* is still one of the rarest swallowtails in the Philippines, second only to *P. carolinensis* (7). The holotype was captured at Sitio Katigahan (450 m) on the outskirts of the village of Catman near St Bernard in southern Leyte (3). In Mindanao the locality is not precisely given, but appears to be inland from Lianga, on the north-east coast (7). Further studies may extend the known distribution into the forests of central and southern Mindanao.

The *veiovis* species-group is believed to have had its origins on the Asiatic mainland and the isolation of the present distribution of *P. osmana* (and *P. veiovis* in Sulawesi and *P. carolinensis* in Mindanao) is somewhat enigmatic. Some Philippine forms have only survived the retreat of the ice caps by adopting montane habits (e.g. *P. chikae, P. benguetanus*), but *P. osmana, P. carolinensis* and *P. veiovis* are lowland species. The distribution of *P. osmana* and *P. carolinensis* may prove to be larger than is presently known, but the reason for their absence from Luzon remains a mystery (4).

Habitat and Ecology Collecting records for this very rare species are limited to April, May, June and July and there is presumed to be only one brood per year (3, 7). The larval foodplant and early stages are unknown (3). Mature larvae of other species in the subgenus are often dark with pale bands or patches and orange segmental spots and spiny tubercles, and are commonly feeding on Lauraceae (2). The type specimens were taken on logging roads in lowland dipterocarp rain forest at altitudes between 150 m and 500 m (3).

Threats *Papilio osmana* and *P. carolinensis* (see next review) are both extremely rare and elusive rain forest swallowtails. The extent and rate of logging, agricultural extension, shifting cultivation and other forms of forest exploitation in the Philippines are such that some degree of threat to both species must be assumed (6). Their ranges are very small and with modern technology the habitat could be virtually destroyed in a matter of decades.

According to a recent FAO/UNEP report the Philippines and Sabah are the tropical countries with the highest intensities of logging and deforestation. The Philippine government has introduced a policy of local processing of logs, which is likely to steady the level of production and exploitation by logging companies, but that level is still extremely high. Estimates for deforestation rates in lowland dipterocarp forest by all causes were 100 000 hectares per year for 1976–80 and are 90 000 hectares per year for 1981–85 (1). This latter figure is a little over 1 per cent of the total area of dipterocarp forest (8.9 million hectares) remaining in the Philippines in 1980. A further 3.4 million hectares are in forest fallow (1).

Although it is difficult to assess the threat from deforestation of the habitat of *P. osmana* and *P. carolinensis*, their main area of distribution, Mindanao, produces three quarters of the total registered commercial log production for the Philippines (1). By implication, the threat to Mindanao forest endemics from logging must be greater than almost anywhere else on earth.

A further threat only recently being fully realized is the serious drought of 1982–3 that has affected eastern Mindanao, eastern Borneo and perhaps other islands. In the wake of the drought, huge fires have destroyed vast areas of forest. In Mindanao whole mountains have been burnt through and both Mt Apo and Mt Katanglad have suffered badly (5). The full extent of defoliation is not yet known and will only be discovered from aerial or satellite survey.

Over the Philippines as a whole logging is not the prime reason for loss of forest. The main cause is the increasing population of shifting cultivators 'kaingineros' and the spread of agriculture (1). Pressure is such that forest regeneration is unable to occur within the cycles imposed by the cultivators, and degeneration of both vegetation and soils follows (1). The number of 'kaingineros' is being swelled by landless tenants, labourers and unemployed rural people who need to find land to grow food for their families. As long as there are thousands of landless people in the basically rural economy of these islands, there will be no end to the destruction of the forest. Southern Leyte and northern Mindanao are also areas of permanent pineapple and coconut plantations, a source of some of the Philippines' most important exports. During 1980–85 the total forest area is projected to decrease by about 1.5 million hectares, about 0.45 million being virgin or near-virgin dipterocarp forest (1).

Conservation Measures Further surveys are needed to assess the distribution of *P. osmana*, particularly within Mindanao and Samar. At present there are no projected or existing national parks in Samar or in eastern Mindanao within the butterfly's known range. Once the butterfly's habitat has been surveyed and its biology studied, areas of forest should be set aside for its protection. Many other endemic species of invertebrates, vertebrates and plants would also benefit from such actions.

References

1. FAO/UNEP (1981). *Forest Resources Assessment Project. Forest Resources of Tropical Asia*. FAO, Rome. 475 pp.

2. Hancock, D.L. (1983). Classification of the Papilionidae (Lepidoptera): a phylogenetic approach. *Smithersia* 2: 1–48.
3. Jumalon, J.N. (1967). Two new papilionids. *Philippine Scientist* 1(4): 114–118.
4. Jumalon, J.N. (1969). Notes on the new range of some Asiatic papilionids in the Philippines. *The Philippine Entomologist* 1(3): 251–257.
5. Lewis, R.E. (1984). *In litt.*, 17 April.
6. Treadaway, C.G. (1984). *In litt.*, 25 May.
7. Tsukada, E. and Nishiyama, Y. (1982). *Butterflies of the South East Asian Islands. Vol. 1 Papilionidae.* (transl. K. Morishita). Plapac Co. Ltd., Tokyo.

Papilio (Chilasa) carolinensis Jumalon, 1967   VULNERABLE

Subfamily PAPILIONINAE   Tribe PAPILIONINI

Summary *Papilio carolinensis* is an extremely rare and little known swallowtail from north-eastern Mindanao in the southern Philippines. A denizen of the rich dipterocarp forests of the island, it is believed to be threatened by rapid deforestation in the region. Surveys of its distribution, biological studies and presence in protected areas are needed.

Description *Papilio carolinensis* is similar in general appearance to *P. osmana* (previous review). The sexes are similar with a forewing length of about 50 mm (2, 5).

Male: differs from that of *P. osmana* in having more extensive black scaling and orange–brown markings. The submarginal and subapical spots are white on the forewing , and orange–yellow on the hindwing. The dark scaling of the hindwing is chocolate brown and the extent of it varies, but it broadly covers the veins. Between the veins is diffuse orange–yellow scaling to the inside of the distal border (1, 2, 5).

Female: similar but may have a lighter ground colour (5).

When first discovered, *P. carolinensis* was presumed to be related to *P. agestor* because of its chocolate brown hindwings (2). However, closer examination of its morphology and genitalia revealed its close association with the other members of the *veiovis* group, *P. osmana* and *P. veiovis* (1, 5). *P. carolinensis* is a danaid-mimic, as distinguished, according to Jumalon, by the rows of white spots on the abdomen (2). Of the Danaidae with which it has been found to associate, it most closely resembles *Danaus luzonensis* (2).

Distribution *Papilio carolinensis* was only discovered relatively recently on Mindanao in the south-eastern Philippines. It is known only from Agusan province in the north-east and Bukidnon province towards the centre of Mindanao (2). Tsukada and Nishiyama refer to it as the rarest Philippine swallowtail, being represented in collections only by five males and one female (5). In north-eastern Mindanao the distribution overlaps with *P. osmana* (5). Both *P. carolinensis* and *P. osmana* are presumed to be relict species of the last glacial era, but their present distribution in lowland tropical habitat is puzzling (3, and see previous review).

Habitat and Ecology All specimens were collected between February and May and the species is presumed to be univoltine (2). Foodplants and early stages are unknown (2). Mature larvae of other species in this subgenus are commonly dark with pale bands or patches and orange segmental spots and spiny tubercles, generally feeding on Lauraceae (1). Tsukada and Nishiyama expect that more specimens will be found in the rain forests of east-central Mindanao (2).

Other species with which *P. carolinensis* has been found to associate include *Atrophaneura mariae*, *Danaus luzonensis* and *D. vitrina* (2). *Graphium idaeoides*, a species which is also considered to be threatened (see separate review), has been found at lower elevations in the same region (2).

Threats Three localities for this species are known so far. The type locality is at 615 m near a small village called Barrio Kitcharao on the east shore of Lake Mainit, north-west of the Diwata Range in Agusan Province (2). Behind the village is a mountain with bare slopes except for relict forest patches in rocky areas. The presumed rain forest habitat of the butterfly, which once covered the mountain, is now almost entirely destroyed. The second locality is in the mountains to the north-east of Malaybalay, the capital of Bukidnan Province, where large remnants of forest still remain. Even so, it seems that suitable habitat is very restricted, most of the Bukidnan Province being dry highland grasslands (2). The third and most southerly locality is Masara Mine in Bukidnon Province. The region is reported to be undergoing development and has poor prospects for this butterfly (2).

The comments on destruction of lowland dipterocarp forest made in the review of *P. osmana* are equally applicable to *P. carolinensis*. Deforestation in the Philippines is proceeding faster than in any other state with the possible exception of Sabah. The main causes are shifting agriculture and logging. The effects of the extensive forest fires that followed the drought of 1982–3 have not yet been assessed. Aerial and satellite surveys are needed (4).

Conservation Measures Further surveys are needed to assess the distribution of *P. carolinensis* and other butterflies on Mindanao. At present there are no projected or existing national parks in eastern Mindanao within the butterfly's range. Biological studies of the butterfly are needed and suitable protected areas should be set up for this and other species native to Mindanao's forests.

References

1. Hancock, D.L. (1983). Classification of the Papilionidae (Lepidoptera): a phylogenetic approach. *Smithersia* 2: 1–48.
2. Jumalon, J.N. (1967). Two new papilionids. *Philippine Scientist* 1(4): 114–118.
3. Jumalon, J.N. (1969). Notes on the new range of some Asiatic papilionids in the Philippines. *The Philippine Entomologist* 1(3): 251–257.
4. Lewis, R.E. (1984). *In litt.*, 17 April.
5. Tsukada, E. and Nishiyama, Y. (1982). *Butterflies of the South East Asian Islands. Vol. 1 Papilionidae*. (transl. K. Morishita). Plapac Co. Ltd., Tokyo. 457 pp.

Papilio (Chilasa) toboroi Ribbe, 1907   RARE

Subfamily   PAPILIONINAE   Tribe   PAPILIONNI

Summary *Papilio toboroi* occurs on Bougainville (Papua New Guinea), Holibara, Santa Isabel and Malaita (Solomon Islands). Although locally common on Bougainville and apparently adaptable to human interference, it is little collected and very poorly known. More data on its ecology, the extent of its distribution, and threats to its survival are required.

Description This species is a large and beautiful short-tailed butterfly in the *laglaizei* group (7). The forewing length is about 60 mm in the female, 56 mm in the male, and the wingspan from apex to apex is about 110 mm. The sexes are similar, though the male tends to be smaller (1, 11, 12). Two subspecies have been described (10).

UFW ground colour black–blue with a blue silky sheen and two narrow, curved green–blue bands (1, 11).

UHW with a short tail, a slightly scalloped outer margin and a ground colour the same as the UFW. The curved blue–green median band is continuous with the inner band of the UFW, and there is a yellow spot at the anal angle (1, 11).

LFW/LHW ground colour duller than the upper surface, but otherwise broadly similar on the forewing. Basal third of the LHW grey with blue–black veins. A postdiscal band is composed of seven large spots merged together; the inner five yellow, the rest pale grey (1, 11).

The final instar larva is brown with black on the head, legs, prothorax, anal segment and prolegs (5, 13). White spots are present on thorax and abdomen and the tubercles are 8–9 mm long, stiff and black with white spots at the base (5, 13). The pupa is yellow with black on the underside and laterally on some abdominal segments (5, 13).

Distribution *Papilio toboroi* is confined to the eastern side of the Solomon Archipelago in the south-west Pacific Ocean 91, 2, 3, 4, 10, 11). The nominate subspecies appears to be widespread on Bougainville (Papua New Guinea) (1, 2, 4, 8, 10, 11), and a second subspecies, *P. t. straatmani* Racheli, occurs further to the south-east in the Solomon Islands (9, 10). It has been recorded from Holibara in south Santa Isabel, and also from Malaita (9, 10). It may possibly have been overlooked on Choiseul Island.

Habitat and Ecology *Papilio toborio* lays its eggs in large masses, as do the other members of the group (13). The larvae are strongly gregarious, the largest recorded group consisting of 700 individuals (13). Feeding is at night (13) on the leaves of its main food plant *Litsea irianensis* (Lauraceae), or alternatively *Flindersia* sp. (Rutaceae) (4). Straatman recorded caterpillars on unidentified host plants on high hills at about 1300 m above sea level (13). The location was in secondary growth far from water but near village gardens where the adults were feeding on flowers. The larvae are heavily predated by toads, frogs, geckos, spiders and ants, only a fraction reaching a spot suitable for pupation. Of these, about 20 per cent have been observed to produce parasitic species of Tachinidae (Diptera) and Braconidae (Hymenoptera) (13).

Threats The most significant threat to *Papilio toboroi* is likely to be forest destruction, but it is difficult to be precise about the extent of that threat. Hundreds of square kilometres of potential habitat have never been searched. In addition, Straatman's notes (12) indicate a degree of adaptability to human interference of its habitat.

On Bougainville *P. toboroi* is said to be fairly widespread, and common in certain places (4). However, the fifth instar caterpillars are highly sought after as a gourmet food item. They are easily collected since 300 or more larvae collect in forks of the food plant (4). *P. laglaizei* is used for the same purpose (4). This practice is unlikely to constitute a serious threat to the species unless it is carried out on a commercial scale, for which there is no evidence at the moment. Natural losses of larvae are anyway known to be high, mainly to reptiles, amphibians and invertebrate predators (12).

Conservation Measures The Solomon Archipelago has an interesting endemic insect fauna, of which *Papilio toboroi* is just one example. Given more information on status and distribution, there is potential for farming of selected rarities for the butterfly market and for release into the wild. Occasional specimens are already collected for sale (6, 8). A survey of the Solomon Islands insects was proposed some years ago, but so far no funding has been available (3). *Papilio toboroi* may be more widespread than is apparent from its known distribution and presence in collections. The Solomon Islands Government requires a permit for export of butterflies, but this is difficult to enforce.

References

1. D'Abrera, B. (1971). *Butterflies of the Australian Region*. Lansdowne Press, Melbourne. 415 pp.
2. Fenner, T.L. (1983). *In litt.*, 15 March.
3. Holloway, J.D. (1983). *In litt.*, 28 February.
4. Hutton, A.F. (1983). *In litt.*, 28 March.
5. Igarashi, S. (1979). *Papilionidae and Their Early Stages*. Vol 1: 219 pp. Vol 2: 102 pp. of plates. Kodansha, Tokyo. (In Japanese).
6. Insect Farming and Trading Agency (1979, 1980). Special butterfly price list. Div. of Wildlife, Bulolo, Morobe Province, Papua New Guinea.
7. Munroe, E. (1961). The classification of the Papilionidae (Lepidoptera). *Canadian Entomologist* Supplement 17: 1–51.
8. Parsons, M.J. (1983). *In litt.*, 2 March.
9. Racheli, T. (1979). New subspecies of *Papilio* and *Graphium* from the Solomon Islands, with observations on *Graphium codrus* (Lepidoptera, Papilionidae). *Zoologische Mededelingen* 54(15): 237–240.
10. Racheli, T. (1980). A list of the Papilionidae (Lepidoptera) of the Solomon Islands, with notes on their geographical distribution. *Australian Entomologists' Magazine*. 7: 45–59.
11. Ribbe, C. (1907). Zwei neuen Papilioformen von der Salomo–Insel Bougainville. *Deutsche Entomologische Zeitschrift "Iris"* 20: 59–63.
12. Smart, P. (1975). *The Illustrated Encyclopedia of the Butterfly World*. Hamlyn, London. 275 pp.
13. Straatman, R. (1975). Notes on the biologies of *Papilio laglaizei* and *P. toboroi* (Papilionidae). *Journal of the Lepidopterists' Society* 29: 180–187.

Papilio (Chilasa) moerneri Aurivillius, 1919   VULNERABLE

Subfamily   PAPILIONINAE   Tribe   PAPILIONNI

Summary *Papilio moerneri* is an extremely rare species known only from New Ireland, although it may occur in unexplored regions of other islands in the Bismarck Archipelago (Papua New Guinea). The butterfly has not been seen since 1924 and surveys are needed to assess its present range. If it could be found and its conservation assured, it has great potential for farming or ranching.

Description This elusive butterfly in the *laglaizei* group has only been recorded from the male which has a wingspan of about 108 mm (3, 7) (Plate 7.1).

UFW ground colour black with a dark blue sheen to all but the costal and outer margins. A broad, relatively straight, grey–green diagonal band runs from the inner margin to the middle of the cell apex (1, 3).

UHW upper surface is all black with a dark blue sheen and a very short tail (1, 3).

LFW black with a yellow submarginal band that curves inwards near the apex and crosses the subapical region to the costal margin (1, 3).

LHW grey–yellow with a broad, black discal band, black veins, and a wavy line inside the outer margin. There is also an orange–yellow streak in the abdominal region (1, 3).

Distribution *Papilio moerneri* is only known from New Ireland in the Bismarck Archipelago (Papua New Guinea), north-east of mainland Papua New Guinea. The species may also occur on the nearby islands of New Britain, and New Hanover, large areas of which are still unstudied (5) and very poorly known for butterflies. In 1939 Bang–Haas described a race of *P. moerneri* from New Britain and called it *mayrhoferi*, but it has not been seen since (2).

Habitat and Ecology Very little is known about the biology and habitat of *Papilio moerneri*. In addition to the 1919 and 1924 specimens mentioned by D'Abrera (3), Straatman notes that there is one specimen in the Australian National Collection, Canberra, and that a few more were taken in 1968 (10). It has not been reported by lepidopterists who have visited the region in more recent years (5, 6, 8, 9). The natural vegetation of the Bismarck Archipelago is rain forest, much of it in hilly or mountainous country, and the butterfly is presumed to inhabit these forests. Members of the *laglaizei* group are mimetic of Uraniidae (7). In other species the larvae are black with orange tubercles or have orange segmental bands and yellow spots (7). The foodplant is unknown, possibly lauraceous (7), and in other members of the group the eggs are laid in large masses and the larvae are gregarious (10) (see review of *Papilio toboroi*).

Threats The distribution of *P. moerneri* is too poorly known to be sure of the major threats to its survival. Its extreme rarity even early in this century implies that timber exploitation and other forms of habitat destruction may not be a prime threat. Nevertheless, logging is now extensive in New Ireland and must represent an encroachment on what is almost certainly a denizen of the primary forest. The Bismarck Archipelago, with recently latent but still active volcanic phenomena, has some highly fertile soils and the islands are responsible for most of the agricultural production of the country (4). Many parts of New Britain and New Ireland have been severely altered (9).

Conservation Measures A survey of potential *Papilio moerneri* habitat is needed in order to assess its conservation status thoroughly. The Whiteman Range and the high interior of the Gazelle Peninsula on New Britain, and the south-eastern lobe of New Ireland, have wilderness areas which may reward a careful search (9). This species is certainly one of the rarest and least known of all Papua New Guinea Papilionidae and would be worthy of international conservation funding. Once re-located, proposals for its management and protection would be required. Once its conservation is assured there would be great potential for farming, and studies of foodplants and other requirements would be quickly repaid. World demand for farmed specimens of this attractive but little-known species would certainly be high. The Insect Farming and Trading Agency at Bulolo already has the experience required to market insects without threatening their wild populations.

References

1. Aurivillius, C. (1919). Eine neue Papilio–Art. *Entomologisk Tidskrift* 1919: 177–178.
2. Bang–Haas, O. (1939). Neubeschreibungen und Berichtigungen der Exotischen Macrolepidopteren fauna II. *Entomologische Zeitschrift* 52(39): 301–302.
3. D'Abrera, B. (1971). *Butterflies of the Australian Region*. Lansdowne Press, Melbourne. 415 pp.
4. FAO/UNEP (1981). *Tropical Forest Resources Assessment Project. Forest Resources of Tropical Asia*. FAO, Rome, 475 pp.
5. Fenner, T.L. (1983). *In litt.*, 15 March.
6. Hutton, A.F. (1983). *In litt.*, 28 March.
7. Munroe, E. (1961). The classification of the Papilionidae (Lepidoptera). *Canadian Entomologist* Supplement 17: 1–51.
8. Parsons, M.J. (1983). *In litt.*, 2 March.
9. Pyle, R.M. and Hughes, S.A. (1978). Conservation and utilisation of the insect resources of Papua New Guinea. Report of a consultancy to the Wildlife Branch, Dept. of Nature Resources, Independent State of Papua New Guinea. 157 pp. unpublished.
10. Straatman, R. (1975). Notes on the biologies of *Papilio laglaizei* and *P. toboroi* (Papilionidae). *Journal of the Lepidopterists' Society* 29: 180–187.

Papilio (Papilio) hospiton Guenée, 1839   ENDANGERED

Subfamily   PAPILIONINAE   Tribe   PAPILIONNI

Summary *Papilio hospiton* is usually known as the Corsican Swallowtail, despite the fact that it occurs on the island of Sardinia (Italy) as well as Corsica (France). It is found at altitudes above 600 m in the mountains and has declined dramatically through the impact of habitat destruction, commercial collecting and destruction of its foodplants, which are poisonous to sheep. Protection of habitat, surveys of distribution and detailed biological studies are urgently required. There is reason to believe that this is one of Europe's most seriously endangered butterflies.

Description *Papilio hospiton* (known as the Corsican swallowtail, le Porte–Queue de Corse or the Korsicher Schwalbenschwanz) is a short-tailed black and yellow swallowtail with blue and red markings. The male and female are similar in appearance, with a wingspan of 72–76 mm (6) (Plate 7.2).

UFW light yellow with black markings dusted with yellow scales. Black areas include the basal region, veins, large spots inside and above the cell, and a distal band that has rows of yellow submarginal and fringe spots.

UHW short-tailed and pale yellow with a black distal band, basal region, and inner margin. Markings include yellow outer marginal fringe crescents and submarginal lunules, blue postdiscal spots, and a small red eye spot.

LFW differs from the upper surface in that the postdiscal band is composed of black-bordered grey lunules, and the narrow submarginal band has a zig-zag edge.

LHW slightly paler than the upper surface and perhaps with a little more red on the hindwings, but otherwise very similar (6).

Distribution This butterfly is restricted to the islands of Corsica (France) and Sardinia (Italy) (1, 6), where it is extremely localized.

Habitat and Ecology The habitat of *Papilio hospiton* is open, mountainous country between 600 and 1500 m above sea level (4, 6). Although there is a considerable area of mountainous country on the two islands, the breeding sites for *Papilio hospiton* are extremely localized on Corsica (3), and presumably also on Sardinia. Precise localities are often kept secret. The larval foodplants are Umbelliferae, but there is some dispute as to which species are utilised. Fennel (*Foeniculum vulgare*) is commonly cited (6), although two authors give Giant fennel (*Foeniculum (Ferula) communis*) (8, 9). Others claim that fennel (and carrot) leaves are refused by caterpillars in captivity (3). Fausser rejects claims that both *Foeniculum communis* and *Ruta corsica* are used, but cites *Peucedanum paniculatum* as the true foodplant (3). Information gathered by Heath at first agreed in indicating *Peucedanum* as a foodplant (4), but has since been refuted (5). Clearly there is a need for further research on the subject. Young stages in Corsica may be attacked by the parasitic ichneumonid wasp *Trogus violaceus* (3). Adults generally emerge between May and June and are on the wing until early August. *P. hospiton* is generally believed to be single-brooded. However, Fausser (3) has found numerous fresh adults and even caterpillars in Corsica in early August, and firmly believes that two broods are possible. In support of this contention he describes morphological differences in the appearance of the early and late generations (3). An alternative suggestion is that some over-wintered pupae could emerge very late in the season, but all the evidence indicates that emergence of the over-wintered brood occurs in April, May or June. Hybridisation with the lowland *Papilio machaon* is a fairly regular occurrence, at least in Corsica (3).

Threats The threats to *Papilio hospiton* appear to be similar in Corsica and Sardinia. Habitat destruction is probably primarily responsible for the decline of the species (4). Much of the habitat, and particularly the status of the foodplants, is adversely affected by agricultural practices. The umbelliferous foodplants are believed to be poisonous to sheep and are destroyed by fires started by local people (4). More localized, but also more permanent, damage to habitat is done by developments such as the growth of skiing resorts on Corsica (4). In addition, it is widely believed that excessive collecting for private and commercial purposes is causing severe declines in butterfly numbers. Both adult and immature stages are taken by local and foreign collectors (4, 5) who are aware of the rarity of the species. This even happens on Corsica, where collecting is forbidden by French law.

Conservation Measures The conservation status of *Papilio hospiton* is widely understood and international concern is reflected in a number of laws and reports (4). The Corsican population of *Papilio hospiton* is protected under a French decree published on 22 August 1979. This is the same decree that protects *Papilio phorbanta* on Réunion and, as stated in the review of that species, the legislation draws attention to the species' plight, but fails to have any impact on the protection of its habitat and foodplants. Under the 1980 Endangered Foreign Species Act of the Netherlands, Dutch citizens may not import or purchase *P. hospiton*. In a report on the status of European butterflies commissioned by the Council of Europe and published in 1981, *P. hospiton* was classified as endangered and urgent action called for (4). The Sardinian population is not protected under Italian law and there is an urgent need for regional, if not national, recognition of the problem. Steps taken by the Piemonte Regional Council in protecting *Carabus olympiae* (Coleoptera: Carabidae) and encouraging the preparation of a recovery plan have set an excellent precedent for such an effort (2). No measures have been taken specifically to protect the habitat of the butterfly and the designation of suitable nature reserves on both islands is urgently needed (4). Detailed surveys of the populations on both islands are required as a matter of urgent priority. These should be followed by an immediate ecological study to evaluate foodplants and habitat requirements, with a view to gazetting and management of protected areas and, if necessary, additional protective legislation.

References

1. Bretherton, R.F. and De Worms, C.G. (1963). Butterflies in Corsica 1962. *Entomologists' Record and Journal of Variation* 75: 93–104.
2. Collins, N.M. (1984). Piemonte, Italy, protects a ground beetle. *Oryx* 18: 69.
3. Fausser, J. (1980). Observations concernant *Papilio hospiton* Gené en Haute–Corse. *Bulletin Liaison l'Association Entomologique d'Evreux* 5: 18–19.
4. Heath, J. (1981). *Threatened Rhopalocera (Butterflies) in Europe*. Nature and Environment Series (Council of Europe) No. 23, 157 pp.
5. Heath, J. (1984). Summary of comments received from a number of authorities during preparation of reference (3). *In litt.*, 27 February.
6. Higgins, L.G. and Riley, N.D. (1980). *A Field Guide to the Butterflies of Britain and Europe*. 4th ed., revised. Collins, London. 384 pp., 63 pl., 384 maps.
7. Panchen, A.L. and Panchen, M.D. (1973). Notes on the butterflies of Corsica, 1972. *The*

Entomologist's Record 85: 149–153, 198–202.
8. Watson, A. (1981). *Butterflies*. Kingfisher Books, London.
9. Whalley, P. (1981). *The Mitchell Beazley Guide to Butterflies*. Mitchell Beazley, London. 168 pp.

Papilio (Princeps) benguetanus Joicey and Talbot, 1923   VULNERABLE

Subfamily PAPILIONINAE   Tribe PAPILIONNI

Summary *Papilio benguetanus* is a butterfly which may be locally common, but has a very restricted distribution within the Cordillera Central of northern Luzon, Philippines. There are no known protected areas in this region. Studies are needed to assess the foodplant, limits of distribution, and the extent of threats to the survival of this butterfly.

Description *Papilio benguetanus* (mistakenly called *benguelana* by D'Abrera (2)) is a lovely butterfly with short tails and a forewing length of 44–48 mm (Plate 7.3). The sexes are similar in appearance, but the female is slightly larger (8).

UFW/UHW yellow with wide brown veins in the basal and discal areas, brown in the postdiscal and submarginal areas. A row of submarginal lunules is present on each wing, yellow on the forewings, orange on the hindwings, with a bluish tinge above the lunule on the anal angle. The female has brighter orange lunules on the hindwing and broader, paler brown–black streaks on its wings (8).

LFW/LHW similar to uppersides but paler, with less brown in the postdiscal and submarginal areas.

Distribution *Papilio benguetanus* is known only from northern Luzon in the Philippines (8). Its distribution is essentially the same as *Papilio chikae*, between Bontoc and Baguio in the Cordillera Central. To the east the River Cagayan valley is a great rice-growing region which separates the Cordillera Central from the Sierra Madre. The butterfly is not known from the Sierra Madre, although there could be access at a fairly high level at the southern end. Mt St Thomas and Mt Paoai are the most famous localities for this species.

Habitat and Ecology *Papilio benguetanus* is a relict species of continental origin, as is *Papilio chikae* (see separate review) (8). At times it has been considered an island race of the continental species *P. xuthus*, but it is quite distinct in appearance. Unlike *xuthus*, it has no known seasonal variation (8). On Mt St Thomas and Mt Paoai the most famous localities for the species, the butterflies occur in some numbers from March to June, when many individuals can be found flying along pathways and hill-topping between 7 a.m. and 11 a.m. (7, 8). Males sometimes fight each other quite fiercely in their territorial battles (7). In fine weather it associates with *Parantica phyle* (Danainae), *Appias phoebe* (Pieridae), *Celastrina* spp. (Lycaenidae) and *Hestina* spp. (Nymphalidae) (8). Both *Parantica phyle* and *Hestina* bear a superficial resemblance to *P. benguetanus* on the wing (8). Females are more secretive than males and the hostplant and immature stages are unknown (4, 8). Apart from the nominate species *P. benguetanus* is the only other member of the *xuthus* group (3). *P. xuthus* has a green eye-spotted caterpillar with segmental orange spots, feeding on Rutaceae (5).

Threats This attractive and rare butterfly clearly has a very limited distribution in northern Luzon and although it is rather more common than *Papilio chikae*, it may nevertheless suffer from a similar level of disturbance. It is listed under category Vulnerable in recognition of the human population pressures that are resulting in extensive habitat alteration for agriculture and forestry in Luzon, and also because the species is coveted by collectors (1). There are some indications that *P. benguetanus* may thrive in partially cleared areas and even in areas around vegetable growing operations, possibly even deriving benefit from man's activities (7). In this event its conservation status might require review. As stated in the review of *P. chikae*, the precise extent of habitat disruption in the Cordillera Central remains unknown. Baguio is an important summer resort and the surrounding region may be subject to recreational pressure as well as agricultural and forestry expansion. Much may depend upon the habitat of the foodplant, and its ability to withstand disturbance.

Conservation Measures *Papilio benguetanus* rarely appears on commercial lists but it may be the subject of substantial private collecting operations. It is to be hoped that entomologists visiting Luzon will exercise restraint. Rather than collect numerous specimens, visitors to the region could be encouraged to search for the foodplant and young stages, and to visit remote regions of the Cordillera Central in order to assess the distribution of the species more fully. It has not been possible to assess whether there are any functional national parks within the range of this butterfly, but this seems improbable. The Philippine national parks system is curently under review (6). More data are needed on the extent of habitat disruption at high altitude in the Cordillera Central. This information should be used to propose a site for a new national park, not only as a conservation area for wildlife, but also as a public amenity and for protection of the watersheds.

References

1. Ae, S.A (1983). *In litt.*, March 18.
2. D'Abrera, B. (1982). *Butterflies of the Oriental Region. Part 1. Papilionidae and Pieridae.* Hill House, Victoria, Australia. xxxi + 244 pp.
3. Hancock, D.L. (1983). Classification of the Papilionidae (Lepidoptera): a phylogenetic approach. *Smithersia* 2: 1–48.
4. Iwase, T. (1965). How *Papilio chikae* was found and named. *Tyo To Ga (Transactions of the Lepidopterists' Society of Japan)* 16: 44–47.
5. Munroe, E. (1961). The classification of the Papilionidae (Lepidoptera). *Canadian Entomologist* Supplement 17: 1–51.
6. Pollisco, F.S. (1982). An analysis of the national park system in the Philippines. *Likas–Yaman, Journal of the Natural Resources Management Forum* 3(12): 56 pp.
7. Treadaway, C.G. (1984). *In litt.*, 25 May.
8. Tsukada, E. and Nishiyama, Y. (1982). *Butterflies of the South East Asian Islands. Vol. 1 Papilionidae.* (transl. K. Morishita). Plapac Co. Ltd., Tokyo. 457 pp.

Papilio (Princeps) acheron Grose–Smith, 1887   RARE

Subfamily   PAPILIONINAE   Tribe   PAPILIONINI

Summary *Papilio acheron* is a rare montane species from the northern highlands of Borneo. Its known distribution is from Mt Kinabalu in Sabah to Mt Dulit in Sarawak, but research may reveal more localities further inland. Whilst some national parks appear to guarantee the livelihood of the species, vigilance is needed to ensure the integrity of the park boundaries.

Description *Papilio acheron* is a large, mainly black butterfly with a wingspan of 58–66 mm. Both sexes appear to mimic the sympatric *Atrophaneura nox* (3, 7, 8).

Male: upperside black with no markings but lighter bluish scales scattered over the discal part of the hindwing between the veins (3, 7, 8). Hindwing with a slightly scalloped outer margin but no tail. Underside also black, with a basal red spot on the LHW and a short golden or buff distal band enclosing three black postdiscal spots. Diffuse, blue postdiscal lunules extend from this band to the costal margin (3, 8).

Female: very similar to the male but with a paler ground colour; usually resembling the female of *A. nox* in having pale brown scaling bordering the veins of the upper part of the UFW. Occasionally a uniformly black male-like example is found (3, 8).

Distribution *Papilio acheron* is found in the mountains of Sabah, Sarawak and Brunei in north-western Borneo. It occurs at altitudes over 1000 m on Gunung (Mt) Kinabalu, south along the Crocker Range to Gunung Pagonprick and Bukit Retak in eastern Brunei and into northern Sarawak where it is known from Gunung Mulu and Gunung Dulit (2, 3, 5, 7, 8). It may also occur on other mountains in Sarawak, notably the Tamabo and Penambo Ranges, the Linau Balui Plateau and the Hose Mountains, and further east in the border area of the Eastern Province of Kalimantan (Indonesia), but such localities have not yet been reported.

Habitat and Ecology *Papilio acheron* is a rare and localized species from lower montane forest. Males fly low and rapidly on patrol routes along streams, often stopping to drink (7, 8). Females are much more scarce and the immature stages and foodplants are unknown (8). *P. acheron* is a member of the *memnon* group, in which the foodplant is normally rutaceous and the caterpillars have raised bands on the metathorax and first abdominal segment, the former ending in a prominent eye-spot (4, 6).

Threats Montane habitats in Borneo tend to be rather better protected than lowland ones. Both Kinabalu and Mulu are designated as national parks. At present *Papilio acheron* is not a matter for great conservation concern, but it must be emphasized that even montane habitats are not free of threat. Kinabalu National Park is under pressure from small farmers, copper mining concessions and tea plantations. The cutting of virtually all the forest on the Pinosuk plateau for cattle farming, sawmills and hydro-electric projects was a particularly significant loss (1). In the Eastern Province of Kalimantan where it borders with Sarawak lies the giant recently declared Sungai Kayan—Sungai Mentarang Reserve (1.6 million ha). The reserve is largely hill forest with some lowland and montane forest (elevation range 200–2558 m) and may well include *Papilio acheron*. However, the area has always supported quite a high density of people; their influence can be seen in the extensive grasslands and the effects of hunting (9). (See also the review of *Graphium procles*).

Conservation Measures Further research in the highlands of northern Borneo may extend the known distribution of *P. acheron*. In addition to distributional study, vigilance is needed to ensure that national park boundaries are not violated by farmers, and that parcels of land are not excised for mining, large-scale plantations of cash crops, or extensive tourist facilities, particularly on Kinabalu. In the Kayan–Mentarang Reserve in Kalimantan surveys are needed not only to check for the butterfly, but also with a view to revisions of the boundaries to exclude disturbed areas and include more lowland areas (9).

References

1. Barlow, H.S. (1983). *In litt.*, 29 June.
2. Cassidy, A.C. (1982). An annotated checklist of Brunei butterflies including a new species of *Catapaecilma* (Lycaenidae). *Brunei Museum Journal* 5: 202–272.
3. D'Abrera, B. (1982). *Butterflies of the Oriental Region. Part 1. Papilionidae and Pieridae.* Hill House, Victoria, Australia. xxxi + 244 pp.
4. Hancock, D.L. (1983). Classification of the Papilionidae (Lepidoptera): a phylogenetic approach. *Smithersia* 2: 1–48.
5. Holloway, J.D. (1978). Butterflies and Moths. In *Kinabalu Summit of Borneo* Sabah Society Monograph, 25–278.
6. Munroe, E. (1961). The classification of the Papilionidae (Lepidoptera). *Canadian Entomologist* Supplement 17: 1–51.
7. Robinson, J.C. (1975–6). Swallowtail butterflies of Sabah. *Sabah Society Journal* 6: 5–22.
8. Tsukada, E. and Nishiyama, Y. (1982). *Butterflies of the South East Asian Islands. Vol. 1 Papilionidae.* (transl. K. Morishita). Plapac Co. Ltd., Tokyo. 457 pp.
9. UNDP/FAO National Parks Development Project (1981). *National Conservation Plan for Indonesia Vol. 5: Kalimantan.* FAO, Bogor.

Papilio (Princeps) jordani Fruhstorfer, 1902   RARE

Subfamily PAPILIONINAE   Tribe PAPILIONNI

Summary *Papilio jordani* is a large and striking swallowtail from the Minahassa Peninsula in northern Sulawesi, Indonesia. It is extremely rare and its life-cycle and habits are very poorly known. Hopefully it breeds within the boundaries of the new Dumoga–Bone National Park. Surveys of the region and studies of the butterfly's breeding and feeding habits are needed.

Description *Papilio jordani* is a beautiful, large, tailless butterfly which was placed in the *fuscus* group by Munroe (5) but has since been transferred to the *polytes* group (2, 4). The male differs markedly from the female, which mimics *Idea blanchardii* (4, 8). The forewing length is 75–80 mm (male) or 80–85 mm (female) (1, 8).

Male: almost entirely black–brown with a scalloped outer edge to the hindwing. There is a series of white, outer marginal fringe spots on both wings, those on the forewing becoming larger towards the tornus, and a band of large, elongated, white discal spots on the hindwing (1, 8).

Female: white with black or dark brown scaling over the veins, forming lines or streaks in the cell, and between the veins of the forewing (1, 8). The hindwing has some dark submarginal wedge-shaped spots and some smaller discal ones, with dark scales scattered between them. The forewing also has dark scales scattered over the distal region (1, 8).

Distribution *Papilio jordani* is now found only on the Minahassa Peninsula, which is the northern limb of Sulawesi (Celebes), Indonesia. The two known localities are Tolitoli in the north-west and Doluduo near Manado in the extreme north-east of the island (1). The type locality, however, is apparently in southeastern Sulawesi (3). The reason why the type locality is so far removed from the area where *P. jordani* is now found remains unexplained. Either the species' range has retracted quite drastically in the past 85 years, or the type specimen was a vagrant or mistaken locality.

Habitat and Ecology This extremely rare and local butterfly is very poorly known. Males have been observed to fly slowly and straight along sunny streams keeping to a height of about 3 m (8). Females mimic *Idea* in their flight as well as their appearance, fluttering gently between the banks of streams, but when alarmed can fly off swiftly in the manner of other *Papilio* species (8). *P. jordani* associates with *P. ascalaphus* but whilst *jordani* usually remains near streams in primary forest, *ascalaphus* inhabits areas near villages (8). Tsukada and Nishiyama note that mimics of *Idea* are often rare (8). The mimics *Graphium idaeoides* (see separate review), *Papilio veiovis* and *Elymnias kunstleri* (Satyridae) are certainly rare, but *Graphium delesserti* is often common (8). The larval foodplant of *P. jordani* is unknown and the young stages undescribed. However, other members of the *polytes* group have larvae with metathoracic and abdominal bands present, and vestigial eye-spots (2). The pupa is rough and curved and the foodplants are in the Rutaceae (2). Adults occur all year round, with a peak of emergence in November and February (8).

Threats Sulawesi is well known for its richness and high levels of endemicity in Lepidoptera, but many butterfly habitats are being destroyed by extensive deforestation (7). Large areas stretching between Makassar and the Northern and

Central Districts are already covered with the unpalatable and useless grass *Imperata cylindrica* (lalang), a sure sign of over-exploitation. Sulawesi has 10.4 million people in a land area of 190 000 sq. km, but they are not evenly distributed. Settlement is concentrated on the southern arm and the Minahassa Peninsula (6), where this butterfly lives. Nevertheless, *Papilio acheron* is too poorly known to be sure of the extent to which this concentration threatens its survival. The peninsula is mountainous with a natural vegetation of lowland and montane forest. Logging, shifting cultivation and some irrigated rice-growing in the valley bottoms are the main uses to which the land is put. Irrigation projects in the valleys of the Dumoga and Bone rivers in the centre of the Minahassa Peninsula have received substantial support from the World Bank.

Conservation Measures At the end of the 1970s, a WWF/IUCN project in northern Sulawesi identified the catchment areas of the Dumoga and Bone rivers as the most pertinent location for a large conservation area that would afford protection to the rich and varied flora and fauna of the province (6). This culminated in an agreement between the Government of Indonesia and the World Bank, whereby the latter would give considerable financial support towards the development of the proposed Dumoga–Bone National Park, in co-operation with WWF. The project aims to increase threefold the production of rice on 13 000 ha of prime agricultural land, whilst maintaining the vital watersheds which feed the land by preserving 280 000 ha of hilly country as the new national park. The project began in 1980. WWF personnel were employed to write a management plan and the park has now been properly gazetted. Encroachment, which was at first severe, has now stabilized.

In 1985 the Royal Entomological Society of London is celebrating its 150th anniversary by organizing 'Project Wallace' in this very important park. With over 100 entomologists taking part, *P. jordani* will certainly be tracked down if it occurs there. Once found, it will hopefully be possible fully to elucidate the life-cycle and biology of the species.

References

1. D'Abrera, B. (1982). *Butterflies of the Oriental Region. Part 1. Papilionidae and Pieridae.* Hill House, Victoria, Australia. xxxi + 244 pp.
2. Hancock, D.L. (1983). Classification of the Papilionidae (Lepidoptera): a phylogenetic approach. *Smithersia* 2: 1–48.
3. Haugum, J. (1984). *In litt.*, 14 April.
4. Haugum, J., Ebner, J. and Racheli, T. (1980). The Papilionidae of Celebes (Sulawesi). *Lepidoptera Group of 1968 Supplement* 9: 21 pp, 1 map, 2 pl.
5. Munroe, E. (1961). The classification of the Papilionidae (Lepidoptera). *Canadian Entomologist* Supplement 17: 1–51.
6. National Parks Development Project, UNDP/FAO (1982). *National Conservation Plan for Indonesia. Vol. VI: Sulawesi.* FAO, Bogor.
7. Straatman, R. (1968). On the biology of some species of Papilionidae from the island of Celebes (East-Indonesia). *Entomologische Berichten* 28: 229–233.
8. Tsukada, E. and Nishiyama, Y. (1982). *Butterflies of the South East Asian Islands. Vol. 1 Papilionidae.* (transl. K. Morishita). Plapac Co. Ltd., Tokyo. 457 pp.

Papilio (Princeps) weymeri Niepelt, 1914   RARE

Subfamily PAPILIONINAE   Tribe PAPILIONNI

Summary *Papilio weymeri* is endemic to Los Negros Island and Manus Island, Papua New Guinea. There is little information on its biology and distribution. Further studies and surveys are needed. The island has extensive rain forests and secondary growth. Plans for sustainable development should include conservation measures for this endemic butterfly and other forms of wildlife.

Description *Papilio weymeri* is a large butterfly with a forewing length of about 70 mm (male) or 75 mm (female) (Plate 7.4). The Papuan group of swallow-tails of which this species is a member was described by Munroe as the *aegeus* group (7), but Hancock has made a few changes and renamed it the *gambrisius* group (4). *Papilio weymeri* is used here in preference to the synonym *cartereti* Oberthür, although seniority may be disputed. Both names were published in the first six months of 1914 (2).

Male: UFW/UHW black with a cream postdiscal band of large spots across the forewing, and a discal band of cream streaks over the hindwing. Both wings have white fringe spots. The hindwing is slightly scalloped and has no tail (1, 10).

Female: UFW blackish-brown relieved by a postdiscal band of white streaks. These become very diffuse and are suffused with dark scales in the submarginal region where they terminate as narrow yellow spots. There are also some cream fringe spots (1, 10).

UHW blackish-brown with yellow submarginal lunules, relatively large, iridescent blue postdiscal spots, and an orange–red eyespot. Blue scaling occurs on the inside of the eye-spot, yellow spots inside some of the blue postdiscal spots, and there are cream fringe spots (1, 10).

Distribution *Papilio weymeri* is endemic to Manus and Los Negros Islands of the Admiralty group in the northern part of Papua New Guinea. Manus Island is about 80 km long and 32 km wide, with an approximate area of 2100 sq. km (6). Los Negros is much smaller.

Habitat and Ecology The young stages of *P. weymeri* have been collected and photographed but not described in print (5). The adults apparently have the unique habit of laying their eggs on the upperside of the hostplant's leaves (5). The early stages have been seen on *Micromelum minutum* (Rutaceae) (8), and possibly on *Flindersia* sp. (Rutaceae) (3). Many of Manus Island's forests were disturbed during the Second World War, but have since returned to secondary forest and bush (9). It seems likely that the rutaceous shrubs used by *P. weymeri* can regenerate in secondary growth. It is not clear whether *P. weymeri* is a denizen of rain forest on the island, or of forest edges and areas of regeneration. Females have been recorded flying slowly in small, stunted brush; males fly erratically and never seem to settle (2). Females are rarer than males; June to August are good flight months (2).

Threats Since *Papilio weymeri* has adapted to areas disturbed during wartime (9), it would appear to be a relatively resilient species. The human population of Manus is relatively large, but is concentrated along the shorelines (9). The interior is hundreds of square kilometres of wild and sparsely populated country with large expanses of natural and regenerating vegetation (3, 6, 9). Nevertheless, other endemics on Manus Island are threatened and there is a need for careful monitoring of *Papilio weymeri*. The Manus Green Tree Snail *Papustyla pulcherrima* (Gastropoda: Camaenidae) is endemic to the Manus forests and threatened by logging of the trees upon which it lives, and by large scale collecting (11).

Conservation Measures Basic details of habitat and life cycle are needed for this butterfly. It is known to feed on Rutaceae, but whether disturbance of natural forest is likely to threaten or encourage the species remains unknown. Studies on Manus are a clear priority, followed by monitoring of butterfly populations and human development on the island.

It has been proposed that a Wildlife Management Area could be designated in central Manus, where the rain forest is relatively undisturbed (9). Wildlife Management Areas are an expression of local as well as national concern for wildlife, being set up on privately owned or traditionally held land. Such areas could effectively serve to protect *Papilio weymeri*, the Manus Tree Snail and other wildlife (9). There is already a certain amount of ranching of *P. weymeri* on Manus (9). This could be further encouraged as a measure for both conservation and development. Most of Manus Island is unsuitable for arable or tree crops, or pasture improvement (6). The comprehensive 1980 report by the Papua New Guinea Office of Environment and Conservation on the resources of Manus Province (6) is an ideal starting point for the development of the island's resources in an ecologically sustainable manner.

References

1. D'Abrera, B. (1971). *Butterflies of the Australian Region*. Lansdowne Press, Melbourne. 415 pp.
2. Ebner, J.A. (1971). Some notes on the Papilionidae of Manus Island, New Guinea. *Journal of the Lepidopterists' Society* 25: 73–80.
3. Fenner, T.L. (1983). *In litt.*, 15 March.
4. Hancock, D.L. (1983). Classification of the Papilionidae (Lepidoptera): a phylogenetic approach. *Smithersia* 2: 1–48.
5. Hutton, A.F. (1983). *In litt.*, 28 March.
6. Kisokau, K.M. (1980). Manus Province, a physical resource inventory. Report of the Office of Environment and Conservation, Papua New Guinea. 73 pp.
7. Munroe, E. (1961). The classification of the Papilionidae (Lepidoptera). *Canadian Entomologist* Supplement 17: 1–51.
8. Parsons, M.J. (1983). *In litt.*, 2 March.
9. Pyle, R.M. and Hughes, S.A. (1978). Conservation and utilisation of the insect resources of Papua New Guinea. Report of a consultancy to the Wildlife Branch, Dept. of Nature Resources, Independent State of Papua New Guinea. 157 pp. unpublished.
10. Smart, P. (1975). *The Illustrated Encyclopedia of the Butterfly World*. Hamlyn, London. 275 pp.
11. Wells, S.M., Pyle, R.M. and Collins, N.M. (1983). *The IUCN Invertebrate Red Data Book*. IUCN, Gland. L + 632 pp.

Papilio (Princeps) sjoestedti Aurivillius, 1908   RARE

Subfamily PAPILIONINAE   Tribe PAPILIONNI

Summary *Papilio sjoestedti*, sometimes known as the Kilimanjaro Swallowtail, flies in the montane forests of Mt Meru, Mt Kilimanjaro and Ngorongoro in north-eastern Tanzania. It has a very restricted range but is well protected in national parks.

Description *Papilio sjoestedti* is a relatively large tailless swallowtail of the *cynorta* group (5, 7) with a forewing length of 45–54 mm (Plate 7.5). The female is slightly larger than the male and has a different pattern; like most other members of the *cynorta* group, the female mimics members of the genus *Amauris* (Danaidae) (1, 2, 7, 8).

Male: ground colour blackish-brown with white markings consisting of a straight median band of postdiscal spots and a series of fringe spots on the forewing (1, 3, 8). The hindwing has a slightly broader, unbroken, white discal band and a series of white fringe spots (1, 3, 8).

Female: The forewing has fewer, more widely spaced white postdiscal spots and several discal spots. The hindwing has a series of large fringe spots and a very broad yellow–ochre discal median band (1, 3, 8).

The lower surfaces are darker and mimetic of Acraeinae (5). The female has a truncated orange–brown basal area on the LHW (1).

Distribution *Papilio sjoestedti* occurs as two subspecies in north-eastern Tanzania. The nominate form *P. s. sjoestedti* is found on Mt Meru, the highlands around the rim of Ngorongoro Crater, and further south in the hills near Mbulu (6), while *P. s. atavus* is only known from Mt Kilimanjaro (1, 3, 6, 8).

Habitat and Ecology *Papilio sjoestedti* flies in montane forest at altitudes between 1900 and 2800 m (1, 6). The immature stages and foodplants are unknown, but in other members of the *cynorta* species-group the caterpillars are green when mature, with metathoracic and abdominal bands and no eye-spots, feeding on Rutaceae. Possible foodplant genera include *Calodendron*, *Vepris* and *Teclea* (4). The mountains occupied by *P. sjoestedti* are all volcanic, and have a much more recent origin than the Usambaras, Uzungwas and other smaller massifs of eastern Tanzania.

Threats *Papilio sjoestedti* is listed here because it has a very limited range within an East African vegetation type that is being depleted daily. In general, the main threats are forest clearance for agriculture and plantation forestry. However, in this case it is heartening to note that the range of the species lies almost entirely within the boundaries of protected areas. Highly priced specimens of this species have recently appeared in dealers' lists.

Conservation Measures Mt Meru, Mt Kilimanjaro and the Ngorongoro Crater are all national parks, open to visitors but protected from incursion by agriculturalists and foresters. Although there is no immediate threat to the forested areas in these parks, there is increasing evidence of inadequate management resources within the Tanzanian National Parks authority. This is particularly manifest in the disturbing levels of poaching reported from Ngorongoro Crater, a park not only designated as a Biosphere Reserve under the Man and the Biosphere programme of UNESCO, but also proposed as a World Heritage Site of international importance.

References

1. Carcasson, R.H. (1960). The swallowtail butterflies of East Africa (Lepidoptera. Papilionidae). *Journal of the East African Natural History Society* Special Supplement 6: 33 pp. + 11pl. (Reprinted by E.W. Classey, Faringdon, 1975).
2. Carcasson, R.H. (1980). *Collins Handguide to the Butterflies of Africa*. Collins, London. Hardback edition, 188 pp.
3. D'Abrera, B. (1980). *Butterflies of the Afrotropical Region*. Lansdowne Editions. Melbourne. xx + 593 pp.
4. Hancock, D.L. (1979). Systematic notes on three species of African Papilionidae (Lepidoptera). *Arnoldia* 8(33): 1–6.
5. Hancock, D.L. (1983). The classification of the Papilionidae (Lepidoptera): a phylogenetic approach. *Smithersia* 2: 1–48.
6. Kielland, J. (1983). *In litt.*, 6 March.
7. Munroe, E. (1961). The classification of the Papilionidae (Lepidoptera). *Canadian Entomologist* Supplement 17: 1–51.
8. Williams, J.G. (1969). *A Field Guide to the Butterflies of Africa*. Collins, London 238 pp.

Papilio (Princeps) manlius Fabricius, 1798   INDETERMINATE

Subfamily PAPILIONINAE   Tribe PAPILIONNI

Summary *Papilio manlius* is said to be quite common in Mauritius, despite the widespread loss of native vegetation there. The caterpillars are able to survive on cultivated *Citrus*, but native habitat is virtually restricted to the Black River Gorges.

Description The male of this short-tailed butterfly in the *nireus* species-group (8) is black with bright greenish-blue markings. The female, at first described as a separate species (9), is paler in colour. The forewing length is 42–52 mm, and the wingspan is 70–89 mm.

Male: UFW ground colour black with a median band of iridescent greenish-blue spots and yellow fringe spots.

UHW slightly scalloped with a short tail and similar greenish-blue spots.

LFW/LHW blackish-brown with lighter wing margins and on the hindwing a yellow tinge, a white band conspicuous in flight (15), white submarginal spots and a white lunule (4, 9, 11).

Female: ground colour brown with pale bluish-green spots in a similar pattern (4, 9, 11), lacking the white band found on the underside of the male (15).

Distribution *Papilio manlius* is endemic to Mauritius, in the Mascarene Islands, Indian Ocean. It is a close relative of *P. phorbanta* of Réunion (see separate review), *P. epiphorbas* of Madagascar and the other members of the *nireus* species-group, or 'blue-banded papilios' of the mainland (3, 5).

Habitat and Ecology The vegetation types of Mauritius range from coastal marshes and palm savannas to lowland, upland and dwarf forests (14). The lowland forests were said to be extremely productive and were one of the main reasons for the settlement of the island (10). They once covered much of the island below about 220 m, where the rainfall is 1000–2500 mm per year (10). The upland forests once covered most of the fertile uplands of the Plaine Wilhelms and the south of the island, where the rainfall is heavy (about 3800 mm per year), and plant growth luxuriant, although the soils are poor (10).

Little is known of the habits and biology of *Papilio manlius* in its natural state, mainly because so little of its habitat remains (see Threats). As in other members of the *nireus* species group, the natural foodplant would probably be a wild member of the family Rutaceae, but this has not been ascertained. *P. manlius* caterpillars are bright green at all stages, with a red osmeterium (5). The adults are reported to be common throughout the island and in no danger of extinction because they can breed on *Citrus* (15). However, the populations on natural vegetation are believed to be small and are a matter of conservation concern.

Threats The majority of the natural vegetation of Mauritius has been cut down and the land developed for agricultural purposes, particularly for growing sugar cane. Only 2 per cent of the 1865 sq. km of Mauritius remains under native vegetation (6). The lowland forests are nearly extinct, only a few impoverished remnants remain on the western slopes of the mountains between Mt du Rempart and Chamarel (10). Upland forest is restricted to about 1600 ha in the south-west of the island, around the Black River Gorges (10). These gorges are believed to be one of the few remaining areas of natural forest suitable for the butterfly. Although the gorges are strictly protected (1, 5) they may be doomed because of the encroachment of introduced competitors such as Chinese guava (*Psidium cattleianum*) and Privet (*Ligustrum walkeri*), which strangle the understorey and prevent tree regeneration (5). The major tree species are hardy plants adapted to cyclones. They are long-lived and robust, but reproduce at a low rate and at irregular intervals. As a result they must be considered highly vulnerable to disturbance of this kind (1, 10). The situation is a parallel of the problems of *Papilio aristophontes* with encroaching bananas in the forests of Grande Comore, but far more serious (see separate review).

Because of its ability to breed on cultivated *Citrus*, *P. manlius* itself is seemingly reprieved from extinction, even if its natural habitat disappears completely. Nevertheless, it is reviewed here and given Indeterminate status in order to draw attention to its circumstances. It is certainly undesirable that a species should be threatened with the possibility of being entirely dependent upon man and his crops. There are many pitfalls to such an existence. For example, insecticidal measures are no doubt periodically taken against the more serious pest *P. demodocus* which, however, has an assured existence elsewhere. Developments in the efficacy of insecticides, or of biological control measures, could threaten *P. manlius*. In addition, *P. demodocus* is known to be a competitive and aggressive species, capable of ousting other species. It has been suggested that *P. manlius* could suffer in this way (13). However, records indicate that *P. demodocus* was absent from Mauritius in 1833 (2) and 1866 (12), but present in 1908 (7). During the past three quarters of a century some sort of balance between the species has presumably been struck.

Conservation Measures Because of its ability to feed on *Citrus*, it seems unlikely that *P. manlius* is threatened with extinction. Nevertheless, the likelihood that the remaining naturally-breeding populations are restricted to native forests that are declining in the face of encroachment of exotic species is a matter of concern not only for *P. manlius* but for many other forest endemics as well. Mauritius has a wealth of endemic plants, invertebrates and vertebrates, all of which are under threat through habitat destruction. Three butterfly subspecies, *Hypolimnas dubius drucei*, *Antanartia borbonica mauritiana* and *Salanis angustina vinsoni* are believed to be already extinct on Mauritius, although the first of these still survives on the Comoro Islands (5). *Cyclyrius mandersi*, an endemic lycaenid butterfly of coastal regions, may be threatened with total extinction (5).

It is clear that careful management and recovery plans are needed for the remaining native vegetation of Mauritius. Every encouragement should be given to research programmes that seek further to facilitate the protection of these internationally important localities.

References

1. Barnes, M.J.C. (1982). *In litt.*, 2 August.
2. Boisduval, J.B.A.D. de (1833). *Faune Entomologique de Madagascare, Bourbon et Maurice*. Paris.
3. Carcasson, R.H. (1980). *Collins Handguide to the Butterflies of Africa*. Collins, London. Hardback edition, 188 pp.
4. D'Abrera, B. (1980). *Butterflies of the Afrotropical Region*. Lansdowne Editions, Melbourne. xx + 593 pp.
5. Davis, P.M.H. and Barnes, M.J.C. (in press). The butterflies of Mauritius. *Journal of Research on the Lepidoptera*.

6. Jones, C.G. (1980). Parrot on the way to extinction. *Oryx* 15: 350–354.
7. Manders, N. (1908). The butterflies of Mauritius and Bourbon. *Transactions of the Entomological Society of London* 1907: 429–454.
8. Munroe, E. (1961). The classification of the Papilionidae (Lepidoptera). *Canadian Entomologist* Supplement 17: 1–51.
9. Paulian, R. and Viette, P. (1968). *Faune de Madagascar. XXVII Insectes Lépidoptères Papilionidae*. O.R.S.T.O.M. and C.N.R.S., Paris. 97 pp., 19 pl. (2 col.), 34 figs.
10. Procter, J. and Salm, R. (1974). Conservation in Mauritius. Report to IUCN/WWF.
11. Smart, P. (1975). *The Illustrated Encyclopedia of the Butterfly World*. Hamlyn, London. 275 pp.
12. Trimen, R. (1866). Notes on the butterflies of Mauritius. *Transactions of the Entomological Society of London* 1866: 329–344.
13. Vane–Wright, R.I. (1983). Pers. comm., 9 June.
14. Vaughan, R.E. and Wiehe, P.O. (1937). Studies on the vegetation of Mauritius, I. A preliminary survey of the plant communities. *Journal of Ecology* 25: 289–343.
15. Vinson, J.M. (1938). Catalogue of the Lepidoptera of the Mascarene Islands. *Mauritius Institute Bulletin* December 1938 : 1–69.

Papilio (Princeps) phorbanta Linnaeus, 1771   VULNERABLE

Subfamily PAPILIONINAE   Tribe PAPILIONNI

Summary *Papilio phorbanta*, the Papillon La Pature, is protected by law on the island of Réunion where it is now endemic, having been extinct in the Seychelles since 1890. Without adequate protection of its habitat and food plants the species could still be very seriously threatened, despite the legislation. Surveys of its habitat and designation of protected areas are urgently required.

Description This beautiful short-tailed butterfly in the *nireus* species-group (10) has a forewing length of 40–55 mm. The male is black and blue and the female is brown with white markings (6, 11) (Plates 8.1 and 8.2). This difference between the sexes is unique in the *nireus* species group (4), and may be caused by mimicry of *Euploea goudotii* (Danainae) (7).

Male: UFW/UHW black with a broad median band and blue spots.

LFW/LHW ground colour blackish brown becoming yellowish at the wing margins, with whitish submarginal spots (11).

Female: UFW/UHW red–brown with white submarginal spots (2, 11).

LFW/LHW brown with a pale grey marginal band, black triangular spots and a grey mark on the hindwing (2, 11).

Distribution This species is now endemic to the island of Réunion, a French Overseas Department 900 km east of Madagascar in the Indian Ocean (11). The much smaller subspecies *nana* Oberthür is known only from a single pair from the Seychelles and is believed to be extinct since 1890 (6).

Habitat and Ecology Réunion is a volcanic island with a remarkable topography. The highest point is 3069 m above sea level and 61 per cent of the land surface is at altitudes over 1000 m. Because of the altitude and ruggedness of the terrain, large areas of upland vegetation remain unaffected by man. In the lowlands however, virtually no forest remains. *Papilio phorbanta* is a forest-dwelling butterfly which, like other members of the *nireus* species group, feeds on Rutaceae. The 4 cm long green or yellow–green caterpillars (11) are nowadays usually found on *Citrus* trees, to which the species has apparently adapted. The 3 cm long pupae are known to be collected and raised in captivity for commercial purposes. *P. phorbanta* has long used *Citrus* as a food source, having been recorded in garden fruit trees by Manders in 1908 (7) and by Boisduval as long ago as 1833 (1, quoted in 5). However, the extent of its present colonization of *Citrus* groves has not been reported.

Threats It has been suggested that the pair of *Papilio phorbanta nana* from the Seychelles was either artificially introduced or wind-blown from Réunion (2, 6), but in view of the unusually small size of the pair, such a provenance seems unlikely. Although the origin of the subspecies was probably Réunion, the small size suggests a lengthy period of isolated evolution and natural selection. In the absence of further evidence there is no reason to suppose that *nana* was anything but a rare subspecies of *P. phorbanta*, now extinct for unknown reasons. Such an event may act as a warning to prevent the same fate for the nominate subspecies on Réunion.

Well into the 1950s *Papilio phorbanta* was still common on Réunion (13). However, with the rapidly expanding human population, more and more land has been turned over to agriculture, mainly for sugar (9). Although this has caused severe reductions in the butterfly's range, there are still significant areas of upland forest on Réunion, theoretically protected due to the rugged relief (13). Evergreen rain forest still covers a large belt around Piton des Neiges (3069 m) and Piton de la Fournaise (2631 m) (9). Réunion is the largest and still the least ecologically disturbed of the Mascarene islands, but deforestation continues, often in favour of reafforestation with Japanese red cedar (*Cryptomeria japonica*). A number of vertebrates have become extinct in the 300 years since human settlement (9) and there is no reason to suppose that many invertebrates have not disappeared also. The ability of the larvae of *P. phorbanta* to feed on *Citrus* trees might be expected to ensure the butterfly's survival even in the face of destruction of its natural forest habitats, but three factors mitigate against this. Firstly, feeding on *Citrus* is no insurance against destruction because of the stringent insecticidal measures taken against species that are regarded as pests of that crop. Secondly, there is a suspicion that the introduction of a parasitic tachinid fly 10–15 years ago to control the caterpillars of *P. demodocus* may have also taken a toll of *P. phorbanta* (13). Thirdly, *P. demodocus* itself may competitively exclude *P. phorbanta* from the *Citrus* groves, as it is suspected of doing for *P. manlius* on Mauritius. *Papilio demodocus*, like *P. demoleus*, is an aggressive butterfly that actively chases other species (2). Even were these fears to prove unfounded, it is quite unacceptable for wild species to decline so far that they are only able to survive as a minor pest of man's agricultural labours.

Conservation Measures In recognition of the increasing threat to the species the French Ministers of Environment and Agriculture declared on 22 August 1979 that *P. phorbanta* would henceforth be a protected species (8). The decree outlawed the destruction or removal of eggs, caterpillars, pupae and adults and forbade trade in the species. Such legislation has served to draw attention to the plight of *P. phorbanta*, but failed to have any impact in the more important matter of protecting the butterfly's natural habitat or food plants. In addition, it has not encouraged much-needed biological studies, nor facilitated ranching or farming of specimens for release into the wild. During the five years since the decree was passed, there is no evidence of any conservation effort being made on behalf of this species. There are no recent reports of the butterfly's status, but a new survey is currently under way (3).

References

1. Boisduval, J.B.A.D. de (1833). *Faune Entomologique de Madagascare, Bourbon et Maurice*. Paris.
2. D'Abrera, B. (1980). *Butterflies of the Afrotropical Region*. Lansdowne Editions, Melbourne. xx + 593 pp.
3. Dove, H. (1983). *In litt.*, 18 April.
4. Hancock, D.E. (1983). Classification of the Papilionidae (Lepidoptera): a phylogenetic approach. *Smithersia* 2: 1–48.
5. Hancock, D.E. (1983). *In litt.*, 21 June.
6. Legrand, H. (1959). Note sur la sous-espèce *nana* Ch. Oberthür de *Papilio phorban* Linné des îles Seychelles [Lep., Papilionidae]. *Bulletin de la Société Entomologique de France* 64: 121–123.
7. Manders, N. (1908). The butterflies of Mauritius and Bourbon. *Transactions of the Entomological Society of London* 1907: 429–454.
8. Ministère de l'environnement and Ministère de l'agriculture (1979). Liste des insectes protégés on France. *Journal Officiel* 22 August: 93–94.

9. Moutou, F. (1984). Wildlife on Réunion. *Oryx* 18: 160–162.
10. Munroe, E. (1961). The classification of the Papilionidae (Lepidoptera). *Canadian Entomologist* Supplement 17: 1–51.
11. Paulian, R. and Viette, P. (1968). *Faune de Madagascar. XXVII Insectes Lépidoptères Papilionidae.* O.R.S.T.O.M. and C.N.R.S., Paris. 97 pp.
12. Smart, P. (1975). *The Illustrated Encyclopedia of the Butterfly World.* Hamlyn, London. 275 pp.
13. Viette, P. (1983). *In litt.*, 6 July.

Papilio (Princeps) aristophontes Oberthür, 1897 INDETERMINATE

Subfamily PAPILIONINAE Tribe PAPILIONNI

Summary *Papilio aristophontes* is endemic to the forests of Anjouan, Moheli and Grande Comore, Comoro Islands. So far the impact of subsistence agriculture has been mainly in the lowlands, but rapid population growth requires that forest reserves be designated in order to protect the wildlife, particularly on Grande Comore. The butterflies still survive in upland forests, but research is needed to study the biology and distribution of the species and to locate suitable areas of its habitat for protection.

Description *Papilio aristophontes* was originally described as a full species but was subsequently considered to be a subspecies of *P. (P.) nireus*, albeit very distinct (1, 5). D'Abrera, following Carcasson, recently reinstated it to full species rank (3), a recognition which concurs with the views of lepidopterists familiar with the butterfly in the wild (2, 6). *P. aristophontes* is in the *nireus* group (4), and the male and female differ somewhat in appearance. The male is black with iridescent blue markings and a brown lower surface, whilst the female is brown and olive-green. The forewing length is 48–52 mm, and the wingspan is 80–90 mm. The young stages are unknown.

Male: UFW black with an iridescent blue median band and five blue spots.

UHW also black with an iridescent blue band, a row of blue submarginal spots, a scalloped margin to the hindwing and virtually no tail.

LFW/LHW. The LFW is entirely blackish-brown; LHW red–brown with black veins and a narrow cream submarginal band with a silver reflection.

Female: UFW/UHW brown with a dull olive green median band and a submarginal band of yellowish-olive lunules, continuous over both wings.

LFW/LHW red–brown with a silvery submarginal band continuous over both wings; a fawn brown median band to the hindwing with pinkish-grey scales on the outside, and a white abdominal spot.

Distribution *Papilio aristophontes* is endemic to the forests of the Comoro Islands, between Mozambique and Madagascar. The main stronghold is Grande Comore, but there are also populations on Moheli and Anjouan (2).

Habitat and Ecology The general ecology of the four Comoro Islands has been briefly described in the review of *Graphium levassori*. *Papilio aristophontes* is another inhabitant of the forests of La Grille (1087 m) and Karthala (2560 m), the two volcanic regions of Grande Comore. In 1980 and 1983 it was seen flying at Oussoudjou and other places in the massif of La Grille (7, 8) and at Nioumbadjou (550 m) (8), Hantsongoma (1000 m) and other localities up to 1700 m (6) on Karthala. The species may be locally abundant, flying along the edges of clearings and paths (7). The larval feeding habits have not been documented but, like the other members of the *nireus* species group, the caterpillars probably feed on wild Rutaceae (4). They have also been reported on wild-growing trees of the domesticated lemon (2).

Threats Although slash and burn agriculture is comparatively uncommon in the Comoros, the forests on Karthala and more particularly on La Grille are retreating from a more insidious threat, the prevention of forest regeneration by the planting of a dense understorey of banana trees. Bananas are one of the staple crops of the rapidly growing population of the Comoros. The main agricultural areas are coastal and lowland, but population pressure is forcing agricultural development to altitudes of up to 1200 m (7), where bananas, guavas, vanilla and ylang ylang (*Cananga oderata*) can still be grown. On the smaller islands of Anjouan and Moheli there is much less high ground than on Grande Comore and therefore agriculture is even more extensive. The distinctive population of *P. aristophontes* on Moheli is under serious threat from agricultural conversion and intensification (2).

Conservation Measures Despite increasing modification of its habitat, *P. aristophontes* still survives in good numbers on La Grille and Karthala (Grande Comore). However, the species is more seriously threatened by loss of habitat on Moheli and Anjouan (2). Throughout the Comoros the human population is growing rapidly, and it is inevitable that the threats to this and other endemic forest species will gradually increase. At the moment there are no protected areas on the Comoro Islands and it is clearly necessary to designate representative reserves at the earliest opportunity. Already the forests of La Grille are seriously damaged, and suitable areas of undisturbed middle altitude forest is hard to find on Karthala (8). Nevertheless, some low altitude forest could be set aside for natural regeneration. As proposed in the review of *Graphium levassori*, all land over 1200 m altitude on Karthala might be given protected status. This would inconvenience few people and would ensure the survival of many forest species (8).

Conservation measures specifically for *Papilio aristophontes* should be incorporated into efforts to conserve representative biotopes. Studies of its distribution and reproductive biology are needed.

References

1. Carcasson, R.H. (1960). The swallowtail butterflies of East Africa (Lepidoptera, Papilionidae). *Journal of the East African Natural History Society* Special Supplement 6: 33 pp. + 11 pl. (Reprinted by E.W. Classey, Faringdon, 1975).
2. Collins, S. (1983). *In litt.*, 12 July.
3. D'Abrera, B. (1980). *Butterflies of the Afrotropical Region.* Lansdowne Editions, Melbourne. xx + 593 pp.
4. Munroe, E. (1961). The classification of the Papilionidae (Lepidoptera). *Canadian Entomologist* Supplement 17: 1–51.
5. Paulian, R. and Viette, P. (1968). *Faune de Madagascar. XXVII Insectes Lépidoptères Papilionidae.* O.R.S.T.O.M. and C.N.R.S., Paris. 97 pp., 19 pl. (2 col.), 34 figs.
6. Turlin, B. (1983). *In litt.*, 1 July.
7. Turlin, B. (1983). *In litt.*, 15 September.
8. Viette, P. (1980). Mission lépidoptérologique à la Grande Comore (Océan Indien occidental). *Bulletin de la Société Entomologique de France* 85: 226–235.

Papilio (Princeps) desmondi teita van Someren, 1960   ENDANGERED

Subfamily   PAPILIONINAE   Tribe   PAPILIONINI

Summary The Taita (Teita) Hills in southern Kenya contain a number of endemic plants, birds, reptiles, amphibians and insects, including *Papilio desmondi teita*, and are the type localities for many more. During this century deforestation has been extensive due to increasing population pressure, and only relict patches of natural forest remain. *P. d. teita* and many other forest species are confined to these relict forests, which should be protected from further interference.

Description When it was first found, this species in the *nireus* species group (13) was thought to be a race of *Papilio brontes desmondi* (17), now known simply as *Papilio desmondi*. Later, when more material from the Usambara Mountains and the Chyulu Hills was available, the Taita specimens were noted to be distinctive in wing pattern, although close to *P. desmondi* in the form of the genitalia (18). After much confusion, it was considered sufficiently different from its neighbours to be described as a full species, *Papilio teita*. Subsequently the species was questioned as being similar to *P. nireus* (15), but that superficial examination took no account of the totally different genitalia. In recent studies, D'Abrera (5) and Carcasson (1) reduced *teita* to a subspecies of *Papilio desmondi*, a course which is followed here.

As in other blue-banded papilios, the overall wing ground colour is black, with a broad median band of blue across the UFW and UHW and rows of blue submarginal spots, larger on the UHW (Plate 7.6). The lower side bears a characteristic continuous cream submarginal band on the LHW, fading towards the leading edge of the LFW. Also characteristic are the squat forewings and broad-base triangular UFW median area. Length of forewing 42 mm, hindwing 40 mm (18). The larvae have not been described but are probably green in the later stages.

Distribution *Papilio desmondi teita* is confined to the Taita (Teita) Hills in Taveta District, coast Province, Kenya. It has been found in forest on Mt Mbololo, Ngangao, Ronge, Chawia and Bura Bluff (4, 5, 14, 16). It was also once recorded 12 km away on Mt Sagala (18), but this observation has not been repeated. The forest on Ngulia Hill in Tsavo West was visited by N.M. Collins and M.P. Clifton in July 1983 and appeared to be largely unsuitable for the butterfly.

Habitat and Ecology The Taita Hills consist of an almost circular inselberg of 27 km diameter with three main peaks, Vuria (2228 m), Ngangao (2149 m, also recorded as Ngao Ngao or Ngoa Ngoa (5)) and Mbololo (2209 m) (also known as Mraru). The hills lie about 110 km east-south-east of Mt Kilimanjaro and 25 km west of Voi, overlooking to the north, east and west the plains of Tsavo West National Park 1000 m below. As Moreau (12) has pointed out, the name Taita Hills is a ridiculous diminutive for such an imposing massif.

The importance of these hills lies in the probability that montane rainforest has grown on them for tens of thousands of years. During the last Ice Age, when eastern Africa was probably (but not certainly (12)) much drier than it is today (10), all forests were severely reduced in size. However, relict patches remained and acted as vital refugia for forest species of fauna and flora. Such refugia were probably found on a string of mountains through the Usambaras and Ulugurus to Mt Mlanje in Malawi

(21). Each montane forest block has therefore tended to evolve a unique flora and fauna, isolated as they are from other patches of forest.

The Taita Hills include both upper montane evergreen rain forest and lower montane woodland (6), although their extent is now severely limited (see Threats). Only the former is suitable habitat for the butterfly, consisting of closed canopy forest with camphorwood (*Ocotea usambarensis*) as a common tree (6). A list of trees is given by Dale (6), who does not, however, mention the two endemics of the Taita Hills, *Memecylon teitense* (Melastomataceae) and *Millettia oblata teitensis* (Papilionaceae), nor the tree *Craibia zimmermannii* (Papilionaceae) for which the Taitas are the only Kenyan locality (4). The Taita African violet *Saintpaulia teitensis* is also endemic to the hills, as is *Psychotria petiti* (also on Sagala), a shrub in the Rubiaceae. The colourless saprophyte *Gymnosiphon usambaricus* (Burmanniaceae) is also found on the Usambaras, and for the giant orchid *Angraecum giriamae* (Orchidaceae), the Taita Hills are one of few Kenyan localities (4).

Papilio desmondi teita has been reared on the Rutaceae *Vepris eugeniifolia* Verdorn, and *Teclea* sp. (19) but there is some doubt as to its natural foodplant. Clifton (2) considers it possible that the caterpillars use *Diphasiopsis fadeni* Kokwaro, another rutaceous plant which, unlike the widespread *V. eugeniifolia*, is confined to the Taita, Chyulu and Ngulia Hills, and Kasigau. The general habits of the adults have not been well recorded, but are probably similar to those of *Papilio desmondi*. The males are mainly seen patrolling forest margins where nectar plants (e.g. *Pentas* (Rubiaceae), *Vernonia* (Compositae) and the introduced but now ubiquitous *Lantana* (Verbenaceae)) occur. Females are more retiring, being found deeper in the forest along forest tracks or in the forest undergrowth (14).

P. d. teita shares the Taita Hills with a number of other endemic and rare animals. The Taita Hills are the type locality for three small mammals, a shrew *Sincus lixus aequatorialis*, a white-toothed shrew *Crocidura cyanea parvipes*, and an elephant shrew or East African rock jumper, *Petrodromus tetradactylus sangi*. Endemic species and subspecies include the nymphalid butterflies *Cymothoe teita* and *Charaxes xiphares desmondi*, one snake (*Amblyodipsas teitana*), the caecilian *Afrocaecilia teitana* (which has adapted to banana plantations (8)) and three birds, including the endangered Taita thrush, *Turdus helleri* (3, 9, 16, 18).

Threats The steep slopes of the Taita Hills were probably originally completely forested, but have been almost entirely cleared for forestry and agriculture. The hills are believed to have been the first Kenyan settlement of Bantu people, who spread eastwards from central and western Africa 1000–500 years ago (20). These slopes have therefore probably been cultivated longer than any other part of Kenya. At the beginning of this century the forests were probably already reduced to small patches and a combination of maps and records reveals seven main areas: Chawia (including Bura Bluff), Ngangao, Mbololo, Mwaganini, Ronge, Kinyesha Mvua and Choke (4, 7). The National Forest of Chawia was the type locality for *P. d. teita*, but is no longer in existence except as tiny degraded patches. The forest was clear-felled for timber and other forest products several decades ago and replaced largely by plantations of exotic fast-growing trees like *Pinus caribaea*, *Cupressus lusitanica* and *Eucalyptus*. These trees matured and have now also been felled (4). The forests of Mwaganini, Choke and Kinyesha Mvua are now believed to consist entirely of plantations of introduced tree species (4). Ronge was cleared for agriculture in about 1945, found to be unsuitable, and was reafforested with *Pinus caribaea*. Ronge has an area of just over 3 sq. km (7), perhaps 20 per cent of which is still natural forest, mostly tiny relicts in ravines and gullies (4). The butterfly has been seen there, but such broken patches of forest of uncertain future are considered inadequate for long-term survival. The Ngangao forest is little more than 0.5 sq. km in extent (7), most of it on steeply sloping ground. The butterfly has been seen there as recently as July 1983 (4), and on numerous earlier occasions (9, 17). Access to Ngangao is relatively easy and there is evidence of minor encroachment into the forest. The seventh forest is on Mbololo Hill, sometimes known as Mraru. There, along the ridge-top, lies a strip of natural forest no more than 4 sq. km in extent (4, 7). The butterfly flies there still, and was seen in July 1983 (4). Hence the total area of relatively undisturbed forest left in the Taita Hills is less than 5 sq. km (4).

The greatest threat to *P. d. teita* and all the other endemic species in the Taita Hills is that their remaining habitat may be felled. Both Ngangao and Mbololo are currently under the management of the Forest Department and are in the process of being gazetted (11). Although such a move legally protects the forests from incursions by local people and developers, it still leaves the Forest Department free to manage the forest as they please, including felling the forest and planting fast-growing trees. Nevertheless, the Chief Conservator of Forests has stated his intention that part of Mbololo should be gazetted as a Nature Reserve (11).

Conservation Measures In 1981 the Kenya Government prohibited the collection and trade of Lepidoptera and Coleoptera (22). This legislation will do nothing to protect *Papilio desmondi teita* since the subspecies is not threatened by the taking of an occasional specimen, but by the destruction of its habitat. This would almost certainly hold true for all Kenya's insects.

No conservation measures have been taken to protect the biotopes of the Taita Hills. An exploratory visit to the hills by N.M. Collins and M.P. Clifton in 1983 confirmed the delicate situation in which the forest fauna and flora survive (4). Urgent efforts are needed to ensure that all the remaining natural forest on Ngangao and Mbololo be declared protected areas as soon as possible. Mbololo is the more extensive area but Ngangao is important by virtue of the stream that runs through it, providing potential habitat for endemic aquatic species. The Forest Department should be encouraged to adopt protective measures with immediate effect. The forests are too small and the slopes too steep for exploitation to produce any commercial advantage. As they stand they are vital to the maintenance of healthy watersheds, stabilizing the soils and preventing the severe erosion that is in evidence in many other parts of the Taita Hills (4). If further plantations are under consideration for the hills, they should be concentrated on the deforested and derelict lands at lower altitudes. Conservation of the Taita Hills is one of the highest priorities of the National Museums of Kenya, which recognize the need to protect as many endemic species on the hills as possible (16).

References

1. Carcasson, R.H. (1980). *Collins Handguide to the Butterflies of Africa*. Collins, London. Hardback edition, 188 pp.
2. Clifton, M.P. (1983). *In litt.*, 2 June.
3. Collar, N.J. and Stuart, S.N. (1985). *Threatened Birds of Africa and Related Islands. The ICBP/IUCN Red Data Book. Part 1*. ICBP/IUCN, Cambridge. xxxiv + 761 pp.
4. Collins, N.M. and Clifton, M.P. (1984). Threatened wildife in the Taita Hills. *Swara* 7(5): 10–14.
5. D'Abrera, B. (1980). *Butterflies of the Afrotropical Region*. Lansdowne Editions, Melbourne. xx + 593 pp.

6. Dale, I.R. (1939). *The Woody Vegetation of the Coast Province of Kenya*. Imperial Forestry Institute Paper No. 18, pp. 13–14.
7. Doute, R., Ochanda, N. and Epp, H. (1981). *A Forest Inventory of Kenya Using Remote Sensing Techniques*. Kenya Rangeland Ecological Monitoring Unit, Nairobi. 55 pp + 72 maps.
8. Duff–Mackay, A. (1980). *Conservation status report No. 1: Amphibia*. National Museums of Kenya, Nairobi.
9. Hall, B.P. and Moreau, R.E. (1970). *An Atlas of Speciation in African Passerine Birds*. British Museum (Natural History), London. 423 pp.
10. Hamilton, A.C. (1981). The quaternary history of African forests: its relevance to conservation. *African Journal of Ecology* 19: 1–6.
11. Mburu, O.M., Chief Conservator of Forests (1983). *In litt.*, 21 September.
12. Moreau, R.E. (1966). *The Bird Faunas of Africa and its Islands*. Academic Press, New York and London. 424 pp.
13. Munroe, E. (1961). The classification of the Papilionidae (Lepidoptera). *Canadian Entomologist* Supplement 17: 1–51.
14. Rydon, A.H.B. (1983). In litt, 6 June.
15. Smart, P. (1975). *The Illustrated Encyclopedia of the Butterfly World*. Hamlyn, London. 275 pp.
16. Van Someren, G.R.C.–(1982). Review of Habitat status of some improtant biotic communities in Kenya. Report to Natural Sciences Division, National Museums of Kenya, Nairobi.
17. Van Someren, V.G.L. (1939). Butterflies of the Chyulu Range. *Journal of the East Africa and Uganda Natural History Society* 14: 130–151.
18. Van Someren, V.G.L. (1960). Systematic notes on the associated blue-banded black Papilios of the *bromius-brontes-sosia* complex of Kenya and Uganda, with descriptions of two new species. *Boletim do Sociedade de Estudos da Provincia de Mocambique* 123: 65–78, 11 pl., 3 maps.
19. Van Someren, V.G.L. (1974). List of foodplants of some East African Lepidoptera. *Journal of the Lepidopterists' Society* 28: 315–331.
20. Were, G.S. and Wilson, D.A. (1977). *East Africa Through a Thousand Years*. Evans Brothers, London. 372 pp.
21. White, F. (1981). The history of the Afromontane archipelago and the scientific need for its conservation. *African Journal of Ecology* 19: 33–54.
22. Wildlife (Conservation and Management) Act (Cap. 376) Extension of Application of Act. 25 September, 1981. Ministry for Environment and Natural Resources. Government Printer, Nairobi.

Papilio (Princeps) antimachus Drury, 1782   RARE

Subfamily   PAPILIONINAE   Tribe   PAPILIONINI

Summary *Papilio antimachus*, the Giant Papilio or African Giant Swallow-tail is an exceedingly beautiful butterfly with a relatively wide range in Africa's primary rain forests. However, forest destruction and the lack of protected areas in African forests outside Congo and Zaire are matters for concern.

Description *Papilio antimachus* is the largest African butterfly and has a long, narrow wingspan of 20–23 cm and no tails. It is orange to reddish-brown with a complex pattern of black and yellow–ochre markings. The sexes are similar, but, unusually for the Papilionidae, the female is smaller, with more rounded wings (5, 6, 7, 20, 22) (Plate 8.3). The early stages are unknown (1, 15).

UFW black at the base, browner towards the outer margin, with orange to red–brown markings (5, 6, 7, 20, 22).

UHW outer margin tailless and only slightly scalloped; upper surface orange to red–brown with a black submarginal border, a deeply indented inner edge, a number of discal black spots and some small yellow–ochre fringe spots (5, 6, 7, 20, 22).

LFW black with a dark grey apical region, yellow–ochre spots, off-white discal patches and a dark orange streak in the cell (4–7).

LHW pattern with larger fringe spots than the UHW, and the orange colour of the upper surface is replaced with yellow–ochre (4–7).

Distribution *Papilio antimachus* is found in the lowland primary rain forest of Central and West Africa from Uganda west to Sierra Leone, and south to Angola (3, 5, 9, 18, 22). Uganda: recorded from forests in the west (3), including Budonga (12, 17), Kibale (21), Kalinzu (12) and Bwamba (3). Rwanda: recorded from Kayonza forest (3). Congo Republic: recorded from Parc National d'Odzala, Sembé and Dimonika, near Luobomo (16). Central African Republic, Zaire, Gabon, Equatorial Guinea, Cameroon, Fernando Poo: recorded in collections (3, 4, 5, 9, 11, 16). Angola: extremely rare, recorded only from the north-east (Lunda province) and the north-west (Cabinda and Zaire provinces) but not collected in recent years (14). Nigeria: very rare and confined to virgin lowland forests east of the River Niger (13), including Benin State (1) and Bendel State (13); very few records in the past 20 years (19) but a specimen was recently taken near Calabar (13). Benin, Togo: no records. Ghana, Ivory Coast: recorded in collections (5). Liberia: recorded in collections (5, 9) and in Wanau Forest (9); recently found to be still present in the Mt Nimba region and on Mt Tokadeh (10). Guinea: recorded near the southern border with Liberia, between Macenta and 'Nzerekore (9). Sierra Leone: known from the northern and eastern regions (17), these being the western limits of the butterfly's range (5, 9).

Habitat and Ecology *Papilio antimachus* is confined to primary lowland tropical forest, where the males may be observed flying in clearings, rides or forest fringes and drinking at flowering shrubs and muddy puddles (3, 5, 9). The females are much more elusive, confining themselves to the canopy (3, 5, 9). The adults fly with great speed (7) and are apparently on the wing throughout the year (22), but the timing of breeding and appearance of young stages have not been recorded (1). Males have been seen frequenting clearings on hilltops in Liberia and Uganda, defending territories against other males which constantly challenge them (10, 18).

Threats *Papilio antimachus* was noted as an extreme rarity in the 18th century when it was first described (6, 7), but is now known to be widespread. All records indicate that the species is confined to primary rain forest. Sightings of males in glades and clearings are perhaps more closely related to the distribution of observers than of the butterflies themselves, which are probably more common in the canopy. Females tend to stay in the canopy, rarely descending to the ground.

Actual threats to this species are very difficult to assess. The species is widespread, but apparently has thinly spread populations throughout its range and is dependent upon virgin forest, which is fast disappearing. Its status could quickly change for the worst. Tropical forest in Africa is being cleared at a rate of 1.3 million ha per year, or 0.61 per cent per year of the total area (8). Closed canopy broadleaved forest of the type preferred by *P. antimachus* may be broadly divided into an eastern, Cameroon–Congolese block, and a much smaller western block running between Nigeria and Guinea. Deforestation is much more serious in West Africa (4 per cent per year) than in the eastern block (0.2 per cent per year) (8). Over 80 per cent of Africa's broadleaved forest is in the eastern block, 50 per cent in Zaire alone (8). Populations of *P. antimachus* in West Africa are therefore believed to be much more seriously threatened than those in Central Africa.

Conservation Measures Closed canopy broadleaved forests in which logging is not permitted represent only 4.25 per cent of the African total for such forests. Zaire alone includes 2.65 per cent, and that country is exceptional in Africa where practically no significant areas of closed canopy forests have been earmarked as national parks or reserves (8). Of those West African countries where *P. antimachus* occurs, only Ghana and Ivory Coast contain strictly protected reserves. The rain forest and swampland area of Sapo has been proposed as Liberia's first national park. Sierra Leone's last rain forest, Gola North, has been suggested as a reserve, but it is destined to be clear-felled by the state timber company. Of the Central African countries, only Congo and Zaire include adequate protected areas. In Zaire four sites have been proposed as protected areas under the World Heritage Convention, Jalonga, Upemba, Maiko and Kundelungu National Parks. In Angola lowland primary rain forest is found mainly in Cabinda, but with small areas in the northern provinces of Zaire, Uige, Kwanza–Norte, Malanga and Lunda–Norte (2). None of the forests fall within protected areas (2), but shifting cultivation and timber extraction in these areas is considered insufficient to pose a threat to the butterfly (2).

Much more information is needed on the precise distribution of *P. antimachus*, and its vulnerability to local forest exploitation. Rain forest reserves within its range should be checked for the butterfly, and care taken to ensure its continued existence there. Careful studies of the life history and young stages are needed to ensure proper management in such reserves. If *Papilio antimachus* suffers a decline in range, it will be a sure sign that other less spectacular species are also under threat.

References

1. Boorman, J. and Roche, P. (1957). *The Nigerian Butterflies. Part I: Papilionidae.* Ibadan University Press, Ibadan. 7pp., 31 black and white pl.
2. Braga, M.J. (1983). *In litt.*, 3 November.
3. Carcasson, R.H. (1960). The swallowtail butterflies of East Africa (Lepidoptera, Papilioni-

dae). *Journal of the East African Natural History Society* Special Supplement 6: 33 pp. + 11 pl. (Reprinted by E.W. Classey, Faringdon, 1975).

4. Carcasson, R.H. (1980). *Collins Handguide to the Butterflies of Africa*. Collins, London. Hardback edition, 188 pp.
5. D'Abrera, B. (1980). *Butterflies of the Afrotropical Region*. Lansdowne Editions, Melbourne. xx + 593 pp.
6. Drury, D. (1782). *Illustrations of Natural History*. Vol. 3. London. 76 pp + 50 pl.
7. Drury, D. (1837). *Illustrations of Exotic Entomology*. Vol. III. New ed, J.O. Westwood.
8. FAO/UNEP (1981). *Tropical Forest Resources Assessment Project. Forest Resources of Tropical Africa. Part 1: Regional Synthesis*. FAO, Rome. 108 pp.
9. Fox, R.M., Lindsey, Jr., A.W., Clench, H.K. and Miller, L.D. (1965). The butterflies of Liberia. *Memoirs of the American Entomological Society* No. 19, 438 pp.
10. Godfray, H.C.J. (1983). *In litt.*, April 6.
11. Hancock, D.L. (1983). *In litt.*, 3 March.
12. Kingdon, J. (1983). *In litt.*, undated
13. Larsen, T.B. (1983). *In litt.*, 9 March.
14. Luna de Carvalho (1962). Alguns Papilionideos da Lunda. Subsidios para o estudo da biologia na Lunda. *Publicaçoes Culturais da Companhia de Diamantes de Angola* 24: No. 60, 165–169.
15. Munroe, E. (1961). The classification of the Papilionidae (Lepidoptera). *Canadian Entomologist* Supplement 17: 1–51.
16. N'Sosso, D. (1983). *In litt.*, 6 July.
17. Owen, D.F. (1983). *In litt.*, 25 February.
18. Passos de Carvalho, J. (1983). *In litt.*, 4 April.
19. Riley, J. (1983). *In litt.*, 2 September.
20. Smart, P. (1975). *The Illustrated Encyclopedia of the Butterfly World*. Hamlyn, London. 275 pp.
21. Struhsaker, T.T. (1983). *In litt.*, 30 April.
22. Williams, J.G. (1969). *A Field Guide to the Butterflies of Africa*. Collins, London. 228 pp., 24 col. pl.

Papilio (Princeps) morondavana Grose–smith, 1891   VULNERABLE

Subfamily PAPILIONINAE   Tribe PAPILIONINI

Summary *Papilio morondavana*, the Madagascan Emperor Swallowtail, is an attractive swallowtail found in the deciduous forests of western Madagascar. Commercial collecting needs to be monitored, but the main threat is habitat destruction.

Description *Papilio morondavana* is a large brown–black swallowtail with pale yellow markings, and two blue and red eye-spots on the hindwing (7) (Plate 8.4). The female is larger, but has the same coloration as the male. The forewing length is 51–55 mm, and the wingspan is 90–98 mm (6). An unusual feature of this species is its bright orange antennae (1).

UFW brown–black with a pattern of pale yellow markings including a median band of large, irregular spots, a row of small submarginal spots, small fringe spots, three cell spots, and some irregular subapical spots (1, 6).

UHW also brown–black, with a scalloped outer edge and a tail, and yellow markings in the form of a row of fringe spots, a row of submarginal lunules and a continuation of the UFW median band to the inner margin. There is a large yellow–ochre costal eye-spot bordered at the top with blue, in the middle of a large black spot. A second brick-red eye-spot is situated distal to the anal angle, with a yellow spot below it, and black, blue, and brown above it (1, 6).

LFW/LHW similar but may have an incomplete series of blue eyespots across the dark part of the hindwing (1, 6).

Distribution *Papilio morondavana* is confined to the forests of western Madagascar. Its known range extends from the region around the towns of Morondava and Mahabo north towards Mahajanga (Majunga) and Ambato–Boeny and south to Andranovory and Toliara (Tuléar) (3, 6).

Habitat and Ecology *P. morondavana* is recorded from the Ankarafantsika forest in Mahajunga Province to the Andronovory forest in Toliara Province (4, 6). These forests are deciduous and increasingly arid towards the south. Ankarafantsika is a dense, dry forest with the characteristic trees *Dalbergia*, *Commiphora* and *Hildegardia*, numerous Leguminosae and Myrtaceae and many lianas. Some plants are adapted to survive aridity, e.g. *Pachypodium*, Ampelidaceae and Passifloraceae.

Papilio morondavana is in the *demoleus* species group (3), which generally use Rutaceae and Umbelliferae as hosts, rarely other plants. There are no published records of the host-plants of *P. morondavana*, or of the young stages. The butterfly may be locally common, and has its main brood in November (8).

Threats The area occupied by *P. morondavana* is subject to forest destruction for agriculture by local people (2), a process which eliminates a large proportion of the insect fauna (9). In addition, the species is increasingly popular with commercial collectors (4), a situation which requires monitoring and perhaps local control. The Ankarafantsika forest and other western forests are reported to suffer from uncontrolled burning. The lack of resources to ensure the long-term integrity of these areas is a cause for international concern.

Conservation Measures No specific measures have been taken to conserve this butterfly. There are four established reserves in western Madagascar. *P. morondavana* is certainly found in the Réserve Naturelle Integrale de l'Ankarafantsika (60 520 ha), it may occur in the R.N.I. du Tsingy de Namoroka (21 742 ha) and the R.N.I. du Tsingy de Bemaraha (152 000 ha), but is unlikely to be found in the R.N.I. du Lac de Tsimanampetsotsa (43 000 ha, mostly water). These reserves are not open to the public but are closed to any human interference except official scientific activities (5). However, there is some concern that the reserves are not adequately policed.

The level of commercial exploitation may be a matter for concern. It is important that any commercially useful insect should be the subject of a careful biological study. Only then can the species be managed and exploited in a way that will ensure a sustainable local industry without a permanent decline in butterfly populations. Monitoring of the trade in *P. morondavana* would be advisable.

References

1. D'Abrera, B. (1980). *Butterflies of the Afrotropical Region*. Lansdowne Editions, Melbourne. xx + 593 pp.
2. FAO/UNEP (1981). *Tropical Forest Resources Assessment Project. Forest Resources of Tropical Africa. Part 1: Regional Synthesis*. FAO, Rome, 108 pp.
3. Munroe, E. (1961). The classification of the Papilionidae (Lepidoptera). *Canadian Entomologist* Supplement 17: 1–51.
4. Paulian, R. (1983). *In litt.*, 10 May.
5. Paulian, R. (1983). *In litt.*, 1 July.
6. Paulian, R. and Viette, P. (1968). *Faune de Madagascar. XXVII Insectes Lépidoptères Papilionidae*. O.R.S.T.O.M. and C.N.R.S., Paris. 97 pp., 19 pl. (2 col.), 34 figs.
7. Smart, P. (1975). *The Illustrated Encyclopedia of the Butterfly World*. Hamlyn, London. 275 pp.
8. Turlin, B. (1983). *In litt.*, 1 July.
9. Viette, P. (1983). *In litt.*, 22 March.

***Papilio (Princeps) grosesmithi* Rothschild, 1926** RARE

Subfamily PAPILIONINAE Tribe PAPILIONINI

Summary *Papilio grosesmithi* has a slightly wider distribution than *P. morondavana* in the deciduous forests of western Madagascar. The two species are often confused by collectors. Commercial collecting needs to be monitored, but the main threat is habitat destruction.

Description *Papilio grosesmithi* is a large, dark brown and pale yellow swallowtail with two blue and red eye-spots on the hindwing. The forewing length is 51–64 mm, and the wingspan is 90–105 mm. The female is larger than the male but otherwise similar (3).

This butterfly resembles *P. morondavana* (see previous review), from which it differs in being larger and somewhat darker. There is also a black spot between the brick-red anal eye-spot and the yellow spot below it. The lower surface has an incomplete series of blue spots across the discal part, represented in the female by a row of blue lunules on both surfaces (3).

Distribution *Papilio grosesmithi*, like *P. morondavana*, is endemic to western Madagascar, but is found over a slightly wider area. Its known range extends from Mahajanga (Majunga) in the north, to Sakaraha, Toliara (Tuléar) and the Lambomakondro forest in the south.

Habitat and Ecology *P. grosesmithi* was first collected in the deciduous forests of north-western Madagascar and is now known from the Ankarafantsika forest (Majunga Province), the Marofandilia forest near Morondava, the Lambomakondro forest in the Sakaraha region, the Zombitsy Special Reserve, the banks of the Fiherenana, and in the Androvonory forest east of Toliara (2). *P. grosesmithi* is in the *demoleus* group, as is *P. morondavana*. Comments under that species also apply here. In some of its localities *P. grosesmithi* may be seasonally relatively abundant but its globally restricted range in an area subject to extremely rapid alteration by man is cause for concern.

Threats Since the range of *P. grosesmithi* is virtually sympatric with that of *P. morondavana*, the threats from agriculture and forest clearance are similar. The two species are not usually distinguished by local collectors, although they are by commercial outlets, so monitoring and perhaps local control should apply to both species (1).

Conservation Measures No specific measures have been taken to conserve this butterfly and no data are available on the level of commercial collecting of the *demoleus* group in Madagascar. As stated in the review of *P. morondavana*, biological studies are needed to ensure a sustainable yield of these species (1). Protected areas which probably contain this species are listed under *P. morondavana*. It seems that extra resources may be needed to ensure the long-term integrity of these areas.

References

1. Paulian, R. (1983). *In litt.*, 10 May.

2. Paulian, R. and Viette, P. (1968). *Faune de Madagascar. XXVII Insectes Lépidoptères Papilionidae.* O.R.S.T.O.M. and C.N.R.S., Paris. 97 pp.
3. Rothschild, W. (1926). On some African papilios, with descriptions of new forms. *Annals and Magazine of Natural History* (9)17: 112–114.

Papilio (Princeps) leucotaenia Rothschild, 1908   VULNERABLE

Subfamily PAPILIONINAE   Tribe PAPILIONINI

Summary *Papilio leucotaenia*, the Cream-banded Swallowtail, is a large swallowtail restricted to forests in the highlands of south-western Uganda, around Lake Kivu in north-eastern Zaire and Rwanda, and in western Burundi. The small range and very local distribution of this butterfly in a part of Africa that includes areas with very high human populations and the rapid degradation of forests is a matter for concern. The remaining pockets of forest in which it flies are inadequately protected from incursion by local people and commercial logging interests.

Description *Papilio leucotaenia* is a large club-tailed butterfly with a forewing length of 54–60 mm (Plate 8.5). The sexes are similar in general appearance, but the female is paler, with a broader median band (1, 3, 13). *P. leucotaenia* is an unusual species, comprising a species-group of its own (5).

UFW/UHW dark blackish-grey with a broad white median band tapering from the hindwing inner margin to the subapical part of the forewing costal margin. Deeply scalloped outer margins and a long (20–25 mm) club-shaped tail to the hindwing.

LFW/LHW with a complex pattern of various shades of brown, grey and black over the whole of the LHW and the apical region of the LFW. The rest of the LFW resembles the UFW (1, 3, 13).

Distribution *Papilio leucotaenia* is restricted to a small area of highland in central Africa, in the mountains that divide the watersheds of the Nile and the Zaire. It occurs locally in small pockets of forest in the extreme south-west of Uganda (south-western Kigezi District) southwards into adjacent areas of Rwanda and western Burundi. It is also found in the adjoining north-eastern part of Kivu Province, around Lake Kivu, in eastern Zaire (1, 3, 6, 13).

Habitat and Ecology *Papilio leucotaenia* is a very local butterfly, occurring in pockets of montane forest on the Nile/Zaire watershed. The young stages and foodplants are unknown, but adults fly throughout the year, possibly being most numerous during June and July (13). Adults live and breed in the forest but, in common with so many forest butterflies, they like to fly in clearings and along road margins and rides (13). Males can be extremely elusive, diving and zigzagging at high speed through the forest. Adults have been located in a number of forests, including the Mafuga forest and the Bwindi (Impenetrable) forest in south-western Uganda (1, 13), the Nyungwe (or Rugege) forest (970 sq. km) in south-western Rwanda, the forests around Lake Kivu in Zaire, and recently on Mt Teza in Burundi (10). Although many of these forests are not well studied, some information is available for the Nyungwe forest, which contains a mixture of relatively poor soils at high altitudes (2200–2950 m) in the eastern sector and richer soils in the western sector, which has a wide range of altitudes (1650–2500 m) (9). It is in the western forests that the butterfly is presumed to fly, where the canopy reaches 50–60 m in height and is dominated by *Entandophragma* (mahogany), *Newtonia*, *Aningeria* and *Podocarpus* (9). Primates are also abundant here.

Threats Rwanda and Burundi have the highest human population densities in Africa. In Rwanda 4.6–5.1 million people, 90 per cent of them rural, live on

24 950 sq. km of land, a density of 176–193 people per sq. km. In Burundi the pattern is similar: 4.5 million people on 25 800 sq. km, a density of nearly 175 people per sq. km (4). In both countries the birth rate approaches 3 per cent per year. The formidable agricultural and pastoral pressures are particularly severe in the western foothills and mountains and have resulted in very severe degradation and destruction not only of the woody vegetation but also of the soils that support it (4). Threats specific to *Papilio leucotaenia* have not previously been documented, but they are the same as for all other flora and fauna that live in these forests. Their habitat is being rapidly destroyed, first for timber, charcoal and building materials, then for plantations of exotic trees (pine and eucalyptus), tea and coffee.

In Rwanda, sites such as the Gishwati forest have been so heavily cut over that they are not considered viable as biological conservation areas (4). The Nyungwe forest lacks a definitive and coordinated conservation policy and is threatened by local encroachment, illegal logging and even foreign investment and development programmes (4). Despite its importance as a catchment area in the headwaters of the Nile, there have been plans to exploit the forest for tea estates, gold extraction and matchwood production (4). Forest clearance has been proceeding at a rate of over 8 sq. km per year since 1958, when the forest was over 1140 sq. km in extent (8). Nyungwe is famous for its variety and complexity of habitats, ranging from middle altitude to montane dry and wet forest communities, high altitude swamps and heaths (4).

In southern Burundi the forests have disappeared completely except for a few hundred hectares near Bururi, which include some of the finest *Entandophragma* in Africa (12). The forests on Mt Teza, where *P. leucotaenia* is known to fly, are under increasing pressure from tea-planting and other developments. Although the pressures on forests are most extreme in Rwanda and Burundi, similar problems affect the eastern forests of Zaire and the south-western forests of Uganda. The Bwindi forest in Uganda is under increasing pressure from forestry and agriculture and much of its lowland section has already been cleared for agriculture. Most of the mountain slopes along the western shore of Lake Kivu in Zaire have also been totally cleared of their natural vegetation (11).

P. leucotaenia is a rarely seen butterfly with a restricted range and a very patchy and local distribution. Destruction of forest throughout the range of the species is believed to be causing increased disjunction and extinction of its already scattered populations. This species is just one of many forms of wildlife threatened by deforestation and development in this part of Africa. Endemic and rare birds in the region include the African Green Broadbill (*Pseudocalyptomena graueri*), the Papyrus Yellow Warbler (*Chloropeta gracilirostris*), Chapin's Flycatcher (*Muscicapa lendu*), Grauer's Swamp Warbler (*Bradypterus graueri*), the Albertine Owlet (*Glaucidium albertinum*) and the Kungwe Apalis (*Apalis argentea*) (2). A number of primates and other mammals could be added to this list.

Conservation Measures In the extreme south-western corner of Uganda there is a small reserve (about 30 sq. km) called the Gorilla Game Reserve. It lies in a mountainous area at the intersection of the Uganda, Zaire and Rwanda borders and was gazetted in order to protect gorillas (6). It may possibly include habitat for *P. leucotaenia*, although it is contiguous with Rwanda's Volcanoes National Park and the predominantly high montane vegetation may be unsuitable. Known Ugandan localities such as the Mafuga and Bwindi (Impenetrable) forests are not formally protected, but the Bwindi forest is gazetted as a forest reserve.

In Rwanda there is no evidence that the Volcanoes National Park harbours the butterfly, and the only other protected area, the Akagera National Park, is quite unsuitable. Various schemes have been suggested for the protection of the Nyungwe forest in Rwanda, an important locality for *P. leucotaenia*. Unfortunately national park status has so far proved unacceptable to the Ministry of Agriculture to whom the forest is a major supplier of hardwood timber and a reserve of agricultural land. A more feasible alternative suggested in 1981 was the formation of an MAB (Man and Biosphere) reserve comprising a totally protected central core surrounded by a zone with controlled sustainable exploitation, a buffer zone of exotic plantations and finally areas of denser human settlement. The Nyungwe forest is already partly surrounded by a 500–2500 m wide buffer zone of Pine, Eucalyptus and some native trees which is to be extended by Belgian aid to surround the forest completely. Although this will prevent encroachment and firewood cutting by local people, it will not prevent planned or unplanned logging (9). In 1983 a team of six consultants to the Rwandan government, including one from IUCN, recommended that the central core of just over 230 sq. km be totally protected, with an option on a further 167 sq. km (8). The remainder could be exploited in a sustainable manner. The government's response to the recommendations has not yet been received.

Burundi has no national parks or protected areas (7), but the authorities there are aware of the need to protect forest and have ordered a cessation of felling (12). The forest at Bururi apparently has some reserve status (9) and plans to plant a buffer zone of exotic trees around the area have been tabled. The forests on Mt Teza have been proposed as an MAB reserve, as described for the Nyungwe Forest (9).

The Zaire section of the range of *P. leucotaenia* is poorly known, but there are no known protected areas within the butterfly's range.

There have been no conservation measures specifically for this butterfly. Clearly an important priority is the formation of well-protected reserves of forest in Rwanda and Burundi. International support for the conservation of Nyungwe and Teza forests in Rwanda and Bururi forest in Burundi would be in order.

References

1. Carcasson, R.H. (1960). The swallowtail butterflies of East Africa (Lepidoptera, Papilionidae). *Journal of the East African Natural History Society* Special Supplement 6: 33 pp. + 11 pl. (Reprinted by E.W., Classey, faringdon, 1975).
2. Collar, N.J. and Stuart, S.N. (1985). *Threatened Birds of Africa and Related Islands. The ICBP/IUCN Red Data Book. Part 1.* ICBP/IUCN. Cambridge. xxxiv + 761 pp.
3. D'Abrera, B. (1980). *Butterflies of the Afrotropical Region.* Lansdowne Editions, Melbourne. xx + 593 pp.
4. FAO/UNEP (1981). *Tropical Forest Resources Assessment Project. Forest Resources of Tropical Africa. Part II: Country Briefs.* FAO, Rome. 586 pp.
5. Hancock, D.L. (1983). Classification of the Papilionidae (Lepidoptera): A phylogenetic approach. *Smithersia* 2: 1–48.
6. IUCN (1976). Proceedings of a regional meeting on the creation of a coordinated system of national parks and reserves in eastern Africa. *IUCN Publications (NS) Supplementary Paper No. 45*, 205 pp.
7. IUCN/CNPPA (1982). *1982 United Nations List of National Parks and Protected Areas.* IUCN, Gland. 155 pp.
8. Monfort, A. (1983). Nyungwe Forest, Rwanda. *Swara* 6(6): 22.
9. Rodgers, W.A. (1981). Brief report on the conservation status of the Nyungwe forest, Rwanda. University of Dar–es–Salaam. Tanzania. 12 pp.
10. Turlin, B. (1983). *In litt.*, 15 September.
11. Verschuren, J. (1975). Wildlife in Zaire. *Oryx* 13: 25–33.

12. Verschuren, J. (1978). Burundi and wildlife: problems of an overcrowded country. *Oryx* 14: 237–240.
13. Williams, J.G. (1969). *A Field Guide to the butterflies of Africa*. Collins, London. 238 pp.

Papilio (Princeps) mangoura Hewitson, 1875   RARE

Subfamily PAPILIONINAE   Tribe PAPILIONINI

Summary *Papilio mangoura* flies only in the eastern rain forests of Madagascar. Only one quarter of these forests are still untouched and with rapid population growth and deforestation the status of this and many other forest butterflies should be monitored.

Description *Papilio mangoura* is a blue-banded tailed swallowtail with a forewing length of about 45 mm and a wingspan of about 80 mm (1, 7). The female has off-white bands and markings.

Male: UFW/UHW velvety black with a broad blue median band, a variable number of small bluish submarginal spots on the forewing and bluish white lunules on the hind wing (1, 7).

LFW/LHW brownish black, no median band on the forewing, but marked in greyish pink below the cell. LHW with a very straight, and small greyish pink band, often poorly marked marginal spots (1, 7).

Female: Very different from the male. Upperside dark brown with a yellowish white median band, broken into spots on the forewing (1, 7). Submarginal spots on the forewing and marginal spots on the hindwing are of the same colour.

Distribution *Papilio mangoura* is found in eastern Madagascar, from Maroantsetra in the north to Taolanaro in the south. A list of localities is given by Paulian and Viette (7).

Habitat and Ecology This butterfly is a species of the eastern rain forests of Madagascar and is closely related to the much more common *P. delalandei*. No details of their biology have been published and the early stages of both species are unknown (3).

Threats The main threat to this and any other creature endemic to Madagascar's rain forests is the alarming rate at which degradation of the vegetation and soils is occurring (2). Madagascar has 10.3 million hectares of closed canopy broad-leaved forest, but only one quarter of this is believed to be primary (2). The rate of deforestation during the period 1981–85 has been estimated at 150 000 hectares per year, a slight lowering from 165 000 hectares per year in 1976–80 (2). The main pressure is from population growth and the rapid spread of shifting cultivation ('tavy'), but timber exploitation adds to the problem (2). About 650 000 hectares of former forest is now too degraded to be utilized for further timber exploitation (2). Fortunately about 1.75 million hectares of forest are on land too steep to be exploited, and approximately a further one million hectares are for various reasons legally protected from exploitation (2, 6). It therefore seems likely, but by no means certain, that most forest butterflies will survive in protected areas and relict forest patches. However, this is no cause for complacency and does not detract from the enormous difficulties facing conservationists in Madagascar.

Other butterflies possibly threatened in a similar way include the rare danaid *Amauris nossima* (locally common in Montagne d'Ambre (8)), the nymphalids *Euxanthe madagascariensis*, *Charaxes cowani*, *C. phraortes*, *C. andranodorus*, *Neptis decaryi*, *N. metella gratilla*, *N. sextilla*, *Apaturopsis kilusa*, *A. pauliani* and the acraeids *Acraea sambarae* and *A. hova* (5). *Graphium endochus*, another forest papilionid, is apparently well distributed at present but may require monitoring as deforestation progresses. Other families have not been assessed, but would undoubtedly add to this list of threatened butterflies.

Conservation Measures Control of population growth and shifting cultivation are the basic requirements of programmes for rational land use, development and conservation in Madagascar. In addition, the system of national parks and natural reserves in eastern Madagascar is inadequate for the protection of the flora and fauna of lowland forest and requires considerable expansion. The Réserves Naturelles de Tsaratanana, Marojejy and Andringitra include mostly montane vegetation (4), although the latter has some excellent forest at medium altitude (6). However, R.N. de Zahamena has rainfall of 1500–2000 mm per year and includes a fine stand of rich lowland rain forest (8). R.N. d'Andohahela includes an area of forest, but this is virtually the southern extremity of this vegetation type. The Réserve Spéciale de Périnet–Analamazaotra apparently includes good rain forest in which the Indris (*Indri indri*) lives, but the reserve is too small (810 ha) to be significant (4). In addition there are a number of small reserves on the east coast. Possibly the only large expanse of rain forest is in the Masoala peninsula, once a reserve of 76 000 ha, but given over to forest exploitation in 1964. If managed on a sustainable basis this forest could still be an important refuge for wildlife (6).

There is too little information on any of the rain forest butterflies, including *Papilio mangoura*, to make specific conservation recommendations. Clearly, more surveys and a great deal more biological study are needed in order that the very high endemicity of the forest fauna may be properly conserved. On an optimistic note, if butterfly foodplants are able to survive on steep ground then relict forest patches on the eastern slopes of the central massif and in the Masoala peninsula may effectively prevent wholesale extinctions.

References

1. D'Abrera, B. (1980). *Butterflies of the Afrotropical Region.* Lansdowne Editions, Melbourne. xx + 593 pp.
2. FAO/UNEP (1981). *Tropical Forest Resources Assessment Project. Forest Resources of Tropical Africa Part II: Country Briefs.* FAO, Rome. 586 pp.
3. Hancock, D.L. (1983). Classification of the Papilionidae (Lepidoptera): a phylogenetic approach. *Smithersia* 2: 1–48.
4. Oberlé, P. (ed.) (1981). *Madagascar, un Sanctuaire de la Nature.* Le Chevalier, Paris. 118 pp.
5. Paulian, R. (1956). Insectes Lépidoptères Danaidae, Nymphalidae, Acraeidae. *Faune de Madagascar* 2. ORSTOM, CNRS, Paris. 102 pp.
6. Paulian, R. (1984). *In litt.*, 16 February.
7. Paulian, R. and Viette, P. (1968). Insectes Lépidoptères Papilionidae. *Faune de Madagascar* 27. ORSTOM, CNRS, Paris. 97 pp.
8. Viette, P. (1984). *In litt.*, 23 February.

Papilio (Princeps) chikae Igarashi, 1965　　　　　　　ENDANGERED

Subfamily　PAPILIONINAE　　　　　　　　　　Tribe　PAPILIONINI

Summary *Papilio chikae* is a very recent discovery, found in 1965 on Mt St Thomas in the southern Cordillera Central of Luzon, northern Philippines. It has since been found on a number of occasions on peaks near Baguio and Bontoc. Essentially a species of areas above 1500 m in the western montane zone, it is a glacial relict of continental origin. The Baguio area is a popular tourist resort and this naturally rare species is believed to be endangered by over-collecting and probably by habitat alteration.

Description This beautiful green–black, red and purple, long-tailed butterfly, is one of the gloss swallowtails or peacocks. It is relatively large with a forewing length of 55 mm (2, 6, 12) (Plate 8.6). Although *P. chikae* is now widely accepted as a good species, doubts have been raised in the past. In particular, its relationship with *Papilio lorquini* Reakirt, 1864 has not been clarified (8).

Male: UFW with golden–green scales evenly scattered over the black ground colour.

UHW ground colour also black with scattered golden–green scales. A large green submarginal patch in the apical area turns blue in the summer form. The red submarginal lunules may be well developed, especially in the summer form, and have purple inner margins. The tail is relatively long, narrowing at its base, and with dense blue–green scales on its upper surface (6, 12).

LFW sooty black with a broad pale grey band tapering to the inner margin, where it is pale blue (6, 12).

LHW with whitish scales extending outwards from the cell over the black ground colour. The submarginal lunules are enlarged and red–orange, with broader and paler purple inner margins (6, 12).

Female: The female has a longer tail to the hindwing and better developed submarginal lunules. These join with marginal spots to form large red hoops towards the anal angle (12).

Distribution *Papilio chikae* is endemic to Luzon in the northern Philippines. It is only known from the north of the island where it is found in the Baguio and Bontoc regions of the Cordillera Central (12). The type locality is Mt St Thomas (Mt Santo–Tomas), a 2258 m mountain to the south of Baguio city (5). The species has not been found in the nearby Sierra Madre range in north-eastern Luzon much of which may lie at too low an altitude (11). *P. chikae* is a member of the *paris* group, which is mainly distributed in eastern continental Asia. Of the twelve species in the group, only two others reach the South East Asian islands, *paris* is found in Sumatra and Java as well as India, Burma, China and Hong Kong, while *karna* occurs only in Java, Sumatra, Borneo and Palawan.

Habitat and Ecology *Papilio chikae* flies at altitudes above 1500 m (11) and can be found almost all year round, with a short recess from November to January (12). The spring form is found from January to early April and the summer form from mid-April to November (12). The summer form is larger, but the spring form rather brighter in colour (12). Such seasonal variation in a species of the montane subtropical habitat is taken to indicate its status as a relict species of continental origin (1, 12). It probably pushed southwards from China and Taiwan during the last ice age, when water accumulated as ice at the poles and the sea level fell considerably. As the ice retreated and sea levels rose once again *P. chikae* was left behind as an isolated species.

The young stages and foodplant of *Papilio chikae* are unknown, but in other species of the *paris* group the larvae are found on Rutaceae and have thoracic and first abdominal segments with a dorsal, shield-like raised area, no metathoracic band and reduced eye spots (4).

The habitat is broken countryside with open grassy meadows, scattered bushes and small copses in ravines and rifts (11). The most common tree is the Benguet pine (*Pinus insularis*). Associated butterflies include *Papilio benguetanus*, *Pieris canidia*, *Eurema hecabe*, *Vanessa indica* and *Argyreus hyperbius* (5, 7).

Threats This elusive and rare butterfly clearly has a very limited distribution. Even within the Cordillera Central it may be restricted to peaks over 1000 m above sea level. Access to the peaks of the Baguio and Bontoc areas seems to be relatively simple. A motorable track runs the whole way up to the type locality on Mt Saint Thomas, but mist often reduces visibility and activity of the butterflies. This and other species of gloss swallowtails are regarded by collectors as great prizes (9). The fact that the sale of just a few specimens of *P. chikae* on the international market would at one time have paid for a return trip from Japan was a matter for concern. As J.N. Jumalon has put it, *P. chikae* "is like a criminal with a price on its head" (9). No doubt most collectors exercise restraint, but with easy access and high prices, the temptation to over-collect must be great. Tsukada and Nishiyama explain how easily the butterfly can be attracted by decoys and note that its flight is so slow that it can be captured almost without fail (12). It has been noted that employees in radio stations on the mountains all keep butterfly nets, presumably for commercial collecting (3). D'Abrera records that some Japanese collectors have been known to offer 35 mm cameras to native collectors in exchange for specimens (2). Although it seems that the market in *Papilio chikae* has decreased considerably in recent years, there is still a great need for restraint in collecting.

It has not so far been possible to assess the extent of habitat disruption within the limits of distribution of *P. chikae*. Baguio is noted as being the summer capital of the Philippines (5), and a summer resort which swells to several times its normal population in April and May (12). One might therefore expect fairly heavy recreational pressure on the habitat from human use at certain times of year. Even so, many slopes, gullies and other potential breeding areas for the butterflies may remain uneconomical for agriculture and forestry and may therefore be effectively protected from these more significant threats to wildlife.

Population growth and the development of tourism are said to be the reason for new roads being built to improve the accessibility of villages in the hills of the Cordillera Central (9). This will inevitably result in some destruction of the alpine vegetation. Other forms of development add to the problem: mining has devastated large areas in the north of Luzon. Disturbance by fire is a further possible threat.

Development and population growth are certainly reducing the range of available habitat in the Cordillera Central, but over-collecting does seem to be the most serious threat (11), an unusual situation for a butterfly. It is clear that considerably more information on *Papilio chikae* is needed and that this should be obtained as expeditiously as possible.

Conservation Measures The most immediate requirement is a survey of the Cordillera Central in the appropriate season to assess the distributional limits of the species. At the same time observations on young stages and larval foodplants should be made. This would permit an assessment of the potential for farming the butterfly. The high prices commanded on the international market indicate the demand for the species and a controlled breeding and farming programme would serve to meet the demand, provide local employment and conserve the wild populations of the butterfly. Monitoring and perhaps even legislative control of collecting might be required.

It appears that there are no functional protected areas in the Cordillera Central of Luzon, although the Philippine national parks system is currently under review (10). To judge from the popularity of Baguio in the summer, protected recreational wilderness areas would be an asset to the local people and visitors as well as serving to protect the wildlife of the region. However, great efforts will be needed to convince the many disadvantaged people of the Philippines that conservation will have long-term benefits.

References

1. Ae, S.A (1983). *In litt.*, March 18.
2. D'Abrera, B. (1982). *Butterflies of the Oriental Region. Part 1. Papilionidae and Pieridae.* Hill House, Victoria, Australia. xxxi + 244 pp.
3. Dacasin, G. (1984). *In litt.*, 25 April.
4. Hancock, D.L. (1983). Classification of the Papilionidae (Lepidoptera): a phylogenetic approach. *Smithersia* 2: 1–48.
5. Harada, M. (1965). The capture of *Papilio chikae. Tyo To Ga (Transactions of the Lepidopterists' Society of Japan)* 16: 48–49.
6. Igarashi, S. (1965). *Papilio chikae*, an unrecorded Papilionid butterfly from Luzon island, the Philippines. *Tyo To Ga (Transactions of the Lepidopterists' Society of Japan)* 16: 41–49.
7. Iwase, T. (1965). How *Papilio chikae* was found and named. *Tyo To Ga (Transactions of the Lepidopterists' Society of Japan)* 16: 44–47.
8. Jumalon, J.N. (1969). Notes on the new range of some Asiatic papilionids in the Philippines. *The Philippine Entomologist* 1(3): 251–257.
9. Jumalon, J.N. (1984). *In litt.*, 10 July.
10. Pollisco, F.S. (1982). An analysis of the national park system in the Philippines. *Likas-Yaman, Journal of the Natural Resources Management Forum* 3(12): 56 pp.
11. Treadaway, C.G. (1984). *In litt.*, 25 May.
12. Tsukada, E. and Nishiyama, Y. (1982). *Butterflies of the South East Asian Islands Vol. 1 Papilionidae.* (transl. K. Morishita). Plapac Co. Ltd., Tokyo. 457 pp.

Papilio (Princeps) neumoegeni Honrath, 1890   VULNERABLE

Subfamily PAPILIONINAE   Tribe PAPILIONINI

Summary *Papilio neumoegeni* is a very local species restricted to relict patches of forest in Sumba, in the Lesser Sunda Islands of Indonesia. Man's influence in Sumba has been extensive, with habitat destruction resulting from deforestation, burning and grazing. The proposed Gunung Wanggamati reserve may prove to be suitable for the butterfly and surveys there are needed.

Description *Papilio neumoegeni* is a beautiful, medium-sized swallowtail with a forewing length of 43–48 mm (1, 5, 8) (Plate 8.7). It is a species in the *peranthus* group, whose members extend from the Lesser Sundas and Sulawesi to New Guinea (3, 7). There are four species in the group, none of their young stages have been described.

Male: UFW black with metallic, golden–green scales scattered over all but the outer and costal margins. An entirely metallic, golden–green median band passes from the costal passing outside the cell apex to just inside the tornus, interrupted in the lower part by a large brown–black sex brand (1, 8).

UHW with a scalloped outer edge and a long, spatulate tail which may have green scaling over the tips. The black ground colour has metallic golden–green scales scattered over the basal third of the wing. A broad emerald–green discal band extends inwards over the cell apex but not over the black hindwing apex, and a series of large emerald–green lunules or spots may merge into the band subapically (1, 8).

LFW/LHW blackish-brown with a few golden scales scattered over all but the distal part which is lighter. The LFW has an oblique, pale grey postdiscal band which tapers towards the tornus. The discal region of the LHW is bordered by a series of black lunules and a black eye-spot, all edged with orange and pale blue (1, 8).

Female: slightly larger than the male, differing in the lack of a sex brand on the UFW. She may be browner and has an ochreous subapical spot on the UHW (1, 8).

Distribution *Papilio neumoegeni* is found only on the island of Sumba situated south-east of Sumbawa and south of Flores in the Lesser Sunda Islands of southern Indonesia. Honrath's original description of the species gives the type locality as "Sambawa" (5), although only Sumba is given as a locality in the recent literature (1, 8). It is not clear whether "Sambawa" means Sumbawa or Sumba, or whether *P. neumoegeni* is endemic to Sumba or occasionally found on Sumbawa.

Habitat and Ecology Adults are known to fly all year round and are particularly evident along paths or in glades in forest (8). They fly actively between 10 a.m. and 11 a.m., and rest in bushes during the afternoon (8). Associated species include *P. polytes* and *Atrophaneura oreon* (8). The female flies slowly and generally high up in the forest canopy (8). The larval foodplant has not been identified.

Sumba is one of the most southerly of the Indonesian islands, straddling the 10° southern parallel of latitude. The climate is fairly arid (6), being in the rainshadow of Australia during the south-east monsoon. The natural vegetation ranges from evergreen forest in the higher and wetter areas and evergreen gallery forest where groundwater is present, to monsoon forests and savanna woodlands.

Threats The species occurs only locally in Sumba, the small hills around Lewa in the north being a well known collecting site (8). Most of northern Sumba is open savanna, but forests survive on these hills (8). A handbook to Indonesia describes Sumba as relatively bare, but with widespread savannas of *Eucalyptus* and large tracts of grassland for grazing cattle and other livestock (2). Human population pressure throughout the Lesser Sundas, including Sumba, is such that dry season fires and deforestation for grazing land and firewood is causing habitat destruction in all but the most remote and inaccessible areas. Degraded areas of grassland are spreading every year and there can be little doubt that this is a serious threat to *P. neumoegeni*. In addition, collectors have been reported to return from Sumba with hundreds of specimens (4), a practice which is unlikely to be of advantage to an already local species.

Conservation Measures A survey and report of the distribution and status of *Papilio neumoegeni* is required, together with an assessment of threats. It seems likely that other species of animals and plants may also be threatened by destruction of relict forest patches, including a number of endemic birds (6).

Gunung (Mt) Wanggamati (1225 m, c. 6000 ha) is the most valuable of several proposed reserves on Sumba, and will protect the widest range of habitat types including Sumba's highest mountain (9). The bedrock is limestone and some of Sumba's best and least disturbed moist forest is found there. Most of Sumba's endemic birds are thought to occur in the forest, which may also prove to be a haven for endemic butterflies such as *P. neumoegeni*. There are population pressures around the periphery of the reserve and survey and guard work needs to be stepped up (9).

References

1. D'Abrera, B. (1982). *Butterflies of the Oriental Region. Part 1. Papilionidae and Pieridae.* Hill House, Victoria, Australia. xxxi + 244 pp.
2. Dalton, B. (1978). *Indonesia Handbook*. Moon Publications, Rutland, Vermont, U.S.A.
3. Hancock, D.L. (1983). Classification of the Papilionidae (Lepidoptera): a phylogenetic approach. *Smithersia* 2: 1–48.
4. Haugum, J. (1983). *In litt.*, 2 June 1983.
5. Honrath, E.G. (1890). Diagnosen von zwei neuen Rhopaloceren. *Entomologische Nach-richten*. 16: 127.
6. Mayr, E. (1944). The birds of Timor and Sumba. *Bulletin of the American Museum of Natural History* 83(2): 127–194.
7. Munroe, E. (1961). The classification of the Papilionidae (Lepidoptera). *Canadian Entomologist* Supplement 17: 1–51.
8. Tsukada, E. and Nishiyama, Y. (1982). *Butterflies of the South East Asian Islands Vol. 1 Papilionidae*. (transl. K. Morishita). Plapac Co. Ltd., Tokyo. 457 pp.
9. UNDP/FAO National Parks Development Project (1982). *National Conservation Plan for Indonesia Vol. 4: Lesser Sundas*. FAO Bogor, Indonesia.

Appendix A: List of Papilionidae in threatened categories

This is a list of threatened Papilionidae sorted according to their IUCN status. These categories are constantly reappraised and adapted to changing conditions. New information that may affect the category is solicited from readers. Definitions for the status categories are given in section 1.

ENDANGERED SPECIES

Ornithoptera alexandrae	Papua New Guinea (PNG)
Papilio homerus	Jamaica
Papilio hospiton	Corsica (France), Sardinia (Italy)
Papilio chikae	Luzon (Philippines)

ENDANGERED SUBSPECIES

Eurytides lysithous harrisianus	Brazil
Papilio aristodemus ponceanus	Florida (USA)
Papilio desmondi teita	Kenya

VULNERABLE

Luehdorfia japonica	Japan
Eurytides marcellinus	Jamaica
Eurytides iphitas	Brazil
Graphium levassori	Comoro Is
Graphium sandawanum	Mindanao (Philippines)
Battus zetides	Hispaniola (Haiti, Dominican Rep.)
Parides ascanius	Brazil
Parides burchellanus	Brazil
Atrophaneura jophon	Sri Lanka
Atrophaneura schadenbergi	Luzon, Babuyan (Philippines)
Troides dohertyi	Talaud Is, Sangihe (Indonesia)
Ornithoptera croesus	Moluccas (Indonesia)
Ornithoptera meridionalis	New Guinea (PNG, Irian Jaya)
Papilio esperanza	Mexico
Papilio himeros	Brazil, Argentina?
Papilio maraho	Taiwan
Papilio osmana	Leyte, Mindanao (Philippines)
Papilio carolinensis	Mindanao (Philippines)
Papilio moerneri	Bismarck Archipelago (PNG)
Papilio benguetanus	Luzon (Philippines)
Papilio phorbanta	Réunion
Papilio morondavana	Madagascar
Papilio leucotaenia	Zaire, Uganda, Burundi, Rwanda
Papilio neumoegeni	Sumba (Indonesia)

RARE

Baronia brevicornis	Mexico
Parnassius autocrator	USSR, Afghanistan
Parnassius apollo	Palaearctic
Bhutanitis mansfieldi	China
Bhutanitis thaidina	China
Teinopalpus imperialis	Nepal, India, Bhutan, Burma, China
Graphium idaeoides	Philippines
Graphium meeki	Bougainville (PNG), Solomons
Graphium stresemanni	Seram (Indonesia)
Graphium mendana	Bougainville (PNG), Solomons
Parides hahneli	Brazil
Atrophaneura luchti	Java (Indonesia)
Papilio toboroi	Bougainville (PNG), Solomons
Papilio acheron	Sabah, Sarawak, Brunei
Papilio jordani	Sulawesi (Indonesia)
Papilio weymeri	Bismarck Archipelago (PNG)
Papilio sjoestedti	Tanzania
Papilio antimachus	West and Central Africa
Papilio grosesmithi	Madagascar
Papilio mangoura	Madagascar

INDETERMINATE

Graphium megaera	Palawan (Philippines)
Graphium procles	Sabah (Malaysia)
Atrophaneura atropos	Palawan (Philippines)
Troides andromache	Sabah,Sarawak (Malaysia)
Troides prattorum	Buru (Indonesia)
Ornithoptera rothschildi	Irian Jaya (Indonesia)
Ornithoptera chimaera	New Guinea (PNG, Irian Jaya)
Ornithoptera paradisea	New Guinea (PNG, Irian Jaya)
Ornithoptera aesacus	Obi (Moluccas, Indonesia)
Papilio caiguanabus	Cuba
Papilio aristor	Hispaniola (Haiti, Dominican Rep.)
Papilio manlius	Mauritius
Papilio aristophontes	Comoro Is

INSUFFICIENTLY KNOWN

Bhutanitis ludlowi	Bhutan
Luehdorfia chinensis	China
Teinopalpus aureus	China
Graphium epaminondas	Andaman Is (India)
Graphium aurivilliusi	Zaire
Graphium weberi	Cameroon
Parides pizarro	Peru, Brazil
Parides steinbachi	Bolivia
Parides coelus	French Guiana
Parides klagesi	Venezuela
Atrophaneura palu	Sulawesi (Indonesia)
Ornithoptera tithonus	Irian Jaya (Indonesia)
Papilio garleppi	Bolivia, Brazil, French Guiana, Guyana, Peru, Surinam
Papilio maroni	French Guiana

Appendix B: Papilionidae that require further monitoring and research.

In addition to the 14 swallowtails that have been fully reviewed under the category Insufficiently Known there is a wide range of uncategorized species for which further data are still required. For a variety of reasons it would not have been fruitful to review all of these species in full. They are listed here in order to simplify cross-reference to the species list in section 3, where notes will be found on the type of data that are required. Some examples may serve to illustrate the variety of reasons for which this list has been developed.

Species marked with an asterisk are reasonably well known and not immediately threatened, but require continuous monitoring. Usually this is because they are found in areas where habitat destruction is accelerating, e.g. *Graphium gudenusi* in the forests of Central Africa, *Atrophaneura antenor* in Madagascar, *Ornithoptera euphorion* in Queensland forests, *Papilio thaiwanus* in Taiwan, *Papilio thuraui* in East African montane forests and *Papilio montrouzieri* in New Caledonia. Other species are well known and widespread, yet numerous local threats are gradually building up to a disturbing general picture; e.g. *Allancastria cerisy* in southern Europe and the Middle East, *Zerynthia polyxena* in Europe, *Bhutanitis lidderdalei* in Asia and *Ornithoptera richmondia* in eastern Australia. *Sericinus montela* from eastern Asia and *Protographium leosthenes* from Australia are of particular interest as primitive members of their tribes. A number of species have an uncertain taxonomic position; the *Ornithoptera* species of Australia were until recently considered to be subspecies of *O. priamus*; *Graphium pelopidas* may be only a subspecies of *G. leonidas*. Several species are very restricted in their ranges but not definitely known to be threatened, possibly only because they have never been investigated. In this category we include *Battus eracon* from western Mexico, *Atrophaneura febanus* and *A. horishanus* from Taiwan, *A. rhodifer* and *Papilio mayo* from the Andaman Islands, *Papilio biseriatus* from Timor and *P. godeffroyi* from Western Samoa. However, the majority of the species listed below are simply very poorly known, in some cases only from a handful of specimens. *Graphium mandarinus*, *G. tamerlanus*, *G. phidias* and *G. olbrechtsi* are in this group, as are all the listed species of *Parides*, *Papilio heringi*, *P. maesseni* and many other species of *Papilio*.

Our long-term aim is to reduce this list by assigning proper categories to the species. This can only be achieved by an increased level of professional and amateur research into the world's swallowtails.

Parnassius delphius	Afghanistan, USSR, Pakistan, India, Bhutan, China
Parnassius stoliczkanus	Afghanistan, India, Pakistan, China

Parnassius inopinatus	Afghanistan
Parnassius loxias	USSR, China
Parnassius acco	Pakistan, India, Nepal, China
Parnassius hannyngtoni	India, China, Bhutan(?)
Parnassius felderi	USSR
Parnassius bremeri	USSR, China, N. Korea, S. Korea, Japan, Mongolia(?)
Parnassius actius	USSR, Afghanistan, Pakistan, China
Parnassius tianschanicus	USSR, Afghanistan, Pakistan, India, China
Sericinus montela	USSR, China, N. Korea, S. Korea
*Allancastria cerisy**	Balkans and Middle East
Allancastria caucasica	USSR, Turkey
Allancastria louristana	Iran
*Zerynthia polyxena**	S. Europe to Balkans and USSR
*Bhutanitis lidderdalei**	Bhutan, India, Thailand, China
*Luehdorfia puziloi**	USSR, China, N. Korea, S. Korea, Japan
Meandrusa sciron	China, India, Bhutan, Thailand, Burma
Eurytides xanticles	Colombia, Panama
Eurytides bellerophon	Argentina, Brazil, Paraguay(?)
Eurytides earis	Ecuador, Brazil
Eurytides pausanias	Costa Rica to Brazil
*Protographium leosthenes**	Australia
Graphium mandarinus	China, Burma, Nepal
Graphium alebion	China, Taiwan
Graphium tamerlanus	China
Graphium dorcus	Sulawesi (Indonesia)
Graphium phidias	Vietnam, Laos(?)
Graphium olbrechtsi	Zaire
Graphium auriger	Gabon
Graphium pelopidas	Tanzania (Pemba I.)
Graphium nigrescens	Cameroon, Gabon, Zaire
*Graphium gudenusi**	Zaire, Uganda, Rwanda, Burundi
*Battus devilliers**	Cuba, Bahamas
Battus eracon	Mexico
*Parides gundlachianus**	Cuba
Parides phalaecus	Ecuador, Peru
Parides mithras	Guyana, Surinam, Fr. Guiana, Brazil
Parides chabrias	Peru, Ecuador
Parides quadratus	Brazil, Peru
Parides orellana	Brazil, Peru
Parides cutorina	Peru, Ecuador
Parides phosphorus	Colombia, Venezuela(?), Guyana, Brazil, Ecuador, Peru
Parides castilhoi	Brazil
*Atrophaneura antenor**	Madagascar
Atrophaneura daemonius	China
*Atrophaneura plutonius**	China, India, Nepal, Bhutan
Atrophaneura polla	India, Burma, China(?), Indochina(?)
Atrophaneura crassipes	India, Burma, China(?), Indochina
Atrophaneura adamsoni	Burma, Thailand
Atrophaneura laos	Thailand, Laos
Atrophaneura mencius	China

Atrophaneura impediens	China, Taiwan
*Atrophaneura febanus**	Taiwan
Atrophaneura hedistus	China
Atrophaneura kuehni	Sulawesi (Indonesia)
Atrophaneura priapus	Java, Sumatra (Indonesia)
*Atrophaneura horishanus**	Taiwan
Atrophaneura dixoni	Sulawesi (Indonesia)
Atrophaneura rhodifer	Andaman Is
Atrophaneura oreon	Lesser Sundas (Indonesia)
*Troides minos**	India
*Troides riedeli**	Tanimbar Is (Indonesia)
*Troides plato**	Timor (Indonesia)
*Ornithoptera euphorion**	Australia
*Ornithoptera richmondia**	Australia
Papilio ascolius	Central America to Ecuador
Papilio neyi	Ecuador
Papilio xanthopleura	Peru, Ecuador
Papilio euterpinus	Colombia, Ecuador, Peru
*Papilio thersites**	Jamaica
Papilio oxynias	Cuba
Papilio epenetus	Ecuador
Papilio dospassosi	Colombia
Papilio elwesi	China
*Papilio demetrius**	Japan, N. Korea, S. Korea, China
*Papilio lampsacus**	Java (Indonesia)
*Papilio thaiwanus**	Taiwan
Papilio mayo	Andaman Is
Papilio noblei	Burma, Thailand, Laos, Vietnam
Papilio antonio	Philippines
Papilio biseriatus	Timor (Indonesia)
Papilio diophantus	Sumatra (Indonesia)
Papilio mahadeva	Burma, Indochina, Malaysia
Papilio hipponous	Talaud and Sangihe Is (Indonesia)
Papilio heringi	Halmahera (Indonesia)
Papilio godeffroyi	Western Samoa
Papilio gambrisius	Seram, Ambon, Buru (Indonesia)
Papilio aethiopsis	Ethiopia, Somalia
*Papilio thuraui**	Tanzania, Malawi, Zambia
Papilio interjecta	Uganda, Kenya
Papilio maesseni	Ghana, Togo
*Papilio blumei**	Sulawesi (Indonesia)
Papilio elephenor	India, Burma, Thailand
Papilio karna	Palawan, Malaysia, Brunei, Indonesia
Papilio hoppo	Taiwan
*Papilio montrouzieri**	New Caledonia

* : Species that require monitoring.

Index

Plates

The following eight plates depict 40 threatened species. The choice has been made with the intention of demonstrating the extraordinary variety of form, pattern and colour to be found within the swallowtail butterfly family. With the exception of the three recently described Philippine endemics *Graphium sandawanum, Papilio osmana* and *Papilio carolinensis,* all Endangered and Vulnerable species are included. Both sexes are illustrated if they differ markedly and in the case of *Battus zetides* the attractive underside is shown. A number of species from other RDB categories are also included, particularly where these add to the variety.

The legends opposite the plates give the species' name, sex, provenance and actual size range. All photographs were made to a standard size in order fully to demonstrate the beauty of each specimen, large or small.

All the photographs are of specimens held in the Entomology Department of the British Museum (Natural History), whose co-operation in the preparation of the plates is gratefully acknowledged.

PLATE 1

1.1 *Baronia brevicornis,* the Baronia, female.
RARE, p. 182. Restricted to a small area of Mexico.
Forewing length: 33 mm.

1.2 *Luehdorfia japonica,* female.
VULNERABLE, p. 198. Restricted to the island of Honshu, Japan.
Forewing length: 30-35 mm.

1.3 *Parnassius autocrator,* male.
RARE, p. 185. Afghanistan and U.S.S.R. (Tadzhikistan).
Forewing length: 29 mm.

1.4 *Parnassius autocrator,* female.
RARE, p. 185.
Forewing length: 34 mm.

1.5 *Bhutanitis mansfieldi,* female, type specimen.
RARE, p. 192. China (Yunnan and Sichuan).
Forewing length: 40 mm. The left antenna is missing in this specimen.

1.6 *Bhutanitis ludlowi,* male, holotype.
INSUFFICIENTLY KNOWN, p. 196. Bhutan (Trashiyangsi valley).
Forewing length: 59 mm.

1.7 *Teinopalpus imperialis,* Kaiser-I-Hind, female.
RARE, p. 200. Nepal and northern India through Bhutan and northern Burma
to China (Hubei and Sichuan).
Forewing length: 60–65 mm.

1.8 *Eurytides marcellinus,* Jamaican Kite, male.
VULNERABLE, p. 206. Restricted to Jamaica.
Forewing length: 30-35 mm.

1.1

1.2

1.3

1.4

1.5

1.6

1.7

1.8

PLATE 2

2.1   *Eurytides lysithous harrisianus,* Harris' Mimic Swallowtail, male.
ENDANGERED, p. 208. Brazil: mimics *Parides ascanius* (2.2).
Forewing length: 40 mm.

2.2   *Parides ascanius,* Ascanius or Fluminense Swallowtail, male.
VULNERABLE, p. 240. Brazil (Rio de Janeiro state).
Forewing length: 46 mm.

2.3   *Eurytides iphitas,* Yellow Kite, male.
VULNERABLE, p. 211. Brazil.
Forewing length 50 mm.

2.4   *Graphium idaeoides,* male.
RARE, p. 215. Philippines. Mimics the danaine *Idea leuconoe.*
Forewing length: 77 mm.

2.5   *Graphium levassori,* female.
VULNERABLE, p. 222. Grande Comoro only.
Forewing length: 43-55 mm.

2.6   *Battus zetides,* Zetides Swallowtail, male.
VULNERABLE, p. 236. Haiti and Dominican Republic.
Forewing length: 35-40 mm.

2.7   *Battus zetides,* Zetides Swallowtail, male underside.
VULNERABLE, p. 236.

2.1

2.2

2.3

2.4

2.5

2.6

2.7

PLATE 3

3.1   *Atrophaneura jophon,* Ceylon Rose, female.
VULNERABLE, p. 258. Restricted to Sri Lanka.
Forewing length: 60-64 mm.

3.2   *Atrophaneura schadenbergi,* male.
VULNERABLE, p. 261. Philippines (Luzon and Babuyan Is only).
Forewing length: 50-60 mm. The left wings of this specimen are damaged.

3.3   *Troides dohertyi,* Talaud Black Birdwing, male.
VULNERABLE, p. 271. Talaud and Sangihe Is only (Indonesia).
Forewing length: 73 mm.

3.4   *Troides dohertyi,* Talaud Black Birdwing, female.
VULNERABLE, p. 271.
Forewing length: 82 mm.

3.5   *Ornithoptera tithonus,* male.
INSUFFICIENTLY KNOWN, p. 273. Indonesia (Irian Jaya).
Forewing length: 70-85 mm.

3.6   *Ornithoptera tithonus,* female.
INSUFFICIENTLY KNOWN, p. 273.
Forewing length 105-110 mm.

3.1 3.2

3.3 3.4

3.5 3.6

PLATE 4

4.1   *Ornithoptera rothschildi,* Rothschild's Birdwing, male.
      INDETERMINATE, p. 276. Indonesia (Arfak mountains, Irian Jaya).
      Forewing length: 63-80 mm.

4.2   *Ornithoptera rothschildi,* Rothschild's Birdwing, female.
      INDETERMINATE, p. 276
      Forewing length: 80-93 mm.

4.3   *Ornithoptera chimaera,* Chimaera Birdwing, male.
      INDETERMINATE, p. 278. Papua New Guinea and Indonesia (Irian Jaya).
      Forewing length: 85-95 mm.

4.4   *Ornithoptera chimaera,* Chimaera Birdwing, female.
      INDETERMINATE, p. 278.
      Forewing length: 110-115 mm.

4.5   *Ornithoptera paradisea,* Paradise or Tailed Birdwing, male.
      INDETERMINATE, p. 281. Papua New Guinea and Indonesia (Irian Jaya).
      Forewing length: 70-80 mm.

4.6   *Ornithoptera paradisea,* Paradise or Tailed Birdwing, female.
      INDETERMINATE, p. 281.
      Forewing length: 90-100 mm.

4.1

4.2

4.3

4.4

4.5

4.6

PLATE 5

5.1   *Ornithoptera meridionalis,* male.
      VULNERABLE, p. 284. Papua New Guinea and Indonesia (Irian Jaya).
      Forewing length: 60-65 mm.

5.2   *Ornithoptera meridionalis,* female.
      VULNERABLE, p. 284.
      Forewing length: 80-85 mm.

5.3   *Ornithoptera alexandrae,* Queen Alexandra's Birdwing, male.
      ENDANGERED, p. 288. South-eastern Papua New Guinea.
      Forewing length: 100-110 mm.

5.4   *Ornithoptera alexandrae,* Queen Alexandra's Birdwing, female.
      ENDANGERED, p. 288.
      Forewing length: 125-140 mm; the female of this species is the world's largest butterfly.

5.5   *Ornithoptera aesacus,* male.
      INDETERMINATE, p. 292. Indonesia (restricted to Obi, Moluccas).
      Forewing length: 75-85 mm. The hindwings of this specimen are somewhat stained.

5.6   *Ornithoptera aesacus,* female.
      INDETERMINATE, p. 292.
      Forewing length: 95-105 mm.

5.1

5.2

5.3

5.4

5.5

5.6

PLATE 6

6.1 *Ornithoptera croesus lydius,* male.
VULNERABLE, p. 294. Indonesia (Moluccas).
Forewing length: 80-90 mm.

6.2 *Ornithoptera croesus lydius,* female.
VULNERABLE, p. 294
Forewing length: 100-110 mm. The female of this subspecies is a mimic of
certain Danainae.

6.3 *Papilio homerus,* Homerus Swallowtail, male.
ENDANGERED, p. 297. Restricted to Jamaica.
Forewing length: ca 75 mm; the largest American swallowtail.

6.4 *Papilio esperanza,* male.
VULNERABLE, p. 299. Restricted to Oaxaca State, Mexico.
Forewing length: 51-52 mm. The right hindwing has a damaged tail in this
specimen.

6.5 *Papilio aristodemus ponceanus,* Schaus' Swallowtail, female.
ENDANGERED, p. 301. Restricted to Florida (U.S.A.).
Forewing length: 47-49 mm.

6.6 *Papilio himeros,* male.
VULNERABLE, p. 311. Brazil and possibly also northern Argentina.
Forewing length: 44-50 mm.

6.7 *Papilio himeros,* female, type specimen.
VULNERABLE, p. 311.
Forewing length: 44-50 mm.

6.1

6.2

6.3

6.4

6.5

6.6

6.7

PLATE 7

7.1 *Papilio moerneri,* male.
VULNERABLE, p. 324. Papua New Guinea (Bismark Archipelago only).
Forewing length: 58 mm. The antennae are missing and the wings are badly damaged in this specimen.

7.2 *Papilio hospiton,* Corsican Swallowtail, female.
ENDANGERED, p. 326. Corsica and Sardinia.
Forewing length: ca 38 mm.

7.3 *Papilio benguetanus,* female.
VULNERABLE, p. 329. Philippines (restricted to Luzon).
Forewing length: 44-48 mm.

7.4 *Papilio weymeri,* female.
RARE, p. 335. Papua New Guinea (restricted to Admiralty Is).
Forewing length: ca 75 mm.

7.5 *Papilio sjoestedti,* Kilimanjaro Swallowtail, male.
RARE, p. 337. Northern Tanzania.
Forewing length: 45-54 mm. Slight damage to the left forewing in this specimen.

7.6 *Papilio desmondi teita,* Taita Blue-banded Swallowtail, male.
ENDANGERED, p. 347. Taita Hills only, south-eastern Kenya.
Forewing length: ca 42 mm.

7.1　　　　　　　7.2

7.3　　　　　　　7.4

7.5　　　　　　　7.6

PLATE 8

8.1 *Papilio phorbanta,* Papillon La Pature, male.
VULNERABLE, p. 342. Endemic to Réunion.
Forewing length: 40-55 mm.

8.2 *Papilio phorbanta,* Papillon La Pature, female.
VULNERABLE, p. 342.
Forewing length: 40-55 mm.

8.3 *Papilio antimachus,* African Giant Swallowtail or Papilio, female.
RARE, p. 351. Central and western Africa from Guinea through Cameroon
south to Uganda and Angola.
Forewing length: ca 75 mm; the male is the largest African swallowtail with a
forewing length of up to 115 mm.

8.4 *Papilio morondavana,* Madagascan Emperor Swallowtail, male.
VULNERABLE, p. 354. Madagascar.
Forewing length: 51-55 mm.

8.5 *Papilio leucotaenia,* Cream-banded Swallowtail, male.
VULNERABLE, p. 358. South-western Uganda, Rwanda, Burundi and north-
eastern Zaire.
Forewing length: 54-60 mm.

8.6 *Papilio chikae,* male.
ENDANGERED, p. 364. Philippines (restricted to northern Luzon).
Forewing length: ca 55 mm. In this specimen the right forewing is damaged
and the left antenna is missing.

8.7 *Papilio neumoegeni,* male.
VULNERABLE, p. 367. Indonesia (Sumba only, Lesser Sunda Is).
Forewing length: 43-48 mm.

8.1

8.2

8.3

8.4

8.5

8.6

8.7